D1693331

Posterior Cruciate Ligament Injuries

Springer
*New York
Berlin
Heidelberg
Barcelona
Hong Kong
London
Milan
Paris
Singapore
Tokyo*

Gregory C. Fanelli, MD
Chief, Sports Injury Clinic and Arthroscopic Surgery
Department of Orthopaedic Surgery
Geisinger Medical Center
Danville, Pennsylvania

Editor

Posterior Cruciate Ligament Injuries

A Practical Guide to Management

With a Foreword by M. Mike Malek, MD, FACS

With Illustrations by Joel Herring

With 234 Illustrations, 31 in Full Color

Springer

Gregory C. Fanelli, MD
Chief, Sports Injury Clinic and Arthroscopic Surgery
Department of Orthopaedic Surgery
Geisinger Medical Center
Danville, PA 17822-2130
USA

Library of Congress Cataloging-in-Publication Data
Posterior cruciate ligament injuries : a practical guide to management / editor, Gregory C. Fanelli
   p. ; cm.
   Includes bibliographical references and index.
   ISBN 0-387-98573-5 (hardcover : alk. paper)
   1. Cruciate ligaments—Wounds and injuries—Treatment. 2. Knee—Wounds and injuries—Treatment. 3. Knee—Surgery. I. Fanelli, Gregory C.
   [DNLM: 1. Posterior Cruciate Ligament—injuries. 2. Knee Injuries—therapy. 3. Posterior Cruciate Ligament—surgery. 4. Tendon Injuries—therapy. WE 870 P857 2000]
RD561 .P676 2000
617.5′82044—dc21   99-046324

Printed on acid-free paper.

© 2001 Springer-Verlag New York, Inc.
All rights reserved. This work may not be translated or copied in whole or in part without the written permission of the publisher (Springer-Verlag New York, Inc., 175 Fifth Avenue, New York, NY 10010, USA), except for brief excerpts in connection with reviews or scholarly analysis. Use in connection with any form of information storage and retrieval, electronic adaptation, computer software, or by similar or dissimilar methodology now known or hereafter developed is forbidden.
The use of general descriptive names, trade names, trademarks, etc., in this publication, even if the former are not especially identified, is not to be taken as a sign that such names, as understood by the Trade Marks and Merchandise Marks Act, may accordingly be used freely by anyone.
While the advice and information in this book are believed to be true and accurate at the date of going to press, neither the authors nor the editors nor the publisher can accept any legal responsibility for any errors or omissions that may be made. The publisher makes no warranty, express or implied, with respect to the material contained herein.

Production coordinated by Matrix Publishing Services, Inc., and managed by Terry Kornak; manufacturing supervised by Jeffrey Taub.
Typeset by Matrix Publishing Services, Inc., York, PA.
Printed and bound in China by Everbest Printing Company, through Four Colour Imports, Louisville, KY.

9 8 7 6 5 4 3 2 1

ISBN 0-387-98573-5   SPIN 10682854

Springer-Verlag   New York Berlin Heidelberg
A member of BertelsmannSpringer Science+Business Media GmbH

To my wife Lori, and my children Matthew, David, and Megan whose love, support, and patience have been unwavering during this project.

Foreword

In the 1980s and 1990s several excellent volumes were published dealing with the pathology, recognition, and treatment of anterior cruciate ligament injuries. The idea of treating anterior cruciate ligament injuries nonoperatively changed to aggressive surgical treatment of anterior cruciate–deficient knees in the belief that this provides the best chance for a normal functioning knee, preventing damage to the secondary restraints and development of osteoarthritis. However, this concept has been challenged by many and the controversy continues.

As the new millennium starts we are focusing our attention on a new frontier—the posterior cruciate ligament. This is the strongest ligament in the knee, and yet it has been ignored for a long time. We seem to be revisiting the issues we faced with the anterior cruciate ligament in the 1970s and in the ensuing 25 years. Now we must learn how to manage and successfully treat injuries to the posterior cruciate ligament.

Pioneers of sports medicine and ligament surgery such as O'Donaghue, Hughston, Slocum, Nicholas, Marshall, and others have taught us the principles of knee function and biomechanics, and the basics of ligament surgery. It is up to the younger generation of surgeons and scientists to continue in the footsteps of these men and provide us with a follow-up to the work done by them. I am pleased that Dr. Fanelli has brought together an outstanding group of contributors, compiled an excellent volume on posterior cruciate ligament injuries, and provided a practical guide for management of these injuries.

In this volume, anatomy and biomechanics of the posterior cruciate ligament and the posterolateral corner are well defined. Clinical examination of the posterior cruciate deficient knee as well as diagnostic testing is clearly described. Nonoperative and operative treatment is discussed, including the latest methods of surgical reconstruction. Sections are devoted to rehabilitation as well as complications of posterior cruciate ligament surgery. The authors are the pioneers of our time, and their devotion to this subject is clear.

I am greatly honored to have been asked to provide the Foreword to such an outstanding collection of material on injuries of the posterior cruciate ligament.

M. Mike Malek, MD, FACS
Washington Orthopaedic & Knee Clinic, Inc.
Viena, Virginia

Preface

The pathways that our individual orthopaedic careers take are largely determined by our practice environment. It has been my good fortune to be in a position that enabled me to expand my research interests and surgical techniques in the assessment and treatment of posterior cruciate ligament (PCL) and related injuries. I believe the same situation exists for the other contributors to this book. We all share a passion and a commitment to the treatment of complex instabilities about the knee involving the PCL. The purpose of this book is to share our knowledge and experience in the assessment and treatment of the PCL and related injuries. It is my personal hope that this book will serve as a launch pad for new ideas to further develop treatment plans and surgical techniques for PCL injuries.

Acknowledgments

Many people have influenced my career, and subsequently the development of this book. M. Mike Malek, M.D., with whom I did my fellowship, has been a mentor, a great friend, and an example for the pursuit of excellence in arthroscopic surgery. The faculty and residents in the Orthopaedic Surgery Department at Geisinger Medical Center all support the PCL research program: Craig J. Edson, M.S., P.T./A.T.C., friend, colleague, and research associate is an outstanding resource for our patients and myself and the operating room and clinic orthopaedic teams at Geisinger Medical Center, are the best orthopaedic support staff in the world. Their dedication and drive for excellence are unequaled. Arthrotek, Inc., a division of Biomet, Inc., contributed to the production of this book. Many of my colleagues have taken time out of their busy schedules to contribute chapters to this book. Mary Zubowicz, my secretary, has always been able to keep the most hectic schedules and deadlines organized and moving forward, all with a smile.

Gregory C. Fanelli, MD
Danville, Pennsylvania

Contents

Foreword *M. Mike Malek*	vii
Preface	ix
Contributors	xiii

PART I: ANATOMY AND BIOMECHANICS

1. Anatomy and Biomechanics of the Posterior Cruciate Ligament — 3
   Christopher D. Harner, Tracy M. Vogrin, and Savio L.-Y. Woo

2. Anatomy and Biomechanics of the Posterolateral Aspect of the Knee — 23
   Robert F. LaPrade and Timothy S. Bollom

3. Anatomy and Biomechanics of the Posteromedial Aspect of the Knee — 47
   Fred Flandry and Christian C. Perry

PART II: DIAGNOSIS AND EVALUATION

4. Clinical Examination of the Posterior Cruciate Ligament–Deficient Knee — 65
   Darren L. Johnson

5. Imaging of the Posterior Cruciate Ligament — 77
   D.C. Peterson, L.M.F. Thain, and P.J. Fowler

6. Measurement of the Posterior Cruciate Ligament and Posterolateral Corner — 87
   Don Johnson

7. Arthroscopic Evaluation of the Posterior Cruciate Ligament and Posterolateral Corner — 95
   Gregory C. Fanelli

PART III: NONOPERATIVE TREATMENT

8. The Natural History of the Posterior Cruciate Ligament–Deficient Knee — 109
   Bradley F. Giannotti

9. Nonoperative Treatment of Posterior Cruciate Ligament Injuries — 117
   John A. Bergfeld, Manuel Leyes, and Gary J. Calabrese

PART IV: SURGICAL TREATMENT

10. Graft Selection in Posterior Cruciate Ligament Surgery — 135
    Walter R. Shelton

11. Arthroscopically Assisted Posterior Cruciate Ligament Reconstruction: Transtibial Tunnel Technique — 141
    Gregory C. Fanelli

12. Arthroscopically Assisted Posterior Cruciate Ligament Reconstruction: Tibial Inlay Technique — 157
    Richard D. Parker, John A. Bergfeld, David R. McAllister, and Gary J. Calabrese

13. Arthroscopically Assisted Posterior Cruciate Ligament Reconstruction: Double Femoral Tunnel Technique — 175
    Michael H. Metcalf and Roger V. Larson

14. Cruciate Ligament Reconstruction with Synthetics — 189
    Don Johnson and J.P. Laboureau

15. Arthroscopically Assisted Combined Anterior Cruciate Ligament/Posterior Cruciate Ligament Reconstruction — 215
    Gregory C. Fanelli and Daniel D. Feldmann

16. Surgical Treatment of Posterolateral Instability — 237
    Roger V. Larson and Michael H. Metcalf

17. Surgical Treatment of Medial Ligament Injuries Associated with Posterior Cruciate Ligament Tears — 249
    Fred Flandry and Jack C. Hughston

18. Rehabilitation of Posterior Cruciate Ligament Injuries — 267
    Craig J. Edson and Daniel D. Feldmann

19. Complications in Posterior Cruciate Ligament Surgery — 291
    Gregory C. Fanelli and Timothy J. Monahan

Index — 303

Contributors

John A. Bergfeld, MD
Section of Sports Medicine
Department of Orthopaedics
The Cleveland Clinic Foundation
Cleveland, OH 44131-2143, USA

Timothy S. Bollom, MD
Sports Medicine and Shoulder Divisions
Department of Orthopaedic Surgery
University of Minnesota
Minneapolis, MN 55455-0100, USA

Gary J. Calabrese, PT
Sports Orthopaedic Rehabilitation
Section of Sports Medicine
Department of Orthopaedics
The Cleveland Clinic Foundation
Cleveland, OH 44131-2143, USA

Craig J. Edson, MHS, PT, ATC
PennState Geisinger Center for Rehabilitation
An Affiliate of State HealthSouth
Danville, PA 17822-2130, USA

Gregory C. Fanelli, MD
Sports Injury Clinic and Arthroscopic Surgery
Department of Orthopaedic Surgery
Geisinger Medical Center
Danville, PA 17822-2130, USA

Daniel D. Feldmann, MD
Department of Orthopaedic Surgery
Geisinger Medical Center
Danville, PA 17822-2130, USA

Fred Flandry, MD, FACS
The Hughston Clinic
Department of Orthopaedic Surgery
Tulane University School of Medicine
Columbus, GA 31908, USA

P.J. Fowler, MD, FRCS (C)
Department of Orthopaedic Surgery
University of Western Ontario
London Health Sciences Centre
Department of Orthopaedic Surgery
Fowler Kennedy Sport Medicine Clinic
London, ON N6A 3K7, Canada

Bradley F. Giannotti, MD
Charles Cole Memorial Hospital
Department of Orthopaedics and Sports Medicine
Coudersport, PA 16915, USA

Christopher D. Harner, MD
University of Pittsburgh
Musculoskeletal Research Center
Department of Orthopaedic Surgery
Pittsburgh, PA 15213, USA

Jack C. Hughston, MD
The Hughston Clinic
Columbus, GA 31908, USA

Darren L. Johnson, MD
Department of Orthopaedic Surgery
Section of Sports Medicine
University of Kentucky School of Medicine
Department of Orthopaedic Surgery
Lexington, KY 40536, USA

Don Johnson, MD
Department of Orthopaedic Surgery
University of Ottawa
Sports Medicine Clinic
Carleton University
Ottawa, ON K1N 6N5, Canada

J.P. Laboureau, MD
Orthopaedic and Sports Medicine Center
Chenove 21300, France

Robert F. LaPrade, MD
Sports Medicine and Shoulder Services
University of Minnesota
Department of Orthopaedic Surgery
Minneapolis, MN 55455, USA

Roger V. Larson, MD
Department of Orthopaedic Surgery
University of Washington
Seattle, WA 98195-0005, USA

Manuel Leyes, MD, PhD
Section of Sports Medicine
Department of Orthopaedics
The Cleveland Clinic Foundation
Cleveland, OH 44131-2143, USA

David R. McAllister, MD
Division of Sports Medicine
Department of Orthopaedic Surgery
University of California Los Angeles
Los Angeles, CA 90095-8362, USA

Michael H. Metcalf, MD
Department of Orthopaedic Surgery
University of Washington
Seattle, WA 98195-0005, USA

Timothy J. Monahan, MD
Department of Orthopaedic Surgery
Geisinger Medical Center
Danville, PA 17822-2130, USA

Richard D. Parker, MD
Sports Orthopaedic Rehabilitation
Section of Sports Medicine
Department of Orthopaedics
The Cleveland Clinic Foundation
Cleveland, OH 44195, USA

Christian C. Perry, PA-C
The Hughston Clinic
Columbus, GA 31908, USA

D.C. Peterson, MD, FRCS (C)
University of Western Ontario
London Health Sciences Centre
Department of Orthopaedic Surgery
Fowler Kennedy Sport Medicine Clinic
London, ON N6A 3K7, Canada

Walter R. Shelton, MD
Mississippi Sports Medicine and Orthopaedic Center
Jackson, MS 39202, USA

L.M.F. Thain, AB, MD, FRCPC
Department of Radiology and Surgery
University of Western Ontario
London Health Sciences Centre
Diagnostic Radiology
London, ON N6A 3K7, Canada

Tracy M. Vogrin, MS
Musculoskeletal Research Center
Department of Orthopaedic Surgery
University of Pittsburgh
Pittsburgh, PA 15238-2887, USA

Savio L.-Y. Woo, PhD, Dsc
Musculoskeletal Research Center
Department of Orthopaedic Surgery
University of Pittsburgh
Pittsburgh, PA 15238-2887, USA

I

Anatomy and Biomechanics

Chapter One

Anatomy and Biomechanics of the Posterior Cruciate Ligament

Christopher D. Harner, Tracy M. Vogrin, and Savio L-Y. Woo

Numerous clinical, biomechanical, and anatomic studies have been devoted to the investigation of the function of the anterior cruciate ligament (ACL) of the knee and the treatment of ACL injuries. These studies have resulted in a better understanding of ACL function and improved surgical outcomes of ACL reconstructions. The success rate of ACL reconstruction is now as high as 80% to 90%.[1-3] In contrast, the posterior cruciate ligament (PCL) has received much less attention regarding its anatomic and biomechanical roles in knee function. The lack of scientific data may in part explain the poorer clinical outcomes following PCL injury and surgery, compared to those for the ACL.

In the general population, PCL injuries account for 3% to 23% of all knee ligament injuries,[4-8] though this number may be as high as 40% in an emergency room setting.[9] PCL injuries are generally classified as isolated (PCL only) or combined (in conjunction with one or more ligaments). Isolated injuries are often considered to have a relatively benign clinical course and have traditionally been treated nonoperatively.[10-13] However, numerous studies have shown that patients treated nonoperatively can develop increased instability and arthritic changes over time.[12-15] Unfortunately, surgical management of isolated PCL injuries has also been problematic, with a high number of patients experiencing residual knee laxity.[5,16-22] Combined PCL injuries are more common, occurring in more than 60% of PCL injuries.[9,23] The most frequent associated injury involves the posterolateral structures (PLS), which is reported to occur in up to 60% of all combined PCL injuries.[9] Nonsurgical management of these injuries has universally failed, while surgical management of these and other combined PCL injuries has been variable.[20,22,24,25] As a result, clinical outcomes of PCL reconstructions remain inconsistent, and the long-term prognosis of surgically treated PCL injuries is significantly worse than that for other knee ligaments.[25]

To improve treatment and outcomes of knee injuries involving the PCL, we suggest that basic science research is necessary to develop new ideas and techniques for PCL reconstruction. This chapter presents data on PCL anatomy and biomechanics. We first address the anatomy of the PCL, including its functional bundles, bony insertions, and microstructure quantitatively. We then describe the structural properties of the femur-PCL-tibia complex, as well as the mechanical properties of the midsubstance of the PCL. Finally, we address the functional biomechanics of the PCL, including measurement of in situ force in the PCL, its contribution to knee kinematics, and PCL reconstruction.

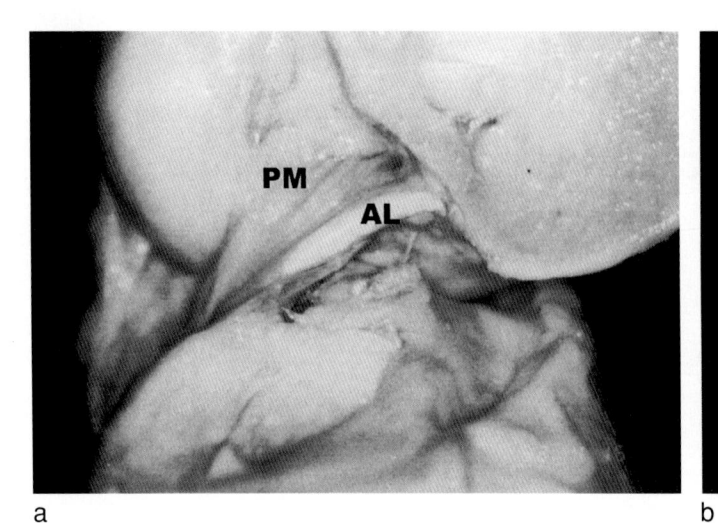

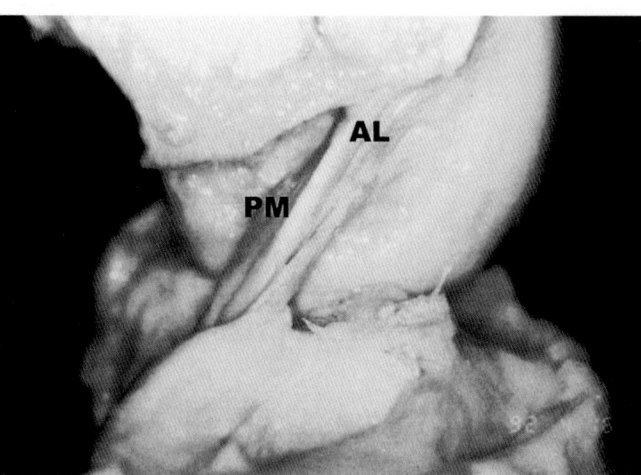

Fig. 1.1. Lateral view of the right knee with the posterior cruciate ligament (PCL) and its components with the lateral femoral condyle removed. (a) Knee in extension. (b) Knee at 80 degree of flexion. Anterolateral (AL) bundle and posteromedial (PM) bundle. (From Harner et al.,[7] with permission.)

Anatomy of the PCL

The PCL extends from the lateral surface of the medial femoral condyle to its insertion on the posterior aspect of the tibia, for which it receives its name.[26,27] It is an intraarticular but extrasynovial ligament and is covered by a sheath of synovium that is reflected from the posterior capsule and covers the ligament on its medial, lateral, and anterior aspects. The posterior aspect of the PCL blends into the posterior capsule and the periosteum.[26,27] The PCL is oriented almost vertically when the knee is extended and becomes more horizontal as the knee is flexed. In 1975, Girgis et al.[27] found the PCL to average 38 mm in length and 13 mm in width as measured by calipers.

The PCL is not a single band, but a complex structure consisting of various components that are most often referred to as the anterolateral (AL) and posteromedial (PM) bundles.[28–31] During passive flexion-extension of the knee (i.e., with no external forces acting on the knee), it is reported that the AL bundle is taut in flexion while the PM bundle is taut in extension[26] (Fig. 1.1). Each of these two bundles has distinct and consistent insertions onto the tibia and femur (Fig. 1.2).[32] The meniscofemoral ligaments (MFLs) are also considered part of the PCL complex. These ligaments originate from the lateral meniscus and insert anterior and posterior to the PCL on the medial femoral condyle. The anterior MFL is known as the ligament of Humphry and the posterior MFL is the ligament of Wrisberg. The incidence of appearance of these ligaments is variable, however,

Fig. 1.2. Schematic of insertions of AL and PM bundles of the PCL onto the tibia and femur. (From Harner et al.,[30] with permission.)

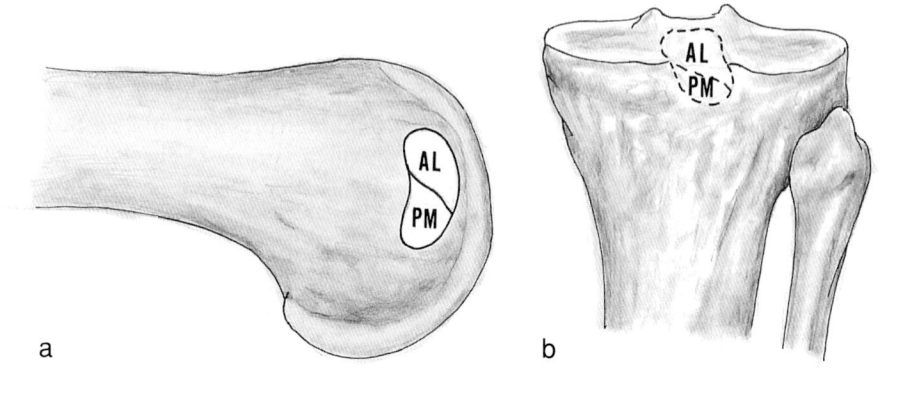

with some knees having one MFL, both, or none. The importance of the MFLs has not been well characterized, but they are believed to help stabilize the lateral meniscus.[34]

While the PLS are not considered part of the PCL complex, they have been shown to work closely with the PCL to provide posterior knee stability.[34–37] There are two primary components to the PLS: the lateral collateral ligament (LCL); and the popliteus complex, which includes the popliteus muscle-tendon unit, the popliteofibular ligament, and various popliteotibial and popliteomeniscal fascicles.[38–41] The anatomy of the PLS is described in detail in Chapter 2.

Determination of Cross-Sectional Area

While there have been numerous qualitative descriptions of the gross anatomy of the PCL,[26,27] few quantitative descriptions exist. This information is important in determining ideal graft substitutes and graft placement. Several quantitative studies have reported cruciate ligament cross-sectional area, measuring a length and width and assuming a rectangular cross section.[45–47] Other contact methods have been used as well, such as compressing the ligament into a slot of known dimensions,[47–50] or taking molds of the ligament and sectioning the molds to determine the area.[51]

To avoid errors that may occur as a result of contacting the soft tissue, several noncontact methods have been developed based on the concept of shadow projections.[49,52–54] In our research center, we used a laser micrometer system to quantitatively determine the cross-sectional shape and area of the PCL.[30,55,56] This method requires no contact with the ligament and is accurate to within 3% for ligaments of various cross-sectional shapes.[56]

The cross-sectional areas of the PCL, ACL, and MFL from eight cadaveric knees were measured using this system to determine the effects of both position along the ligament axis as well as knee flexion angle.[30,124] The femur-PCL-tibia complex was mounted in the assembly such that the beam of the laser was perpendicular to the long axis of the ligament. The specimen was then rotated incrementally while the profile widths of the ligament were recorded and the cross-sectional area of the PCL was calculated. This process was performed at four other levels along the ligament for knee flexion angles of 0, 30, 60, and 90 degrees of flexion. An identical test was then performed for the ACL. The MFLs were tested as isolated specimens, and thus the effect of flexion angle was not determined for that ligament.

Figure 1.3 shows typical cross-sectional areas of the PCL, ACL, and MFL at 30 degrees of flexion. Each structure had a unique and consistent cross-sectional shape. While the cross-sectional shape of both the ACL and PCL was observed to vary with knee flexion, no significant changes in its area occurred. The cross-sectional area of the PCL tended to increase from distal (tibia) to proximal (femur), while the opposite trend was observed for the ACL. The PCL was significantly larger than the ACL at its midsubstance and proximal levels, ranging in size from 120% to 150% of the ACL. The MFLs were approximately 22% of the size of the PCL.

PCL Insertion Site Anatomy

The majority of studies on cruciate ligament reconstruction have focused on appropriate graft materials, placement, and fixation. Fewer anatomic studies have investigated the actual insertion sites of the PCL and ACL. De-

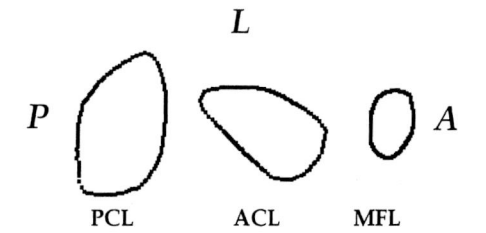

Fig. 1.3. Typical cross-sectional shapes and areas for the PCL, anterior cruciate ligament (ACL), and meniscofemoral ligament (MFL) (Wrisberg) at the midsubstance level for a left knee flexed 30 degrees. (From Harner et al.,[124] with permission.)

Fig. 1.4. Bar graph demonstrating the ratio of ligament insertion cross-sectional area to ligament midsubstance (mean ± standard error of the mean [SEM], $n = 5$). (From Harner et al.,[32] with permission.)

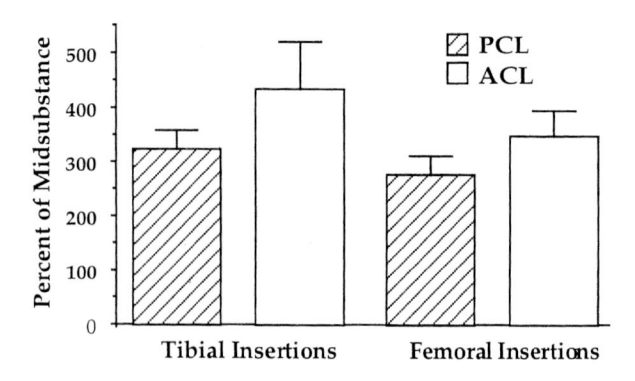

tailed knowledge of insertion site anatomy is essential for the development of new reconstructive techniques, as it guides the basic scientist in the investigation of mechanics and biology, and is critical to the surgeon in developing "maps" for accurate reconstruction. In 1975, Girgis et al.[26] used calipers to measure the length and width of PCL insertions. Morgan et al.[57,58] also used calipers to measure the distance from the center of the insertions to anatomic landmarks. In our research center, we further quantified the insertions of the PCL and ACL and their functional bundles using a three-dimensional digitizing system (Ascension Technologies, Colchester, VT).[32]

The x, y, and z coordinates of the insertions of the PCL and ACL were digitized (accurate to 0.8 mm) and the points were loaded into a program that projected the insertion to a series of different plane views. The largest projected area was taken to be the surface area of the ligament insertion. The insertions of the PCL were approximately 120% of those of the ACL, though this effect was not significant. Relative to the ligament midsubstance, the insertions were found to be 3.5 and 3 times larger for the PCL and ACL, respectively (Fig. 1.4). In the second part of the study, the insertion areas of the two bundles of the PCL (anterolateral and posteromedial) and ACL (anteromedial and posterolateral) were measured. Typical results for the PCL are shown in Fig. 1.5. For both the ACL and the PCL, each component had specific and consistent insertion locations that averaged approximately 50% of the total ligament cross-sectional area.

PCL Ultrastructure

For a complete understanding of the anatomy and function of the PCL, its ultrastructure must be studied as well. Several reports have suggested that a correlation exists between the mechanical properties of soft tissue and the diameter of its collagen fibrils, with areas subjected to greater stresses exhibiting larger fibrils.[59,60] Other factors, such as aging,[61,62] exercise,[61] immobilization,[63,64] and healing,[64] have also been linked to collagen fibril diameters. While the distribution of fibril diameters is well known for the ACL,[61,65,66] fewer studies have investigated the PCL or MFL. Therefore, we quantified the diameters of the collagen fibrils of the PCL, ACL, and MFL at three different locations along their lengths.[67]

The PCL, ACL, and MFL from four young human knees (range 17 to 22 years) were harvested and a 1-mm-diameter core portion of each ligament was prepared for transmission electron microscopy (TEM) using standard procedures. Samples were examined from proximal, middle, and distal sections for each ligament and micrographs were printed at a final magnification of 75,000×. A minimum of 500 fibrils were randomly sam-

1. Anatomy and Biomechanics of the Posterior Cruciate Ligament

Fig. 1.5. Tibial (a) and femoral (b) insertions of the anterolateral (AL) and posteromedial (PM) bundles of the PCL and the mensicofemoral (MFL) ligament of Humphry. (From Harner et al.,[32] with permission.)

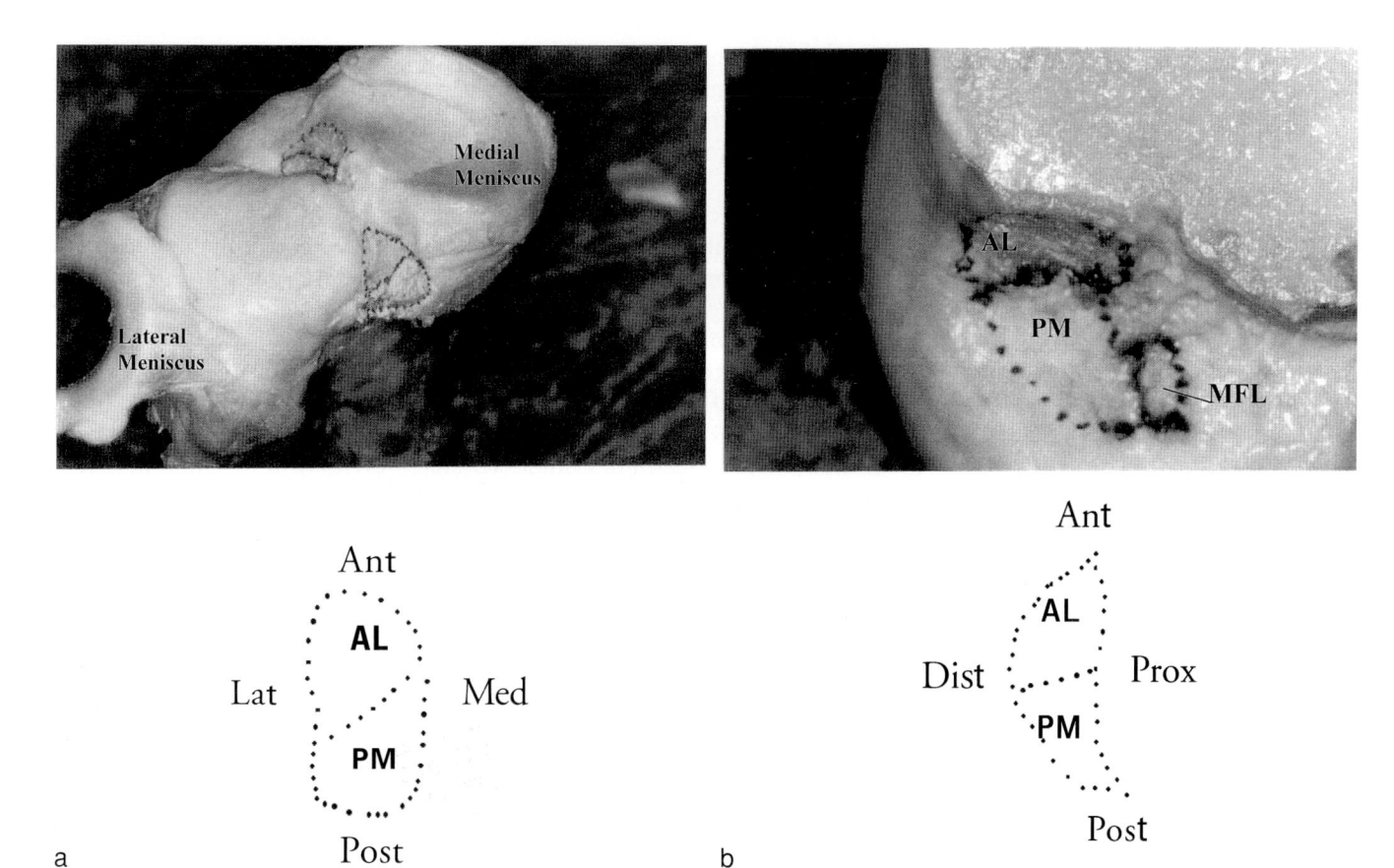

pled from each micrograph and analyzed using an image analysis program to measure the diameters of the collagen fibrils.

The location of the ligament from which the fibrils were sampled (proximal, middle, distal) was found to significantly affect the collagen fibril diameter. The fibrils for the PCL were largest in the proximal region and decreased in the distal direction, while the reverse trend was observed for the ACL. The fibrils of the MFL were largest in the middle region. No significant differences were observed between ligaments, however.

Biomechanical Properties of the PCL

Several authors have attributed the relatively low incidence of PCL injuries, compared to injuries of the ACL or MCL, to the considerable strength of the PCL. Kennedy and colleagues[68] were among the first to measure the tensile strength of the human PCL, reporting that it was twice as strong as the ACL. This study, however, was performed using specimens excised from the knee. The femur-PCL-tibia complex of cadaveric knees at 45 degrees of flexion was tested by Prietto et al.,[69] who reported an ultimate load of $1,627 \pm 491$ N and a linear stiffness of 204 ± 49 N/mm.

Recently, in our research center and others, the structural properties of the AL and PM bundles of the human PCL have been determined.[30,31] Fourteen fresh-frozen cadaveric specimens (mean donor age 52 years) were studied. The specimens were first thawed at room temperature and the tensioning patterns of the PCL observed through several ranges of flexion-extension. The AL bundle was identified on the basis that it was taut in flexion while the PM bundle was taut in extension. The bundles were then separated to their insertion sites, and the insertions of the AL, PM, and MFLs were marked on the bone. The femoral insertions of each component were then separated into bone blocks and potted in polymethylmethacrylate for fixation while the tibia and lateral meniscus were left intact.

The cross-sectional areas of the three components were measured using a laser micrometer.[55,56] The tibia was then placed in an aluminum clamp that allowed movement in five degrees of freedom. Each bone-ligament-bone complex was aligned such that a uniaxial tension could be applied along the anatomic axis of the PCL. For the MFL, the lateral meniscus was clamped such that a uniaxial tension could be applied. Four elastin markers were placed on each component to enable determination of tissue strain using a motion analysis system (Motion Analysis, Santa Rosa, CA). Each component was then preconditioned for 10 cycles of 0 to 2 mm of extension at 20 mm/min and tensile tested to failure at 200 mm/min.

The data obtained consisted of the structural properties of each component from the load-elongation curve of the bone-ligament-bone complex (Fig. 1.6). Stress-strain curves were also plotted and the elastic modulus of the ligament midsubstance was calculated. The cross-sectional area of the AL bundle was found to be two times as large as the PM bundle at their respective midsubstances on average. The linear stiffness of the AL bundle (120 ± 37 N/mm) was also significantly larger than that of the PM bundle (2.6 times) and the MFL (2.7 times) (Fig. 1.7). The ultimate load of the AL bundle (1,120 ± 362 N) was more than three times as large as that of the PM bundle and five times larger than that of the MFL. The primary mode of failure for the young specimens (49 years of age or less) was a midsubstance rupture, while two-thirds of the older specimens failed at the ligament insertion. The elastic modulus of each bundle was also determined, with the AL and PM bundles having moduli of 294 ± 115 megapascal (MPa) and 150 ± 69 MPa, respectively. The modulus of the MFL was not significantly different from that of the AL bundle.

In summary, current quantitative anatomic and biomechanical studies on the human PCL indicate that the AL bundle of the PCL is the largest and the strongest structurally of the PCL complex (AL, PM, MFL). Based on these findings, we conclude that the primary focus of single-bundle reconstructive procedures should be to reproduce the AL component.

Fig. 1.6. Typical load-elongation curves for the AL, PM, and MFL components of the posterior complex from the same knee. (From Harner et al.,[30] with permission.)

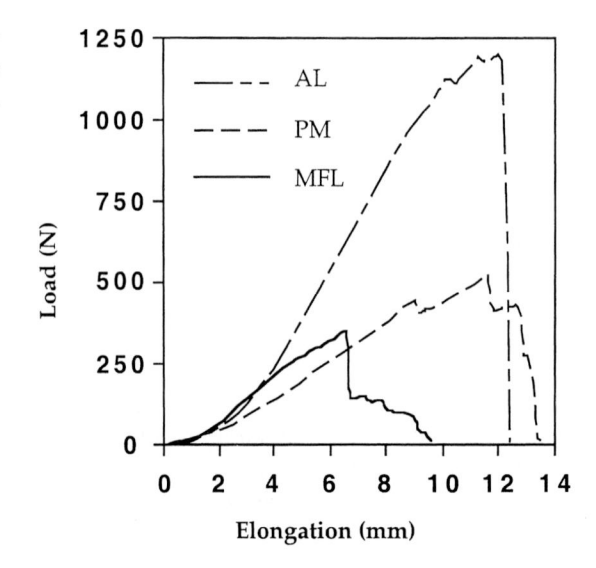

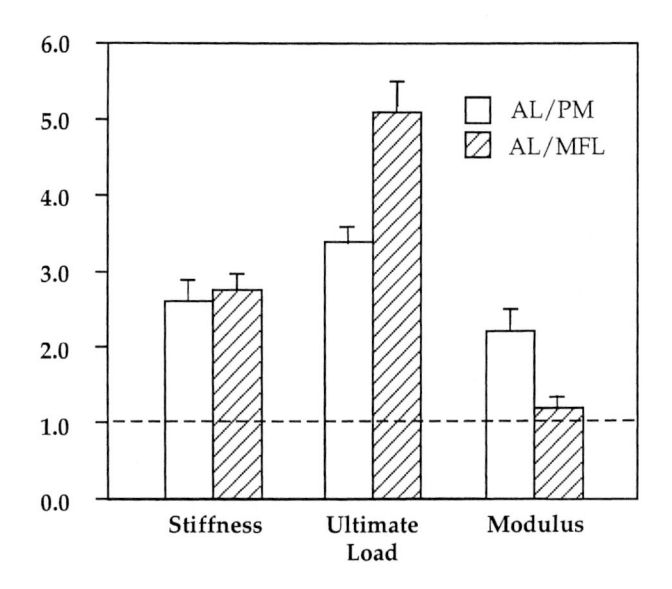

Fig. 1.7. Bar graph detailing the mean (± SEM) AL/PM and AL/MFL intraspecimen ratios of stiffness, ultimate load, and modulus. (From Harner et al.,[30] with permission.)

Biomechanical Function of the PCL and PCL Reconstructions

Characterizing the structural properties of the bone-ligament-bone complex and the mechanical properties of the ligament or tendon substance can aid in the understanding of the functional role of these tissues. However, the function of the PCL can be further characterized by determining its contribution to joint kinematics and its forces in situ, when external loads are applied to the joint, as well as its interaction with other structures of the knee.

Contribution of the PCL to Knee Kinematics

Cutting studies have been used to investigate the individual contributions of the PCL and its associated structures in providing knee stability.[21,34,35,37,70] In general, a device is used that allows multi–degree of freedom (DOF) motion of the tibia with respect to the femur. Known external loads, as determined by load and torque cells, are applied to the tibia, while linear and rotational transducers or six-DOF goniometers measure the resulting knee kinematics in each degree of freedom. Subsequently, a ligament is sectioned and the external loads are again applied to the tibia. By comparing the changes in knee kinematics, the kinematic contribution of that structure can be determined for the given loading condition. The kinematics of the intact knee can also serve as the "gold standard" for the evaluation of reconstructive techniques for the PCL. It is important to recognize when performing kinematic studies, however, that constraining joint motion in one or more DOF can yield vastly different data on joint kinematics.[72,73]

These cutting studies have demonstrated that the primary function of the PCL is to restrain posterior tibial translation.[21,34,35,37,70] Under posterior tibial loading, a 1 to 11 mm increase in posterior tibial translation is observed when the PCL is sectioned, and this effect increases with knee flexion angle. The PCL has also been shown to work in conjunction with the PLS to provide knee stability.[34–37] In response to a posterior tibial load, isolated sectioning of the PLS resulted in less than a 3-mm increase in posterior tibial translation, while isolated sectioning of the PCL resulted in up to 10 mm of posterior tibial translation. However, after combined sectioning of the PCL and PLS, posterior tibial translation was as high as 25 to 30 mm. The PCL and PLS also work together to restrain external and varus rotations as well. In response to varus and external tibial moments, the PLS provided the majority of the restraint in the intact knee, while only small increases in rotation occurred after sectioning the PCL alone. However, large increases in both rotations (up to 7 degrees of varus and 14 degrees of external rotation) were observed with combined deficiency of both the PCL and PLS.[35,36] These results suggest that the PCL and PLS work synergistically in restraining these knee motions.

For each of the loading conditions, the PCL provided more restraint to both posterior translation and varus and external rotations at high flexion angles, while the PLS was important near full extension.[35] These findings are consistent with anatomic observations of the PCL and PLS, as the PCL is more taut with the knee in flexion, while the PLS is more taut with the knee near extension, in response to posterior tibial loads.[26–28] Veltri et al.[37] found no significant contributions of either the PCL or PLS in restraining anterior tibial loads or internal or valgus rotations.

Joint Contact Forces

While deficiency of the PCL significantly affects knee kinematics, biomechanical studies have addressed other implications of PCL and PLS deficiency as well. In 1993, Skyhar et al.[73] investigated the effects of sectioning the PCL on articular contact pressures within the knee. Ten human cadaveric knees were tested in response to a non–weight-bearing rehabilitation exercise loading model. The resulting contact pressures in the patellofemoral and medial compartments were measured using a pressure-sensitive film. It was found that patellofemoral pressures were significantly elevated when the PCL and PLS were deficient, while medial compartment pressures increased significantly with isolated PCL deficiency. These results are in agreement with several clinical studies that have reported that patients with nonoperatively treated PCL injuries develop osteoarthritis of the patellofemoral and medial compartments.[5,13,14]

Measurement of In Situ Forces

While cutting studies and strain measurements provide important information on the function of the PCL, they do not provide data on the force, or tension, in the PCL when external loads are applied to the knee. These data are of great interest, particularly during ligament reconstructions, as the surgeon wants to replicate the condition of the intact knee as much as possible. As a result, several studies have focused not only on kinematic changes, but also the in situ forces in the PCL, in response to externally applied loads.[74–76,80–82]

A number of methods have been employed over the past 15 years to determine the in situ force in the PCL. These include directly implanted devices, such as buckle transducers[74,79] and pressure probes.[77,82] The limitations of direct methods, however, are that contact with the ligament is required, and that forces can be measured only in a portion of the ligament and not the ligament as a whole. Other investigators have utilized indirect methods to measure forces in ligaments in which an external load is applied to the tibia, while the lengths of various portions of the ligament are measured.[78,83] The force in the ligament is then calculated based on the load-elongation curve of the ligament. However, ligament forces that oc-

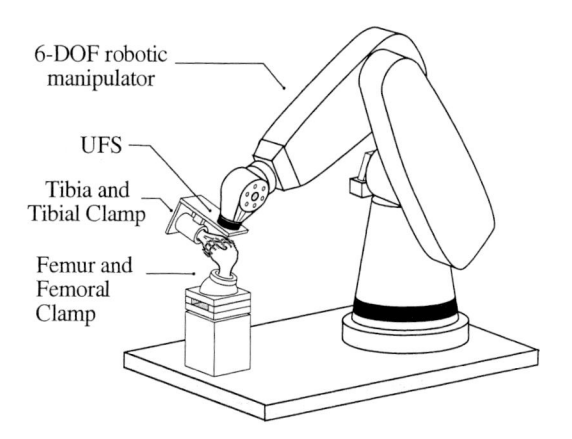

Fig. 1.8. Schematic of the robotic/Universal Force-Moment Sensor (UFS) testing system with a specimen in place.[72] DOF, degrees of freedom. (From Harner et al.,[72] with permission.)

cur in situ may not be repeated as the ligament is tensile tested. More recently, Markolf and colleagues[80,81,84] have developed a method to determine forces in the PCL in which a load cell is rigidly fixed to the subchondral bone at the femoral insertion of the PCL.

Robotics/UFS Testing System

In our research center, we have developed an approach that combines a Universal Force-Moment Sensor (UFS, JR[3], Woodland Hills, CA) with a six-DOF articulated robotic manipulator, called the robotic/UFS testing system (Fig. 1.8).[85–87] This system provides a direct, noncontact method for the determination of in situ force in the entire ligament. The robot is capable of controlling joint motion in six DOF, while the UFS can measure three orthogonal forces and moments, and provides force-moment feedback to the robot. In combination, this system can measure multiple DOF knee kinematics and in situ forces in the PCL in response to various externally applied loads.[87]

To determine the in situ force in the PCL, an external load is first applied to the joint at chosen knee flexion angles while the robot measures the resulting kinematics. The PCL is then sectioned arthroscopically. The previously determined kinematics are repeated by the robot while the UFS measures the resulting forces and moments. Because identical knee positions are repeated before and after sectioning the PCL, the vector decrease in forces measured by the UFS yields the in situ force in the ligament using the principle of superposition.[87] The direction and point of application of the force can also be determined. An advantage of this system is that all effects, such as that of sectioning a ligament or of a reconstruction, can be tested in the same knee. This minimizes interspecimen variability, making statistical comparisons more powerful and reducing the number of specimens needed.

In Situ Force in the PCL

Using the robotic/UFS testing system, we have measured the in situ force in the PCL during passive flexion-extension of the knee, as well as in response to various external loading conditions. During passive flexion-extension of the knee, in situ forces in the PCL ranged from 6 ± 5 N at 30 degrees of flexion to 15 ± 3 N at 90 degrees of flexion (mean $\pm$ standard deviation [SD], $n = 9$). No significant effect of knee flexion on these forces could be demonstrated ($p > .05$).[76]

We also determined the in situ force in the PCL and the distribution between its AL and PM bundles in response to a 110-N posterior tibial load in nine human cadaveric knees.[75] It was found that the in situ force in the

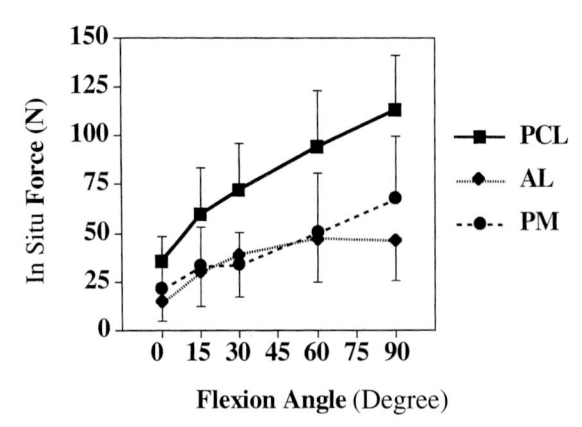

Fig. 1.9. In situ force in the PCL and in the AL and PM bundles under 110 N posterior tibial load with respect to flexion angle. (From Harner et al.,[75] with permission.)

Fig. 1.10. In situ force in the PCL (mean ± standard deviation [SD]) in response to isolated hamstrings and combined hamstrings/quadriceps muscle loads. (From Höher et al.,[76] with permission.)

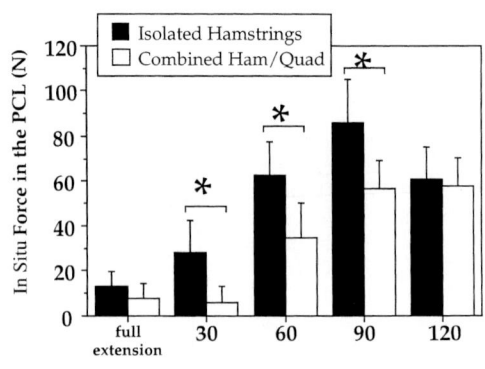

PCL increases with knee flexion, ranging from 36 N at 0 degrees to 112 N at 90 degrees of flexion. However, there were no significant differences between in situ forces in the AL and PM bundles at any flexion angle (Fig. 1.9). These findings are consistent with those of other investigators.[80–82] Markolf et al.[80] also found a significant increase in PCL force with knee flexion in response to several combined external loading conditions (e.g., posterior tibial load and internal/external or varus valgus moments).

Using the robotic/UFS testing system, we also investigated the effects of hamstrings and quadriceps muscle loads on the in situ force in the PCL.[76] The hamstrings (40 N biceps, 40 N semimembranosus) and quadriceps loads (200 N) were applied via a cable and pulley system. In response to an isolated hamstrings load, the in situ force in the PCL increased with knee flexion, ranging from 12 ± 5 N at 0 degrees to 80 ± 20 N at 90 degrees of flexion. With the addition of a 200 N quadriceps load (i.e., combined hamstrings/quadriceps load), these forces decreased significantly by 23 to 31 N at 30 to 90 degrees of flexion (Fig. 1.10).

Because kinematic studies have suggested that a significant interaction exists between the PCL and PLS, we also investigated, using the robotic/UFS testing system, the importance of the PLS. We first investigated the contribution of the popliteus muscle by determining the effects of a 44-N popliteus muscle load on knee kinematics and in situ forces in the PCL.[89] Under a 110-N posterior tibial load, the popliteus muscle load significantly reduced in situ force in the PCL by 36% to 9% at 30 and 90 degrees of knee flexion, respectively (Fig. 1.11). The popliteus muscle load also significantly reduced posterior tibial translation of the PCL-deficient knee by 1 to 3 mm at all flexion angles tested. In our research center and others,[36,81] the effects of deficiency of the PLS on knee kinematics and in situ forces in the PCL have also been investigated. PLS deficiency resulted in significant increases in external tibial rotation and posterior tibial translation in response to an external tibial moment or posterior tibial load, respectively. Further, in situ forces in the PCL were significantly higher (up to six times) in the PLS-deficient knee compared to the intact knee for each loading condition.

Variables of PCL Reconstruction

Biomechanical studies have also investigated the effects of several important factors of PCL reconstruction on kinematics of the reconstructed knee and in situ forces in the PCL replacement graft. A successful PCL reconstruction should restore the kinematics of the intact knee as well as replicate the in situ forces in the intact PCL. Among the variables of PCL re-

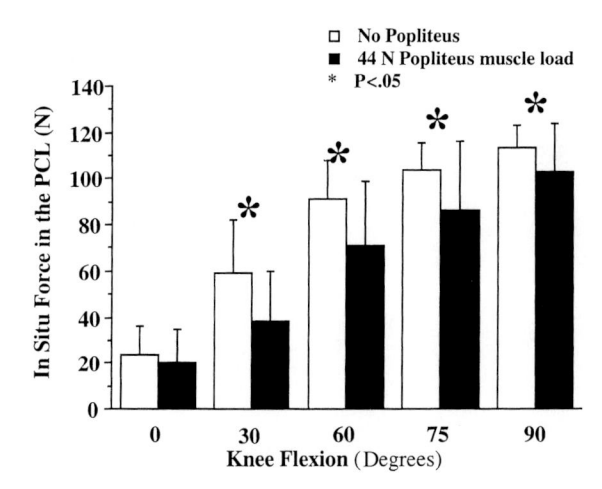

Fig. 1.11. Effects of a 44-N popliteus muscle load on in situ force in the PCL. (From Harner et al.,[88] with permission.)

construction that have been investigated in this manner are tunnel placement,[89–92] knee flexion angle at the time of graft tensioning,[89,93] and deficiency of the PLS.[94]

Effect of Graft Tunnel Placement

The insertion of the PCL onto the femur occupies a large area.[27,33,96] Thus, a number of possibilities exist for the placement of a tunnel 10 mm in diameter within the anatomic insertion of the PCL. While small changes in the placement of the tibial tunnel are not believed to significantly affect the outcome of PCL reconstruction, femoral tunnel placement has been shown to be an important factor.[91,95–98] Some authors have advocated an isometric tunnel placement, i.e., one in which the PCL graft maintains a constant length throughout the range of flexion,[99,100] while others prefer a more anatomic placement, in which the tunnel is drilled through the original tibial and femoral insertion of the AL bundle of the PCL[5,89,91,94,97,101] (Fig. 1.12).

This variable has been addressed in numerous biomechanical studies.[90–92,96,98,99,101–106] Several studies have measured ligament lengths in various regions of the PCL and for different tunnel placements.[90,95,97,100,101] The isometric point of the PCL was found to be located more proximal and posterior than the anatomic insertion,[89–91,95,97,98,103,106–108] and thus an isometric reconstruction would restore the posterior fibers of the PCL.[90,95,97] Ligament length studies have shown, however, that the bulk of the PCL, particularly its large anterior and central portions, is nonisometric.[90,98,103] Other biomechanical studies have varied tunnel placements within the same knee and compared the resulting kinematics to those of the intact knee.[89,91,105] In agreement with the observation that the PCL is a nonisometric structure, isometric reconstruction has been less effective in restoring posterior tibial translation than the anatomically placed graft.[89,91,92,105] It has also been observed that moving the femoral attachment in the proximal-distal direction had a greater effect on knee kinematics and tension in the PCL graft than did variation in the AP direction.[90,96,98,102]

Effect of Graft Tensioning Technique

There are very few data on the optimal position of the knee at the time of graft fixation. This is a critical issue, as we know from the functional anatomy that the PCL is under the greatest tension with the knee in flexion. Several authors have recommended that graft fixation be performed

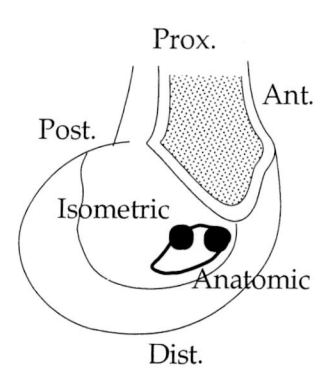

Fig. 1.12. Sagittal view of a medial femoral condyle showing anatomic and isometric tunnel positions for graft fixation. (From Stone et al.,[105] with permission.)

Fig. 1.13. Posterior tibial translation of the PCL-deficient and PCL-reconstructed knees in response to a 134-N posterior tibial load, expressed as a ratio of the reconstructed to the intact knee. (From Harner et al.,[93] with permission.)

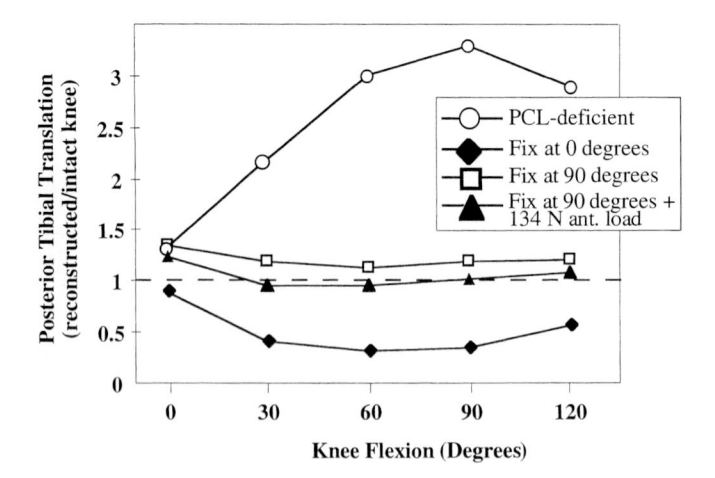

with the knee near full extension[107,109,110] or in mid-flexion[8,20,23,92,111,112] to ensure minimal graft impingement. Others tension the graft with the knee at 90 degrees of flexion[5,19,89,94,113] to more accurately re-create the normal tension in the PCL in flexion.

In our research center, we investigated the effects of the knee flexion angle as well as the application of an anterior drawer at the time of graft fixation using the robotic/UFS testing system. Ten knees were tested in response to a 134-N posterior tibial load at flexion angles of full extension, and 30, 60, 90, and 120 degrees. Reconstruction of the AL bundle of the PCL was performed using an Achilles tendon graft. Three different reconstructions were evaluated within the same knee with the grafts fixed at (1) full extension, (2) 90 degrees of flexion, and (3) 90 degrees of flexion with a 134-N anterior drawer applied. Both knee flexion angle and application of the anterior drawer at the time of graft fixation significantly affected the kinematics of the reconstructed knee (Fig. 1.13). Graft fixation with the knee at full extension resulted in an overconstrained knee, as it significantly reduced posterior tibial translation to 1 to 4 mm less than the intact knee. Further, in situ forces in the graft were significantly higher than those in the intact PCL at all knee flexion angles for this reconstruction. The graft fixed at 90 degrees of flexion with a 134-N anterior load applied to the tibia best restored both knee kinematics and in situ forces in the PCL. This observation is consistent with that of Burns et al.,[89] who also reported that this technique best restored intact knee kinematics.

Effect of PLS Deficiency

Our own clinical observations as well as those of others have noted that a large number of PCL injuries are accompanied by damage to the PLS.[9] To study the effects of PLS deficiency, we evaluated a PCL reconstruction in isolated PCL and combined PCL/PLS injury models.[94] The AL bundle of the PCL was reconstructed using an Achilles tendon graft fixed at 90 degrees of flexion with a 134-N anterior drawer applied. Under a 134-N posterior tibial load, PCL reconstruction reduced posterior tibial translation to within 1 to 2 mm of the intact knee at 30 and 90 degrees of flexion, respectively (Fig. 1.14). After sectioning the PLS, however, posterior tibial translation increased by an additional 4 to 6 mm. The in situ forces in the PCL did not differ significantly from the in situ forces in the PCL graft when the PLS was intact. However, sectioning the PLS significantly increased the in situ force in the PCL graft by 32 to 40 N. Similar results were observed in response to applied external and varus moments.

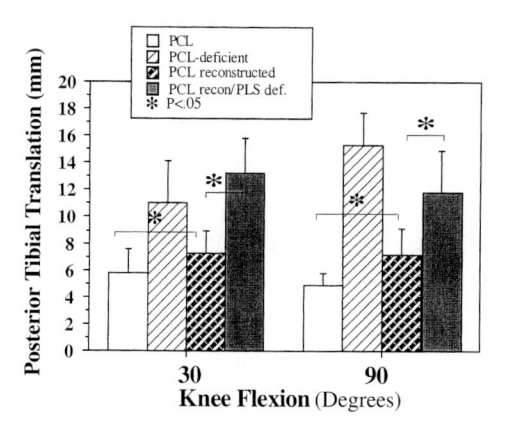

Fig. 1.14. Posterior tibial translation (mean ± SD) in response to a 134-N posterior tibial load in the PCL reconstructed knee. (From Harner et al.,[94] with permission.)

Clinical Relevance

The results of these basic science studies have numerous implications for clinical management of injuries involving the PCL. The anterolateral bundle of the PCL was found to be anatomically and biomechanically more significant,[30,31] thus suggesting that it should be the focus of current single bundle techniques. Further, several studies have revealed that the PCL is not isometric but undergoes increasing in situ force with knee flexion.[75,81] To reproduce in situ forces in the intact PCL, we hypothesized that the replacement graft should be tensioned and fixed with the knee in flexion (70–90 degrees). Our robotic study confirmed that grafts fixed with the knee in extension, as often performed clinically, resulted in an overconstrained knee and increased in situ forces, thus putting the graft at risk for failure.[93]

Biomechanical studies have also clearly demonstrated the importance of the PLS to PCL function. An active popliteus muscle within the PLS induces a load-sharing effect with the PCL and can contribute to knee stability when the PCL is absent.[88] On the other hand, if the PLS is absent, significant increases in posterior and posterolateral rotatory knee laxity will occur, as well as increased in situ forces in the PCL.[36,81]

These findings suggest that the PCL and PLS should both be addressed when combined injury occurs so as to effectively restore intact knee kinematics and forces. Clinically, the diagnosis of a combined PCL/PLS injury can be difficult. Due to the interaction of the PCL and PLS, the posterior drop-back of the tibia can mask increased posterolateral laxity.[18,20,114] Further, due to its complex anatomy, injuries to the PLS are difficult to repair.[38,115–118] However, if surgical treatment is performed and only addresses the PCL deficiency, the graft cannot restore intact knee kinematics and will experience increased in situ forces.[97] This information has guided our clinical approach to PCL injuries to repair and reconstruct all associated ligament injuries, especially those to the PLS.

Finally, various muscle loads have been shown to affect in situ forces in the PCL,[76] as well as strain.[119] Both the strain and in situ force in the PCL increase with knee flexion under hamstrings loads, but the addition of a quadriceps load reduces this effect. These findings support the clinical theory that initial rehabilitation after PCL injury or reconstruction should stress the quadriceps muscle and be performed with the knee near full extension.

The results of these biomechanical studies have greatly influenced our surgical approach to injuries involving the PCL. We prefer to use an Achilles tendon allograft to reconstruct the anterolateral bundle of the PCL (Fig.

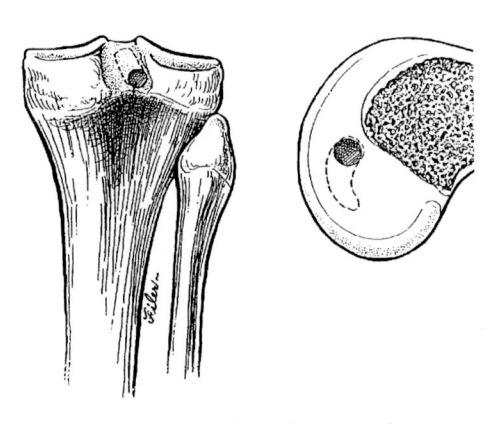

Fig. 1.15. Schematic indicating placement of tunnels within femoral insertion of the PCL. (From Fu et al.,[125] with permission.)

1.15). At the time of graft tensioning and fixation, the knee is placed at 70 to 90 degrees of flexion and an anterior drawer is applied to the tibia. Injuries to the posterolateral structures are repaired or reconstructed at the time of PCL reconstruction.

Future Directions

Knowledge gained on the anatomy and biomechanics of the PCL over the past 10 years has provided valuable information to surgeons treating knee injuries involving the PCL. However, clinical management and treatment of PCL injuries remain inconsistent, and current reconstructive techniques are unable to fully restore the kinematics of the intact knee.[5,7,18,20,22,24,25] These variable outcomes reflect a need for additional information on the biomechanics of the PCL and its functional components.

Recently, several surgeons have performed newer reconstructive techniques, such as double-bundle reconstruction in which both the AL and PM bundles of the PCL are considered. It is believed that this approach will provide additional restraint to posterior tibial translation, particularly when the knee is near extension.[120-122] Another method, the direct or inlay technique, involves direct fixation of the bone block of the graft onto the tibial insertion of the PCL. It is believed that this will alleviate the stresses on the graft that arise as the graft wraps through the tunnel and onto the anterior aspect of the tibia.[123] However, these methods have not yet been the subject of rigorous biomechanical studies. Therefore, additional investigation will need to be performed to answer the question of whether the benefits provided by these new methods merit their increased surgical time and technical difficulty.

Because of the detrimental effect of an associated PLS injury to the PCL replacement graft, methods to reconstruct the PLS must be studied as well. With its complex anatomy and the variable injury patterns for the PLS, the available reconstructive procedures, including tenodesis, advancements, and popliteal bypass,[116] must be subjected to careful biomechanical evaluation to prove their efficacy.

Finally, it is necessary to look into techniques whereby the in vitro data on knee kinematics can be applied to in vivo situations. We believe that it is possible to use the kinematics of the knee in vivo using gait or motion analysis techniques and repeat them on a cadaveric knee using a system such as the robotic/UFS testing system. Thus, the in situ forces in the PCL and PLS during activities of daily living can be obtained, as well as the effects of PCL or PLS deficiency or PCL reconstruction.

Acknowledgments

The authors gratefully acknowledge the financial support of the Whitaker Foundation, the Orthopaedic Research and Education Foundation, and the Musculoskeletal Research Center. We also acknowledge the cast of engineers, students, orthopedic residents, fellows and surgeons, whose skills, ideas, and suggestions have contributed to the work presented in this chapter: Asbjorn Aroen, M.D., Goo Hyun Baek, M.D., Gregory Carlin, M.S., Ross Fox, M.D., Jürgen Höher, M.D., Marsie Janaushek, B.S., Akihiro Kanamori, M.D., Shinji Kashiwaguchi, M.D., Glen Livesay, Ph.D., Benjamin Ma, M.D., Ted Rudy, M.A., Jeffrey Stone, M.D., and Jonathan Woo.

References

1. Harner CD, Olson JJ, Irrgang JJ, Silverstein S, Fu FH, Silbey M. Allograft versus autograft anterior cruciate ligament reconstruction. Clin Orthop 1996;324:134–144.
2. Shelbourne KD, Grey T. Anterior cruciate ligament reconstruction with autogenous patellar tendon graft followed by accelerated rehabilitation. A two- to nine-year follow-up. Am J Sports Med 1997;25:786–795.
3. Shelbourne KD, Nitz P. Accelerated rehabilitation after anterior cruciate ligament reconstruction. Am J Sports Med 1990;18:292–299.
4. Bianchi M. Acute tears of the posterior cruciate ligament: clinical study and results of operative treatment in 27 cases. Am J Sports Med 1983;11:308–314.
5. Clancy WG, Shelborne KD, Zoellner GB, Keene JS, Reider B, Rosenberg TD. Treatment of knee joint instability secondary to rupture of the posterior cruciate ligament. J Bone Joint Surg 1983;65A:310–322.
6. Clendenin MB, DeLee JC, Heckman JD. Interstitial tears of the posterior cruciate ligament of the knee. Orthopaedics 1980;3:764–772.
7. Harner CD, Höher J. Evaluation and treatment of posterior cruciate ligament injuries. Am J Sports Med 1998;26:471–482.
8. Kennedy JC, Grainger RW. The posterior cruciate ligament. J Trauma 1967;7:367–377.
9. Fanelli GC, Edson CJ. Posterior cruciate ligament injuries in trauma patients: part II. Arthroscopy 1995;11:526–529.
10. Cross MJ, Fracs MB, Powell JF. Long-term follow-up of posterior cruciate ligament rupture: a study of 116 cases. Am J Sports Med 1984;12:292–297.
11. Degenhardt TC, Hughston JC. Chronic posterior cruciate instability: nonoperative management. Orthop Trans 1981;5:486–487.
12. Keller PM, Shelbourne DK, McCarroll JR, Retting AC. Nonoperative treated isolated posterior cruciate ligament injuries. Am J Sports Med 1993;21:132–136.
13. Parolie JM, Bergfeld JA. Long-term results of nonoperative treatment of isolated posterior cruciate ligament injuries in the athlete. Am J Sports Med 1986;14:35–38.
14. Dandy DJ, Pusey R. The long-term results of unrepaired tears of the posterior cruciate ligament. J Bone Joint Surg 1982;64B:92–94.
15. Dejour H, Walch G, Peyrot J, Eberhard P. The natural history of rupture of the posterior cruciate ligament. Rev Chir Orthop Repar Appareil Moteur 1988;74:35–43.
16. Barrett GR, Savoie FH. Operative management of acute PCL injuries with associated pathology: long-term results. Orthopaedics 1991;14:687–692.
17. Fleming RE, Blatz DJ, McCarroll JR. Posterior problems in the knee: posterior cruciate insufficiency and posterolateral rotatory insufficiency. Am J Sports Med 1981;9:107–113.
18. Harner CD, Miller M. Isolated PCL reconstruction using fresh-frozen allograft. Paper presented at AOSSM specialty day, New Orleans, LA, 1993.
19. Insall JN, Hood RW. Bone-block transfer of the medial head of the gastrocnemius for posterior cruciate insufficiency. J Bone Joint Surg 1982;64A:691–699.
20. Lipscomb AB Jr, Anderson AF, Norwig ED, Hovis WD, Brown DL. Isolated posterior cruciate ligament reconstruction. Long-term results. Am J Sports Med 1993;21:490–496.
21. Noyes SS, Grood ES, Cummings J, VanGinkel LA. Posterior subluxations of the medial and lateral tibiofemoral compartments: an in vitro ligament sectioning study in cadaveric knees. Am J Sports Med 1993;21:407–414.
22. Torg JS, Barton TM, Pavlov H, Stine R. Natural history of the posterior cruciate ligament-deficient knee. Clin Orthop 1982;164:59–77.
23. Hughston JC, Bowden JA, Andrews JA, Norwood LA. Acute tears of the posterior cruciate ligament. J Bone Joint Surg 1980;62A:438–450.
24. Loos WC, Fox JM, Blazina ME, Del Pizzo W, Friedman MJ. Acute posterior cruciate ligament injuries. Am J Sports Med 1981;9:86–92.
25. Wilk KE. Rehabilitation of isolated and combined posterior cruciate ligament injuries. Clin Sports Med 1994;13:649–677.

26. Girgis FG, Marshall JL, Al Monajem ARS. The cruciate ligaments of the knee joint: anatomical, functional and experimental analysis. Clin Orthop 1975;106:216–231.
27. Van Dommelen BA, Fowler PJ. Anatomy of the posterior cruciate ligament: a review. Am J Sports Med 1989;14:24–29.
28. Barton TM, Torg JS, Das M. Posterior cruciate ligament insufficiency. A review of the literature. Am J Sports Med 1984;12:419–430.
29. Covey DC, Sapega AA. Current concepts review: injuries of the posterior cruciate ligament. J Bone Joint Surg 1993;75A:1376–1386.
30. Harner CD, Xerogeanes JW, Livesay GA, et al. The human posterior cruciate ligament complex: an interdisciplinary study—ligament morphology and biomechanical evaluation. Am J Sports Med 1995;23:736–745.
31. Race A, Amis AA. The mechanical properties of the two bundles of the human posterior cruciate ligament. J Biomech 1994;27:13–24.
32. Harner CD, Baek GH, Vogrin TM, Carlin GJ, Kashiwaguchi S, Woo SL-Y. Quantitative analysis of human cruciate ligament insertions. Arthroscopy 1999;15(7):741–749.
33. Kusayama T, Harner CD, Carlin GJ, Xerogeanes JW, Smith BA. Anatomical and biomechanical characteristics of human meniscofemoral ligaments. Knee Surg Sports Traum Arthrosc 1994;2:234–237.
34. Gollehon DL, Torzilli PA, Warren RF. The role of the posterolateral and cruciate ligaments in the stability of the human knee. J Bone Joint Surg 1987;69A:233–242.
35. Grood ES, Stowers SF, Noyes FR. Limits of movement in the human knee: the effects of sectioning the PCL and posterolateral structures. J Bone Joint Surg 1988;70A:88–97.
36. Vogrin TM, Höher J, Aroen A, Carlin GJ, Woo SL-Y, Harner CD. Effects of sectioning the posterolateral structures on knee kinematics and in situ forces in the posterior cruciate ligament. Knee Surg, Sports Trauma, Arthrosc 2000;8:93–98.
37. Veltri DM, Deng X-H, Torzilli PA, Warren RF, Maynard MJ. The role of the cruciate and posterolateral ligaments in the stability of the knee. Am J Sports Med 1995;23:436–443.
38. Seebacher JR, Inglis AE, Marshall JL, Warren RF. The structure of the posterolateral aspect of the knee. J Bone Joint Surg 1982;64A:536–541.
39. Staubli HU. Posteromedial and posterolateral capsular injuries associated with posterior cruciate ligament insufficiency. Sports Med Arthrosc Rev 1994;2:146–164.
40. Staubli HU, Birrer S. The popliteus tendon and its fascicles at the popliteal hiatus: gross anatomy and functional arthroscopic evaluation with and without anterior cruciate ligament deficiency. J Arthrosc Rel Surg 1990;6:209–219.
41. Watanabe Y, Moriya H, Takahashi K, et al. Functional anatomy of the posterolateral structures of the knee. J Arthrosc Rel Surg 1993;9:57–62.
42. Heller L, Langman J. The meniscofemoral ligaments of the human knee. J Bone Joint Surg 1964;46B:307–313.
43. Odensten M, Gillquist J. Functional anatomy of the anterior cruciate ligament and a rationale for reconstruction. J Bone Joint Surg 1985;67A:257–261.
44. Gratz CM. Tensile strength and elasticity tests on human fascia lata. J Bone Joint Surg 1931;13:334–340.
45. Nunley RL. The ligament flava of the dog. A study of tensile and physical properties. Am J Phys Med 1958;37:256–268.
46. Wright DG, Renneis DC. A study of the elastic properties of plantar fascia. J Bone Joint Surg 1964;46A:482–492.
47. Allard P, Thirty PS, Bourgault A, Drouin G. Pressure dependence of the area micrometer method in evaluation of cruciate ligament cross-section. J Biomed Eng 1979;1:265–267.
48. Butler DL, Grood ES, Noyes FR, Zernicke RF, Brackett K. Effects of structure and strain measurement technique on the material properties of young human tendons and fascia. J Biomech 1984;17:579–596.
49. Ellis DG. A shadow amplitude method for measuring cross sectional area of biological specimens. Annu Conf Eng Med Biol 1986;51:6.
50. Walker LB, Harris EH, Benedict JV. Stress-strain relations in human cadaveric plantaris tendon: a preliminary study. Med Electron Biol Eng 1964;2:31–38.

51. Race A, Amis AA. A molding method to find cross-sections of soft tissue bundles with complex shapes. Trans ORS 1994;19:783.
52. Gupta BN, Subramanian KN, Brinker WO, Gupta AN. Tensile strength of canine cranial cruciate ligaments. Am J Vet Res 1971;32:183–190.
53. Iaconis F, Steindler R, Marinozzi G. Measurements of cross-sectional area of collagen structures (knee ligaments) by means of an optical method. J Biomech 1987;20:1003–1010.
54. Njus GO, Njus NM. A noncontact method for determining cross-sectional area of soft tissues. Trans ORS 1986;11:126.
55. Lee TQ, Woo SL-Y. A new method for determining cross-sectional shape and area of soft tissues. J Biomech Eng 1988;110:110–114.
56. Woo SL-Y, Danto MI, Ohland KJ, Lee TQ, Newton PO. The use of a laser micrometer system to determine the cross-sectional shape and area of ligaments: a comparative study with two existing methods. J Biomech Eng 1990;112:426–431.
57. Morgan CD, Kalman VR, Grawl DM. Definitive landmarks for reproducible tibial tunnel placement in anterior cruciate ligament reconstruction. Arthroscopy 1995;11:275–288.
58. Morgan CD, Kalman VR, Grawl DM. The anatomic origin of the posterior cruciate ligament: where is it? Reference landmarks for PCL reconstruction. Arthroscopy 1997;13:325–331.
59. Craig AS, Eikenberry EF, Parry DAD. Ultrastructural organization of skin: classification on the basis of mechanical role. Connect Tissue Res 1987;16:213–223.
60. Flint MH, Craig AS, Reilly HC, Gillard GC, Parry DAD. Collagen fibril diameters and glycosaminoglycan content of skin: indices of tissue maturity and function. Connect Tissue Res 1984;13:69–81.
61. Oakes BW. Collagen ultrastructure in the normal ACL and ACL graft. In Jackson DW, ed. The Anterior Cruciate Ligament: Current and Future Concepts. New York: Raven Press, 1993:209–217.
62. Parry DAD, Barnes GRG, Craig AS. A comparison of the size distribution of collagen fibrils in connective tissues as a function of age and a possible relation between fibril size distribution and mechanical properties. Proc R Soc Lond Biol Sci 1978;203:293–303.
63. Binkley JM, Peat M. The effect of immobilization on the ultrastructure and mechanical properties of the medial collateral ligaments of rats. Clin Orthop 1986;203:301–308.
64. Christel PS, Gibbons DF. Collagen fiber changes in the exercised, immobilized or injured anterior cruciate ligament. In: Jackson DW, ed. The Anterior Cruciate Ligament: Current and Future Concepts. New York: Raven Press, 1993:195–208.
65. Neurath M, Stofft E, Zschabitz A, Printz H. Comparative microstructural studies on collagen and elastic fiber systems of the cruciate ligaments. Z Unfallchir Versicherungsmed 1991;84:170–176.
66. Strocchi R, Pasquale VD, Gubellini P. The human anterior cruciate ligament: histological and ultrastructural observations. J Anat 1992;180:515–519.
67. Baek GH, Carlin GJ, Vogrin TM, Woo SL-Y, Harner CD. A quantitative analysis of collagen fibrils of the cruciate and meniscofemoral ligaments. Clin Orthop 1998;357:205–211.
68. Kennedy JC, Hawkins RJ, Willis RB, Danylchuk KD. Tension studies of human knee ligaments: yield point, ultimate failure, and disruption of the cruciate and tibial collateral ligaments. J Bone Joint Surg 1976;58A:350–355.
69. Prietto MP, Bain JR, Stonebrook SN, Settlage RA. Tensile strength of the human posterior cruciate ligament (PCL). Trans ORS 1988;13:195.
70. Butler DL, Noyes FR, Grood ES. Ligamentous restraints to anterior-posterior drawer in the human knee. J Bone Joint Surg 1980;62A:259–270.
71. Inoue M, McGurk-Burleson E, Hollis JM, Woo SL-Y. Treatment of the medial collateral ligament injury. I: The importance of anterior cruciate ligament on the varus-valgus knee laxity. Am J Sports Med 1987;15:15–21.
72. Livesay GA, Rudy TW, Woo SL-Y, et al. Evaluation of the effect of joint constraints on the in situ force distribution in the anterior cruciate ligament. J Orthop Res 1997;15:278–284.
73. Skyhar M, Warren R, Ortiz G, Schwartz E, Otis J. The effects of sectioning of the posterior cruciate ligament and the posterolateral complex on the ar-

ticular contact pressures within the knee. J Bone Joint Surg 1993;75A:694–699.
74. Barry D, Ahmed AM. Design and performance of a modified buckle transducer for the measurement of ligament tension. J Biomech Eng 1986;108:149–152.
75. Fox RJ, Harner CD, Sakane M, Carlin GJ, Woo SL-Y. In situ forces in the human posterior cruciate ligament: a cadaveric study. Am J Sports Med 1998;26:395–401.
76. Höher J, Vogrin TM, Woo SL-Y, Carlin GJ, Aroen A, Harner CD. In situ forces in the human posterior cruciate ligament in response to muscle loads: a cadaveric study. J Orthop Res 1999;17:763–768.
77. Holden JP, Grood ES, Korvick DL, Cummings JF, Butler DL, Bylski-Austrow DI. In-vivo forces in the anterior cruciate ligament: direct measurements during walking and trotting in a quadruped. J Biomech 1994;27:517–526.
78. Hollis MJ, Takai S, Adams DJ, Horibe S, Woo SL-Y. The effects of knee motion and external loading on the length of the anterior cruciate ligament: a kinematic study. J Biomech Eng 1991;113:208–214.
79. Lewis JL, Lew WD, Schmidt J. A note on the application and evaluation of the buckle transducer for knee ligament force measurement. J Biomech Eng 1982;104:125–128.
80. Markolf KL, Slauterbeck JR, Armstrong KL, Shapiro MS, Finerman GA. Effects of combined knee loadings on posterior cruciate ligament force generation. J Bone Joint Surg 1996;14:633–638.
81. Markolf KL, Wascher DC, Finerman GAM. Direct in vitro measurement of forces in cruciate ligaments. Part II: the effect of section of the posterolateral structures. J Bone Joint Surg 1993;75A:387–394.
82. Race A, Amis AA. Loading of the two bundles of the posterior cruciate ligament: an analysis of bundle function in A-P drawer. J Biomech 1996;29:873–879.
83. Takai S, Woo SL-Y, Livesay GA, Adams DJ. Determination of load in the human anterior cruciate ligament. Trans ORS 1991;16:235.
84. Markolf KL, Gorek JF, Kabo JM, Shapiro MS. Direct measurement of resultant forces in the anterior cruciate ligament. J Bone Joint Surg 1990;72A:557–567.
85. Fujie H, Livesay GA, Woo SL-Y, Kashiwaguchi S, Blomstrom G. The use of a universal force-moment sensor to determine in-situ forces in ligaments: a new methodology. J Biomech Eng 1995;117:1–19.
86. Fujie H, Mabuchi K, Woo SL-Y, Livesay GA, Arai S, Tsukamoto Y. The use of robotics technology to study human joint kinematics: a new methodology. J Biomech Eng 1993;115:211–217.
87. Rudy TW, Livesay GA, Woo SL-Y, Fu FH. A combined robotic/universal force sensor approach to determine in situ forces of knee ligaments. J Biomech 1996;29:1357–1360.
88. Harner CD, Höher J, Vogrin TM, Carlin GJ, Woo SL-Y. Effects of a popliteus muscle load on in situ forces in the PCL and knee kinematics: a cadaveric study. Am J Sports Med 1998;26(5):669–673.
89. Burns WC, Draganich LF, Pyevich M, Reider B. The effect of femoral tunnel position and graft tensioning technique on posterior laxity of the posterior cruciate ligament-reconstructed knee. Am J Sports Med 1995;23:424–430.
90. Covey DC, Sapega AA, Sherman GM. Testing for isometry during reconstruction of the posterior cruciate ligament. Am J Sports Med 1996;24:740–746.
91. Galloway MT, Grood ES, Mehalik JN, Levy M, Saddler SC, Noyes FR. Posterior cruciate ligament reconstruction: in vitro study of femoral and tibial graft placement. Am J Sports Med 1996;24:437–445.
92. Race A, Amis AA. PCL reconstruction: in vitro biomechanical evaluation of isometric versus single and double bundled anatomic grafts. J Bone Joint Surg 1998;80B:173–179.
93. Harner CD, Janaushek MA, Ma CB, Kanamori A, Vogrin TM, Woo SL-Y. The effects of knee flexion angle and application of an anterior tibial load on the biomechanics of a posterior cruciate ligament-reconstructed knee. Am J Sports Med 2000;28:460–465.
94. Harner CD, Vogrin TM, Höher J, Ma CB, Woo SL-Y. Biomechanical analy-

sis of a posterior cruciate ligament reconstruction: deficiency of the posterolateral structures as a cause of graft failure. Am J Sports Med 2000;28:32–39.
95. Sidles JA, Larson RV, Garbini JL, Downey DJ, Matsen FA III. Ligament length relationships in the moving knee. J Orthop Res 1988;6:593–610.
96. Bach BR, Daluga DJ, Mikosz R, Andriacchi TP, Seidl R. Force displacement characteristics of the posterior cruciate ligament. Am J Sports Med 1992;20:67–72.
97. Grood ES, Hefzy MS, Lindenfield TN. Factors affecting the region of most isometric femoral attachments. Part 1. The posterior cruciate ligament. Am J Sports Med 1989;20:351–355.
98. Hefzy MS, Grood ES, Lindenfield TL. The posterior cruciate ligament: a new look at length patterns. Trans ORS 1986;11:128.
99. Markolf KL, Slauterbeck JR, Armstrong KL, Shapiro MS, Finerman GA. A biomechanical study of replacement of the posterior cruciate ligament with a graft: forces in the graft compared with forces in the intact ligament. J Bone Joint Surg 1997;79A:381–386.
100. Petermann J, Gotzen L, Trus P. Posterior cruciate ligament (PCL) reconstruction—an in vitro study of isometry Part II. Tests using an experimental PCL graft model. Knee Surg Sports Traum Arthrosc 1994;2:104–106.
101. Ogata K, McCarthy JA. Measurements of length and tension patterns during reconstruction of the posterior cruciate ligament. Am J Sports Med 1992;20:351–355.
102. Bomberg BC, Acker JH, Boyle J. The effect of posterior cruciate ligament loss and reconstruction on the knee. Am J Knee Surg 1990;3:85–96.
103. Dorlot JM, Christel P, Sedel L, Witvoet J. The displacement of the bony insertion sites of the cruciate ligaments during flexion of the knee. Trans ORS 1983;8:328.
104. Pearsall AW, Pyevich M, Draganich LF, Larkin JJ, Reider B. In vitro study of knee stability after posterior cruciate ligament reconstruction. Clin Orthop 1996;327:264–271.
105. Stone JD, Carlin GJ, Ishibashi Y, Harner CD. Assessment of PCL graft performance using robotics technology. Am J Sports Med 1996;24:824–828.
106. Friederich NF. Kniegelenksfunktion und kreuzbänder. Biomechanische grundlagen für rekonstruktion und rehabilitation. Orthopäde 1993;22:334–342.
107. Jakob RP, Ruegsegger M. Therapy of posterior and posterolateral knee instability. Orthopaedics 1993;22:405–413.
108. Juergensen K, Edwards JC, Jakob RP. Positioning of the posterior cruciate ligament. Knee Surg Sports Traum Arthrosc 1994;2:133–137.
109. Warren RF, Veltri DF. Arthroscopically assisted posterior cruciate ligament reconstruction. Oper Tech Sports Med 1993;1:136–142.
110. Wirth CJ, Jager M. Dynamic double tendon replacement of the posterior cruciate ligament. Am J Sports Med 1984;12:39–43.
111. Hughston JC, Andrews JA, Cross MJ. Classification of knee ligament instabilities: part II: the lateral compartment. J Bone Joint Surg 1976;58A;173–179.
112. Southmayd WW, Rubin BD. Reconstruction of the posterior cruciate ligament using the semimembranosus tendon. Clin Orthop 1980;150:196–197.
113. Swenson TM, Harner CD, Fu FH. Arthroscopic posterior cruciate ligament reconstruction with allograft. Sports Med Arthrosc Rev 1994;2:120–128.
114. Noyes FR, Barber-Westin SD. Surgical restoration to treat chronic deficiency of the posterolateral complex and cruciate ligaments of the knee joint. Am J Sports Med 1996;24:415–426.
115. Terry GC, LaPrade LF. The posterolateral aspect of the knee: anatomy and surgical approach. Am J Sports Med 1996;24:732–739.
116. Veltri DM, Warren RF. Operative treatment of posterolateral instability of the knee. Clin Sports Med 1994;13:615–627.
117. Veltri DM, Warren RF. Posterolateral instability of the knee. J Bone Joint Surg 1994;64A:460–472.
118. Watanabe Y, Moriya H, Takahashi K, et al. Functional anatomy of the posterolateral structures of the knee. J Arthrosc Rel Surg 1993;9:57–62.
119. Dürselen L, Claes L, Kiefer H. The influence of muscle forces and external loads on cruciate ligament strain. Am J Sports Med 1995;23:129–136.
120. Clancy WG, Bisson LJ. Double bundle technique for reconstruction of the posterior cruciate ligament. Oper Tech Sports Med 1999;7:110–117.

121. Petrie RS, Harner CD. Double bundle posterior cruciate ligament technique: University of Pittsburgh approach. Oper Tech Sports Med 1999;7:118–126.
122. Kimar R, Beacon JP, LaBoreau JP. Two bundle posterior cruciate ligament reconstruction. Paper presented at American Academy of Orthopaedic Surgeons, New Orleans, 1998.
123. Berg EE. Posterior cruciate ligament tibial inlay reconstruction. Arthroscopy 1995;11:69–76.
124. Harner CD, Livesay GA, Kashiwaguchi S, Fujie H, Choi NY, Woo SL-Y. Comparative study of the size and shape of human anterior and posterior cruciate ligaments. J Orthop Res 1995;13:429–434.
125. Fu FH, Harner CD, Vince KG. Knee surgery. Baltimore: Williams & Wilkins, 1994.

Chapter Two

Anatomy and Biomechanics of the Posterolateral Aspect of the Knee

Robert F. LaPrade and Timothy S. Bollom

Because of the developmental phylogeny that has occurred owing to the initial articulation of the fibula with the femur and its eventual descent over time to articulate with the tibia, the anatomy of the posterolateral aspect of the knee is more complex than the medial side. In addition, in lower species, there is a meniscus between the articulation of the femur and the fibular head. It has been speculated that this meniscus may have eventually evolved into the popliteus attachment to the fibular styloid (popliteofibular ligament) or the popliteus tendon. In addition, the popliteus complex and the biceps femoris complex in lower species are noted to have attachments around the knee that are different from those found in humans. These anatomic differences have made the posterolateral aspect of the knee more complex and less thoroughly studied than the medial aspect of the knee in terms of both its anatomy and biomechanics.

Over the past decade, several studies have looked at the fine anatomic details of the posterolateral knee, which has contributed greatly to further interest in this topic. Stäubli et al.[1,2] reported on the details of the popliteomeniscal fascicles and popliteofibular ligament, and Terry et al.[3] reported on the components of the iliotibial band. Terry and LaPrade[4] reported on the individual components of the long and short heads of the biceps femoris about the knee, and the same authors also reported on many of the individual anatomic structures of the posterolateral knee.[5] Veltri et al.[6] also described the anatomy and biomechanical importance of the popliteofibular ligament as well as its biomechanical importance.

Prior to these studies, a number of studies addressed similar or related topics using varying terminology to describe these structures. The variations in terminology used in different reporting institutions over many years created confusion regarding the anatomy of the posterolateral aspect of the knee. This chapter reviews many of these studies, correlates these authors' descriptions with currently utilized terminology (cited in italics) for these posterolateral knee structures, and discusses the anatomy and biomechanics of the posterolateral knee.

Previous Studies on the Anatomy of the Posterolateral Knee

In 1884, Sutton[7] reported that the "external lateral ligament of the knee joint" (*fibular collateral ligament* [*FCL*]) was in reality an extension of the tendon of the peroneus longus muscle. He found this to be true in many species. In 1948, R. J. Last[8] reported on old terminology used to describe structures of the posterolateral aspect of the knee and attempted to replace some of the terminology with the then-current British terminology. Last described the short external lateral ligament of the knee. He reported that it represented part of the true joint capsule on the lateral side of the knee and that it attached proximally to the lateral epicondyle of the femur and distally to the styloid process. It appears that the short external lateral ligament in his description represents a portion of the *fabellofibular ligament* and the *capsular arm of the short head of the biceps femoris*. Last also described the arcuate ligament as a fascial attachment posterior to the popliteus and firmly adherent to the popliteus fascia (possibly part of the *posterior division of the popliteofibular ligament* and *capsular arm of the short head of the biceps femoris*). Last stated that the lateral ligament of the knee joint, in the then-current British terminology, was represented by the long external lateral ligament of old terminology (FCL). It was described to have arisen from the lateral epicondyle of the femur, in continuity with the upper part of the short external lateral ligament (a portion of the *fabellofibular ligament* and *capsular arm of the short head of the biceps femoris*), and to insert on the upper surface of the head of the fibula. Last also observed that the tendon of the biceps femoris wrapped around the posterior aspect of the lateral ligament (FCL) of the knee. He also revealed that the popliteus emerged from underneath the arcuate ligament and that a very substantial portion of the popliteus muscle arose not from the tendon of the popliteus but from the arcuate ligament. He suggested that the attachment of the popliteus to the lateral meniscus (*popliteal aponeurosis*) was the primary factor in protecting the meniscus from injury.

Sneath[9] described the insertions of the biceps femoris in 1955. He stated that the long head of the biceps femoris attached onto the head of the fibula, the lateral ligament of the knee (FCL), and the lateral condyle of the tibia. He also reported that the biceps tendon blended anteriorly with the iliotibial tract and gave off an expansion to the crural fascia (*anterior aponeurosis* and *distal fascial expansion of the long head of the biceps femoris*) covering the anterior, lateral, and posterior compartments of the leg. Sneath said that the remaining portion of the common biceps tendon was derived mainly from the short head of the biceps femoris. He reported that it had three lamina. The superficial lamina (*anterior arm of the long head of the biceps femoris*) passed superficial to the lateral ligament of the knee and inserted into the lateral condyle of the tibia. The intermediate lamina (*lateral aponeurosis of long and short heads of biceps femoris*) blended with the posterior border of the distal third of the lateral ligament. The deep lamina (*anterior arm of the short head of the biceps femoris*) passed deep (medial) to the lateral ligament and inserted onto the lateral condyle of the tibia immediately posterior to the iliotibial band. He observed a constant synovial bursa that separated the lateral ligament from the superficial and deep lamina of the biceps tendon (*FCL–biceps femoris bursa*). Sneath also observed that with active flexion of the knee, the intermediate lamina of the biceps tendon would keep the lateral ligament taut by a bowstring effect.

In 1957, Kaplan[10] presented a surgical approach to the lateral (peroneal) side of the knee joint. His investigation of comparative anatomy used a large number of animal species and the dissection of human knees of various age groups. Kaplan stated that upon making an oblique straight incision on the lateral aspect of the knee, the first fascia layer was represented by the fascia lata (*superficial layer of the iliotibial tract*), which adhered intimately to the anterior tubercle of the tibia, to the lateral tibial tubercle of Gerdy, and to the head of the fibula. He reported that over the head of the fibula, the fascia was connected with the biceps tendon (*direct arms of the long and short heads* and *anterior arm of the long head of the biceps femoris*), which surrounded the fibular portion of the lateral collateral ligament (*FCL*). He also revealed that the tendon of the biceps expanded onto the fascia lata (*reflected arm of the long head of the biceps femoris*), and distally connected into the fascia (*anterior aponeurosis of the long head of the biceps femoris*) covering the muscles of the leg. He also stated that the fascia lata adhered intimately to the lateral supracondylar tubercle and blended into the lateral intermuscular septum (*deep and capsulo-osseous layers of the iliotibial tract*). The iliotibial band was described as a thickening of the fascia lata, running from the base of the greater trochanter to the lateral tubercle of the tibia (Gerdy's tubercle). He reported that the anterior part of the iliotibial band (*iliopatellar band*) curved forward, forming a group of arciform fibers, which blended with the fascia over the patella.

Kaplan stated that the most important structure found under the fascia is the lateral collateral ligament (*FCL*) of the knee. This ligament was found to be between the lateral condylar tubercle (*lateral epicondyle*) of the femur and the middle of the lateral surface of the head of the fibula, about 1.5 cm anterior to the apex of the head. It was observed that the insertion of the biceps femoris muscle enveloped the fibular insertion of the lateral collateral ligament (*FCL*) with a bursa (*FCL–biceps femoris bursa*) between the ligament and the capsule. He also reported that the coronary ligament for the lateral meniscus was found deep to the inferior lateral genicular artery. He stated that the popliteus tendon originated in front and below the origin of the lateral collateral ligament (*FCL*), and although these two structures were in close contact, they were actually separated by the joint capsule. Kaplan said that the popliteus tendon almost never adhered to the meniscus (*popliteomeniscal fascicles*). He did observe that there was an intimate connection of the popliteus tendon with the fibular head (*popliteofibular ligament*). He also reported that this connection sometimes reached the tibiofibular joint (*anterior division of the popliteofibular ligament*), and that this connection between the popliteus and the fibular head is found in other species, especially in the chimpanzee.

Kaplan also stated that the muscular body of the popliteus, in contrast to the popliteus tendon, adhered firmly to the lateral meniscus, the coronary ligament, and the posterior capsule of the knee joint. He reported that this occurred through the means of a thick fascia, which he termed the popliteal fascia (*popliteal aponeurosis*). It was noted that there was relative immobility of the tendon of the popliteus, which he felt made the tendon a stabilizer of the knee in all positions.

Kaplan also observed that the lateral head of the gastrocnemius muscle (*lateral gastrocnemius tendon*) originated from the femoral condyle just above the origin of the lateral collateral ligament (*FCL*), and also from the capsule of the joint (*meniscofemoral portion of the posterior capsule*).

Kaplan[11] described in 1958 the significance of the iliotibial tract. He performed several comparative anatomy studies of the knees of other species to the human knee, focusing on the posterolateral aspect. He found that in most mammals, the biceps femoris formed a wide sheet of insertion that covered the entire lateral aspect of the knee, passing over the head of the fibula and the lateral tuberosity of the tibia, without inserting into either of these two structures. He found that even in higher primates such as the gibbon, orangutan, and gorilla, there is no iliotibial tract. He noted that only in the human lower extremity is there an open space between the biceps femoris tendon and the vastus lateralis muscle at the knee joint, which permits passage of the iliotibial tract for insertion onto Gerdy's tubercle. Kaplan speculated that the iliotibial tract is a structure found only in humans because of erect posture, and it acts as an anterolateral ligament of support for the knee. Kaplan also found, from dissections of stillborn infants, that there are longitudinal fibers from the iliotibial tract that adhere firmly to the lateral intermuscular septum between the biceps and vastus lateralis, and that with flexion and extension of the knee these longitudinal fibers moved anteriorly or posteriorly. Kaplan noted that in the adult, the distal half of the iliotibial tract is intimately connected from the lateral intermuscular septum and the linea aspera to the supracondylar tubercle of the lateral femoral condyle (*deep layer of the iliotibial band*). He reported that the iliotibial tract is connected by curved fibers with the lateral border of the patella anteriorly (*iliopatellar band*) and to the anterior fibers of the biceps femoris tendon posteriorly (confluence with the *capsulo-osseous layer of the iliotibial band*).

Kaplan[12] in 1961 also described the fabellofibular and short lateral ligaments of the knee joint. Kaplan reported that the fabellofibular ligament had been described in 1905 by the French anatomist Dujarier, who stated that the short lateral ligament was the peroneosesamoid ligament of the knee joint. It had also been described as the short external lateral ligament of Vallois. Kaplan had found in his review of much earlier literature that the short lateral ligament was described mostly as a part of the lateral arch of the arcuate ligament, or as a reinforcement of the posterior capsule (*capsular arm of the short head of the biceps femoris*). He noted that there had been multiple names given to this ligament in the past, and he gave his reasons for calling it the fabellofibular ligament. He also reported that the oblique popliteal ligament was always present and that its lateral border arched over the popliteus muscle and constituted the medial arch of the arcuate ligament. He observed that the inferior lateral genicular artery crossed the joint capsule anterior to the lateral arch of the arcuate ligament. He found that the arcuate ligament (*capsular arm of the short head of the biceps femoris*) overlaid the popliteus muscle, which was invested by its own popliteal fascia. He reported that another structure originated near the condylar origin of the lateral gastrocnemius and ran toward the fibular head. He stated in this instance that the inferior lateral genicular artery ran between the capsule of the knee and this ligament structure. He felt that this ligament, which ran from under the gastrocnemius to the fibular head, represented the short lateral ligament (*fabellofibular ligament*). He reported that the medial border of this ligament blended with the lateral arch of the arcuate ligament (*capsular arm of the short head of the biceps femoris*). He felt that the presence of a large fabella was always associated with a notable increase in size of the short lateral ligament. He observed that this ligament ran almost parallel to the lateral collateral ligament (FCL) toward the fibular head, where it inserted posterior to the insertion of the biceps muscle tendon (*direct arm of the short head of the biceps femoris*). Kaplan stated that the fabellofibular ligament was strongest in the kangaroo and in those species where the function and size of the gastrocnemius is important. He observed that in the absence of the fabella, a short lateral ligament represents a homologue of the fabellofibular ligament.

Kaplan[12] also reported that the popliteus muscle originates in lower animal species from the fibular head and has no origin on the femur. He stated that in the course of evolution, the fibular head receded from its contact with the femur to assume its new position at the proximal tibiofibular joint. The disappearance of the femorofibular joint in the process of evolution eliminated the necessity for the femorofibular meniscus and led to its complete transformation. He speculated that the meniscus underwent a change from a fibrocartilaginous structure into a tendon that then formed a continuous structure that extended from the femur to the fibula and into the popliteus muscle. He felt that the structure became the tendon of the popliteus muscle. Kaplan also observed that the lateral arch of the arcuate ligament represented a reinforcement of the posterior capsule of the knee joint where the popliteus muscle is attached to the coronary ligament of the lateral meniscus (*popliteal aponeurosis* and a portion of the *capsular arm of the short head of the biceps femoris*).

In 1962, Kaplan[13] evaluated the functional anatomy of the knee joint. He reported that the popliteus muscle is intimately attached to the posterior capsule (*popliteal aponeurosis*) and the coronary ligament of the lateral meniscus. He stated that in the human knee, the popliteus may have a direct or an indirect connection with the fibular head. He noted that Fürst reported in the German literature in 1903 that the tendon of the popliteus and the muscle are two independent structures that developed independently in the animal world. Kaplan further described the short lateral ligament as the ligament that is present when there is no fabella. When the fabella is present, it is called the fabellofibular ligament. He also observed that it runs almost parallel to the lateral collateral ligament (FCL) to the fibular head, and it reinforced the posterior capsule of the knee (*capsular arm of the short head of the biceps femoris*).

Marshall et al.[14] in 1972 described the structure of the biceps femoris tendon at the knee. In a series of dissections, they found that the fleshy fibers of the long head of the biceps started to form a broad, flat, tendon about 7 to 10 cm above the level of the knee joint. The short head of the biceps femoris joined the tendon of the long head at its undersurface, remaining fleshy almost to the fibular head, at which point the two heads joined to form a short, thick tendon. They found that just before it reached the lateral collateral ligament (FCL), the tendon split into three layers—superficial, middle, and deep. The superficial layer was described to be lateral to the lateral collateral ligament, the middle layer surrounded it, and the deep layer was medial and deep to the lateral collateral ligament. They stated that the superficial layer (*anterior arm and aponeurosis of the long head of the biceps femoris*) formed three expansions. The first of these was an anterior expansion (*anterior aponeurosis of the long head of the biceps femoris*) that was thin and sheet-like and extended distal to the lateral collateral ligament (FCL), and fanned out to blend with the anterior crural fascia. They reported that some of these fibers reached Gerdy's tubercle and that some of the deep fibers of the superficial layer inserted into the lateral aspect of the head of the fibula (*anterior arm of the long head of the biceps femoris*). A posterior expansion extended inferiorly and blended with the fascia over the calf muscles (*distal fascial expansion of the long head of the biceps femoris*).

Marshall et al.[14] also observed that the middle layer of the common biceps tendon was a thin, poorly defined layer that surrounded approximately the distal fourth of the lateral collateral ligament like a sling. They reported that this layer split to surround the lateral collateral ligament and was separated from the ligament by a bursa (*FCL–biceps femoris bursa*). They stated that in every case there was a fibrous attachment into the posterior aspect of the lateral collateral ligament (*lateral aponeurotic attachments of the long and short heads of the biceps femoris*). They also said that the deep layer of the common biceps tendon bifurcated and had a fibular and tibial attachment. The tibial attachment was just posterior to Gerdy's tubercle (*anterior arm of short head of biceps femoris*), and the fibular attachment was into the styloid process of the fibula and into the upper surface of the head, deep to the distal attachment of the lateral collateral ligament (*direct arm insertions of long and short heads of biceps femoris*). An extension of this layer was also reported to cross the upper surface of the tibiofibular syndesmosis, which strengthened its anterior capsule and inserted into the anterolateral aspect of the tibia (felt to be the *anterior division of the popliteofibular ligament*). In addition, they described a well-developed extension that passed to the posterolateral aspect of the knee joint capsule (*capsular arm of the short head of the biceps femoris*). They also stated that at 5 to 7 mm proximal to the knee joint, a distinct broad fiber span sweeping anteriorly was present and was a fibrous attachment between the biceps and the iliotibial band (*confluence with the capsulo-osseous layer of the iliotibial band*).

In 1975, Reis and de Carvalho[15] described the proximal attachments of the human popliteus muscle. They reported that Higgins in 1894 and Fürst in 1903 reported on the popliteus muscle as having two proximal attachments with one on the lateral condyle of the femur and the other on the fibular head. They also reported that a proximal attachment of the popliteus tendon was observed to occur constantly in the depression just below the lateral epicondyle, beneath, anterior, and lateral to the FCL. They observed that the tendon has an oblique direction upward and outward, forming an angle of approximately 45 degrees over the horizontal plane when it passed by the tibial condyles. They also reported that they found expansions to the oblique popliteal ligament in all knees (*popliteal aponeurosis*) as well as an attachment into the head of the fibula (*popliteofibular ligament*), which extended out approximately from the musculotendinous junction of the popliteus muscle. They also found meniscal attachments that united the popliteus tendon to the lateral meniscus (*popliteomeniscal fascicles*). They concluded that the popliteus muscle had four proximal attachments: insertion into the lateral femoral condyle (origin on femur), meniscal attachments (*popliteomeniscal fascicles*), attachments to the posterior capsule (*popliteal aponeurosis*), and the fibular head attachment (*popliteofibular ligament*).

In 1979, Cohn and Mains[16] described the anatomy of the popliteal hiatus. They reported that the anatomy of the popliteus was constant. They observed two distinct ligamentous fascicles—one for the superior surface and one for the inferior surface of the meniscus. These fascicles were noted to make up the floor and the roof of the popliteal hiatus. In their illustrations, the superior fascicle corresponds to the *posterosuperior popliteomeniscal fascicle* and the inferior fascicle corresponds to the *anteroinferior popliteomeniscal fascicle*.

Fabbriciani et al.[17] reported in the Italian literature in 1982 on an anatomic study of the popliteus muscle complex. They observed that the popliteus muscle continues into a complex aponeurosis consisting of popliteocapsular (*popliteal aponeurosis*), popliteofibular, and popliteomeniscal fibers (*popliteomeniscal fascicles*). It then develops into a strong tendon that attaches on the lateral femoral condyle. They stated that there were superior and inferior popliteomeniscal fibers. In addition, they summarized some of the developmental phylogeny literature, which is useful in the understanding of the development of the various components of this muscular complex. They reported that the comparative anatomy studies have demonstrated that in lower order vertebrates, the fibula and the tibia provide articulation with the femur, and each has its respective menisci. It also was noted that in lower species, the fibers of the popliteus muscle attach on the head of the fibula posteriorly. As the fibular head dips in higher species, there is differentiation of the meniscus between the femur and the fibula. In later stages of development, the popliteus muscle was noted to attach to the femorofibular meniscus, which became the popliteus tendon, and the popliteus muscle developed a tendinous attachment with the fibular head (*popliteofibular ligament*). In addition, the popliteus tendon maintained its ties with the lateral meniscus (*popliteal aponeurosis*).

Seebacher et al.[18] in 1982 reported on the anatomy of the posterolateral aspect of the knee. They divided the posterolateral structures of the knee into three layers. Layer I, the most superficial layer, was described as having two parts: the iliotibial tract and its anterior expansion (*iliopatellar band*), and the superficial portion of the biceps femoris and its posterior expansion. Layer II was actually an incomplete layer that consisted of the retinaculum of the quadriceps mechanism and the lateral patellofemoral ligaments. Layer III, the deepest layer, was described as the lateral portion of the joint capsule. It was noted that the coronary ligament is a capsular attachment to the outer edge of the lateral meniscus. It was reported that just posterior to the iliotibial tract, the capsule divided into two laminae. The superficial lamina (*lateral aponeuroses of the long and short heads, capsular arm of the short head of the biceps femoris*) was noted to encompass the lateral collateral ligament (*FCL*) and the fabellofibular ligament. The deeper lamina was noted to pass along the edge of the lateral meniscus where it formed the hiatus for the popliteus tendon and the coronary ligament.

Seebacher et al.[18] described the inner laminae as terminating posteriorly as the Y-shaped arcuate ligament (*popliteofibular ligament*), which was noted to span the junction between the popliteus muscle and its tendon from the fibula to the femur. They stated that the two laminae were always separated from each other by the inferior lateral genicular vessels. They noted that both the arcuate (*popliteofibular ligament*) and fabellofibular ligaments inserted at the apex of the fibular styloid. Both these ligaments then ascended with their respective capsular laminae to the lateral head of the gastrocnemius. They stated that the arcuate ligament (*popliteofibular ligament*) was firmly adherent to the underlying musculotendinous junction of the popliteus. They also stated that in addition to the limb that went from the fibula to the lateral head of the gastrocnemius, the arcuate ligament also fanned out medially over the popliteus muscle (*popliteal aponeurosis*) to join the fibers of the oblique popliteal ligament of Winslow. They also noted that the midlateral part of the capsule (*middle third lateral capsular ligament*) was very strong at the point of fusion of the two laminae and was probably responsible for the avulsion of tibial metaphyseal bone in extreme lateral disruptions of the knee (Segond fractures).

In 1987, Lobenhoffer et al.[19] reported on the distal femoral fixation of the iliotibial tract. They studied the anatomy of the attachments of the iliotibial tract to the distal femur. They described a supracondylar bundle that was fixed to the supracondylar area of the femur. In addition, they noted fibers near the lateral intermuscular septum that had a transverse course between the superficial layer of the iliotibial tract and the dorsolateral femur (*deep layer of the iliotibial tract*). They also reported on retrograde tracks that connected Gerdy's tubercle with the dorsolateral femur and formed an arc bridging the knee joint. They reported that there was an arched fiber arrangement, with these fibers having a reverse course in relation to the other insertions of the iliotibial track to the femur. Thus, they called them "retrograde fiber tracks" (*capsulo-osseous layer of the iliotibial tract*). They also reported that there were connections to the biceps tendon and their dorsal tract. They did observe that the supracondylar insertion did not seem to contribute to lateral knee stability. However, they felt that the "retrograde" fibers seemed to stabilize the lateral side of the knee. There was minimal change in isometry with range of motion of the retrograde fibers. The authors also stated that these fiber tracks corresponded to the ligament femorotibiale laterale anterius described by Werner Müller. The authors felt that reinsertion of the origin of the torn fibers back to the lateral intermuscular septum would assist in lateral reconstructions of the knee.

In 1989, Oransky et al.[20] provided some information on the embryonic development of the posterolateral structures of the knee. They studied human embryos and fetuses and found that the attachments of the lateral meniscus and fibular head to the popliteal tendon and muscle are formed during the process of cavitation, which formed the popliteal hiatus. They reported that cavitation of the popliteal hiatus occurred at the 10th week of development, which separated the popliteus tendon and the lateral meniscus. They found that the popliteofibular ligament, the connection between the distal portion of the popliteus tendon and the fibular head, developed soon after cavitation occurred. By 14 weeks of development, the superolateral attachment of the popliteus tendon to the lateral meniscus (*popliteal aponeurosis*) was thick. By 16 weeks of development, the connections between the popliteus tendon, the lateral meniscus, and the fibular head were fully formed (*popliteomeniscal fascicles* and *popliteofibular ligament*). They verified the location of the popliteofibular ligament in adult human cadavers and reported that it was a strong ligamentous band that attached the posterolateral capsule to the upper fibula via the popliteus tendon. They reported that the popliteofibular ligament was a strong and constantly found structure, 1.5 cm in width and approximately 2 cm in length, which inserted on the posteromedial aspect of the fibular head.

In the early 1990s, Stäubli and coauthors[1,2] reported on the popliteus tendon and its attachments to the lateral meniscus. They identified a constant femoral attachment of the popliteus tendon at the anterior end of the popliteal sulcus. The anteroinferior popliteomeniscal fascicle was always present and blended into the middle third of the lateral meniscus. The posterosuperior popliteomeniscal fascicle was also found to blend into the posterior horn of the lateral meniscus. They also described a popliteofibular fascicle (*popliteofibular ligament*), which consisted of anterior and posterior divisions. They reported that it arose from the proximal tibiofibular joint and from the apex of the fibula at the musculotendinous junction of the popliteus. They stated that the popliteofibular fascicle has a shape of an inverted **Y** and consisted of an anterolateral portion that blends into the anteroinferior popliteomeniscal fascicle and a posterior division that attaches to the posterior aspect of the fibular styloid and is adjacent and adherent to the proximal tibiofibular joint capsule. They also observed that the superficial lateral portion of the popliteus muscle joins with the arcuate popliteal ligament (*fabellofibular ligament* and *capsular arm of long head of biceps femoris*). They said that the inferior lateral genicular artery passes anterior to the arcuate popliteal ligament. They also noted that the femoral attachment of the popliteus tendon is at the anterior end of the popliteus sulcus, anterior and distal to the lateral epicondyle. They also reported that the superoposterior portion of the popliteus tendon blends with the posterior meniscofemoral capsule (*meniscofemoral portion of mid-*

third lateral capsular ligament), which was noted to be adjacent to the lateral gastrocnemius complex. An inferior popliteomeniscal fascicle at the posteroinferior rim of the posterior horn of the lateral meniscus (*posteroinferior popliteomeniscal fascicle*) was also described. It was also noted that the superior popliteomeniscal fascicle blended with the meniscofemoral portion of the posterior capsule.

Watanabe et al.[21] in 1993 reported on the functional anatomy of some of the posterolateral structures of the knee. They found that the popliteal origin from the fibular head (*popliteofibular ligament*) was present in 93% of their specimens. They stated that the arcuate ligament was posterior to the inferior lateral genicular artery and was quite thin in these specimens. They described the arcuate ligament to be a thin, triangular band of capsular fibers that originated from the posterior aspect of the fibula and arched upward immediately over the popliteus tendon and spread out over the posterior capsule (*capsular arm of the short head of the biceps femoris*). They also observed that the fabellofibular ligament ran between the fibula and the fabella and was under the greatest tension when the knee was in full extension but relaxed as the knee flexed.

The literature on the posterolateral knee is valuable but at times confusing and conflicting. The wide variations in terminology make it difficult to compare studies. In the more recent anatomic articles on which we are basing our descriptions in this chapter,[1-6] more standard nomenclature is used for the structures that have been observed to be present in virtually all knees. The major confusion lies with the structures that course from the fibular head and attach to the posterolateral aspect of the knee. The inferior lateral genicular artery is an excellent landmark to use to compare studies and illustrations and to identify the structures being described. The popliteofibular ligament's posterior division is anterior to the inferior lateral genicular artery. This is found to be the only structure from the posterior aspect of the fibular styloid that is anterior to the inferior lateral genicular artery when it crosses at this point. The fabellofibular ligament goes from just lateral to the tip of the styloid to the region of the fabella or fabella analogue. Studies have shown that there is always a cartilaginous fabella when the bony fabella is not present.[22] The large capsular expansion that courses from the fibular styloid in this area to the posterolateral capsule and lateral gastrocnemius tendon is what is termed the capsular arm of the short head of the biceps femoris.[4] It appears that this same structure is what has been called the arcuate ligament in previous studies. The short lateral ligament of other studies[8,12] appears to be the fabellofibular ligament. In fact, one author has tied these two together as being the same structure.[12] While it is recognized that varying terminology will appear based on one's training and the region of the world in which studies are performed, it is also desirable that a standard terminology be developed so that anatomic comparisons can be made, and biomechanical studies and surgical reconstructions can be compared. The terminology used in the next section of this chapter is a synopsis of current studies and current thought, and can serve as the building block of standard nomenclature for anatomic structures in this region.

Anatomy of the Posterolateral Knee

Part of the complexity of the posterolateral aspect of the knee lies in its inherent instability due to its bony anatomy. Patients with posterolateral

Fig. 2.1. Iliotibial band (superficial layer) and iliopatellar band.

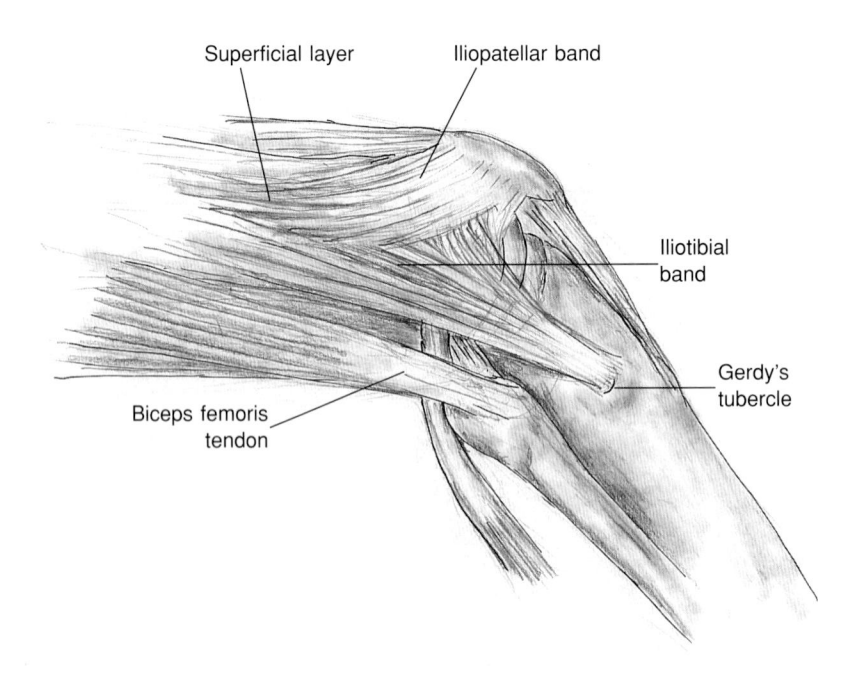

knee injuries can have instability with routine ambulation and this is almost always due to a varus thrust gait. The lateral femoral condyle and the lateral tibial plateau are both relatively round, and if the basic static stabilizers of this side of the knee are injured, the dynamic stabilizers may not be able to exert full control, and there can be either a varus rotational opening or an external rotational opening of the posterolateral aspect of the knee with the initial portion of the stance phase at heel strike. This can contribute to the instability seen with gait in some patients with grade II or III posterolateral rotatory instability of the knee.

Iliotibial Band and Its Components

The iliotibial band is one of the main static stabilizers on the lateral aspect of the knee. It has four main components[3,5] (Fig. 2.1), each of which can be injured with posterolateral knee injuries. The main component of the iliotibial band is the superficial layer, which is the main superficial fascial layer that attaches to the lateral side of the knee. Its distal attachment site is broad based at Gerdy's tubercle. The superficial layer also has a fascial expansion called the iliopatellar band, which attaches to the lateral aspect of the patella. Rarely, the superficial portion of the iliotibial band is injured, and when it is it most commonly involves an avulsion off the tibia at Gerdy's tubercle rather than a midsubstance tear. The other two main components of the iliotibial band are its deep and capsulo-osseous layers. The deep layer of the iliotibial band extends from the superficial layer to the lateral epicondylar flair of the femur. The capsulo-osseous layer initiates from the lateral intermuscular septum posterodistally, where it has a confluence with a fascial sheath that also extends from the short head of the biceps femoris and the lateral gastrocnemius muscle. This complex then extends distally as a fascial sling along the posterior aspect of the lateral femoral condyle and attaches to the tibia approximately 1 cm posterior to Gerdy's tubercle. Its attachment site on the tibia is convergent with the lateral capsular tibial attachment (meniscotibial portion of mid-third lateral capsular ligament) on the lateral side of the knee.

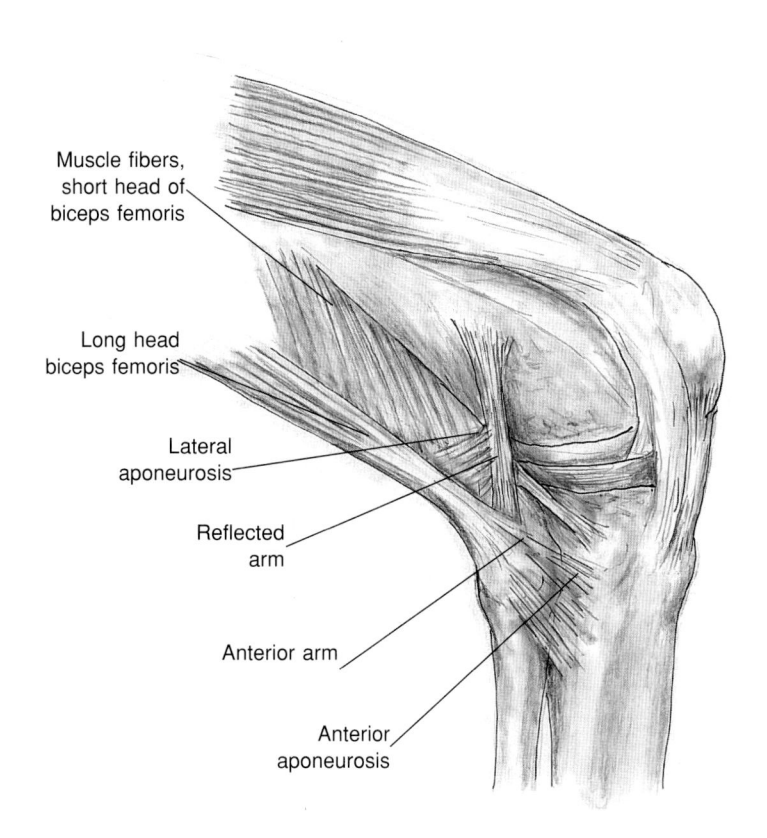

Fig. 2.2. Long head of biceps femoris and its superficial components: anterior arm, reflected arm, and lateral and anterior aponeurosis.

Long Head of the Biceps Femoris

The long head of the biceps femoris has six components at the posterolateral aspect of the knee[4,22] (Fig. 2.2). The direct arm forms a major portion of the tendon of the long head of the biceps femoris and attaches to the midportion of the posterolateral aspect of the fibular styloid. Another tendinous component of the long head of the biceps femoris is the anterior arm, a tendinous expansion that extends from the lateral aspect of the common biceps tendon. The anterior arm partially attaches to the distal and lateral aspect of the fibular styloid, and then extends distally around the lateral aspect of the FCL. A bursa is present over the distal quarter of the FCL where the anterior arm of the long head of the biceps femoris passes the FCL[23] (Fig. 2.3). After the anterior arm of the long head of the biceps femoris extends past the FCL, it forms a fascial sheath, termed the anterior aponeurosis, which extends distally and anterolaterally over the leg.

The long head of the biceps femoris also has a fascial reflection, called the reflected arm, which extends from the anterolateral aspect of the main tendon of the long head of the biceps to the posterior border of the iliotibial band. In addition to the reflected arm, a lateral aponeurosis also extends from the anteromedial aspect of the long head of the biceps femoris and attaches to the posterior and lateral aspect of the FCL proximal to the biceps bursa. Finally, a distal fascial expansion of the long head of the biceps femoris overlies and attaches to the lateral gastrocnemius complex.

Short Head of the Biceps Femoris

The short head of the biceps femoris has five major components at the posterolateral aspect of the knee.[4] The first of these is the main muscle fibers

Fig. 2.3. Fibular collateral ligament–biceps femoris bursa. The anterior arm of the long head of the biceps femoris has been horizontally incised to expose the bursa at the distal aspect of the fibular collateral ligament.

of the short head of the biceps femoris, which extend from the linea aspera of the femur to its main tendinous attachment along the tendon of the long head of the biceps femoris. These muscle fibers remain attached to the posteromedial aspect of the common biceps tendon almost to its fibular styloid insertion. The direct arm attachment of the short head of the biceps femoris is the medial portion of the common biceps tendon, which attaches just lateral and distal to the tip of the posterior aspect of the fibular styloid. In this position on the fibular styloid, the direct arm of the short head of the biceps femoris is slightly proximal and medial to the direct arm of the long head of the biceps femoris (Fig. 2.4). The tendinous anterior arm of the short head of the biceps femoris extends medial to the fibular collateral ligament and attaches on the proximal lateral tibia approximately 1 cm posterior to Gerdy's tubercle (Fig. 2.5). This attachment point on the tibia is along with the attachment of the capsulo-osseous layer of the ili-

Fig. 2.4. Fibular styloid attachments of the biceps femoris tendon (direct arms of short head and long head, anterior arm of long head), fabellofibular ligament, and popliteofibular ligament (posterior division illustrated [posterior view, right knee]).

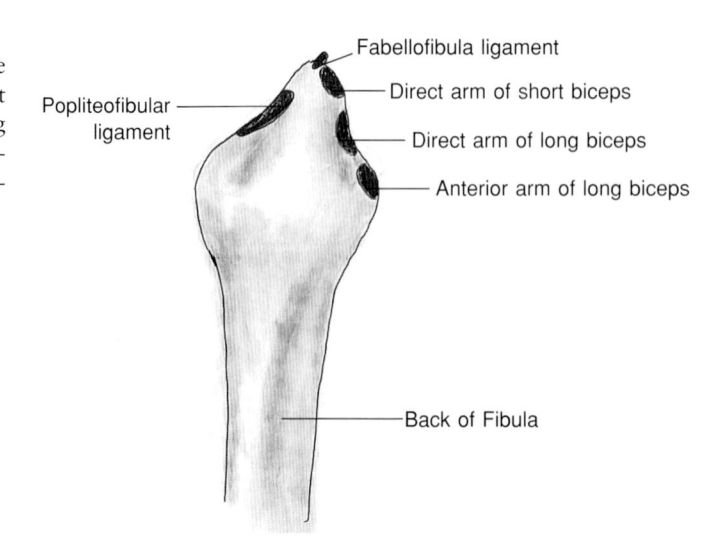

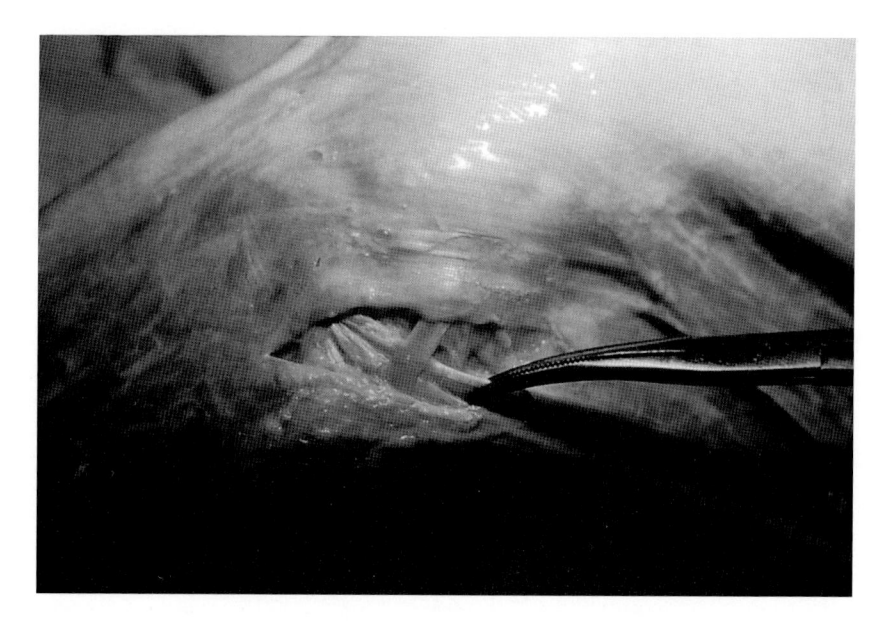

Fig. 2.5. Anterior arm of the short head of the biceps femoris (in hemostat) and its muscle fibers (proximal).

otibial band as well as the meniscotibial portion of the mid-third lateral capsular ligament attachment.

Prior to its direct arm attachment on the fibular styloid, the short head of the biceps femoris has a stout capsular arm attachment to the posterolateral aspect of the lateral joint capsule as well as the lateral gastrocnemius tendon. This capsular arm is located posterior and lateral to the inferior lateral genicular artery as it crosses the posterolateral aspect of the knee. The distal border of the capsular arm of the short head of the biceps femoris extends from just lateral to the tip of the fibular styloid to the fabella or fabella analogue[4,22] and is known as the fabellofibular ligament. This ligament is best seen in vivo or in a fresh frozen cadaveric knee when the knee is close to full extension. Due to the fascial nature of this ligament, it is not well seen during knee dissections or in surgical reconstructions when the knee is flexed past 30 degrees. In addition to the capsular arm fascial attachments, there is also a fascial attachment that forms a confluence with the capsulo-osseous layer of the iliotibial band and also extends to the lateral gastrocnemius tendon. Finally, there is a lateral aponeurosis that is a fine aponeurotic attachment from the main tendinous portion of the short head of the biceps femoris, which attaches onto the posteromedial aspect of the FCL.

The Fibular Collateral Ligament

The FCL also commonly known as the lateral collateral ligament, has a femoral attachment just proximal and posterior to the lateral epicondyle (Fig. 2.6). It is extraarticular and extends distally at a slight caudal angle to the distal femur and primarily attaches to the lateral aspect of the fibular head in a small saddle at a point approximately midway on the fibular head. A portion of the FCL then extends distally as a reinforcement to the fascia over the peroneus longus muscle of the lateral compartment of the leg.[5,7] The FCL appears to be dynamically controlled by the actions of the short and long heads of the biceps femoris and their attachments to the FCL. This can be demonstrated in vivo by applying a muscle stimulator to the muscle fibers of the short head of the biceps femoris and observing during its contraction its bowstring effects on the FCL. The distal quarter of

Fig. 2.6. Fibular collateral ligament. Its attachment on the femur is close to the lateral epicondyle, while its attachment on the fibular is along the lateral aspect of the fibular head.

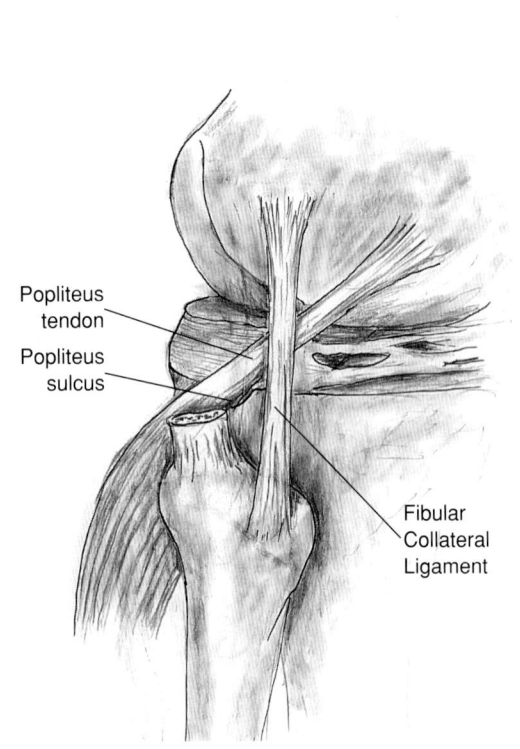

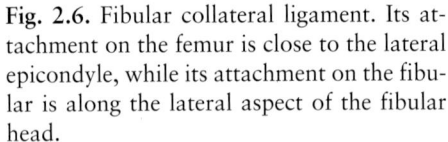

Fig. 2.7. Mid-third lateral capsular ligament. The ligament is composed of two portions—the meniscofemoral and meniscotibial ligaments.

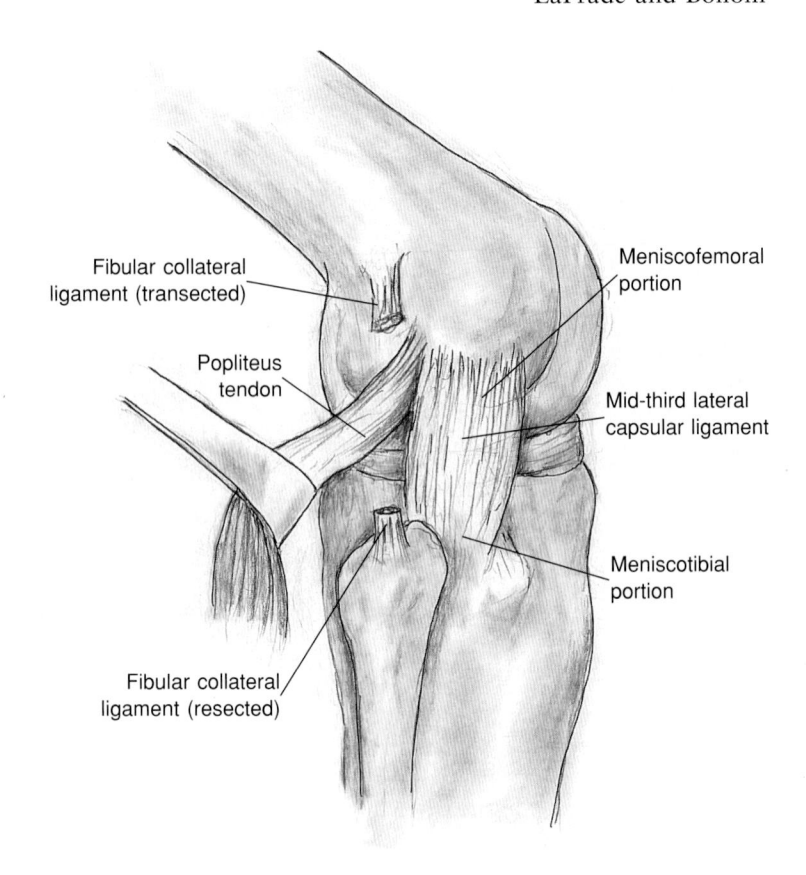

the FCL has a large bursa that covers the ligament as the anterior arm of the long head of the biceps femoris crosses it[23] (Fig. 2.3). The FCL-biceps bursa is a very useful landmark that can be used at the time of surgery to identify the FCL by making a horizontal incision through the anterior arm of the long head of the biceps femoris approximately 1 cm proximal to the lateral aspect of the fibular head.

The Mid-Third Lateral Capsular Ligament

The mid-third lateral capsular ligament is a thickening of the lateral capsule of the knee.[5,22,24,25] It is analogous to the mid-third medial capsular ligament or deep medial collateral ligament (Fig. 2.7). The meniscofemoral ligament portion of the mid-third lateral capsular ligament is that portion of the lateral capsule that extends from the femur, from just anterior to the popliteus tendon attachment on the femur posteriorly to the lateral gastrocnemius tendon attachment on the femur, and to the proximal portion of the lateral aspect of the lateral meniscus. The meniscotibial ligament portion is a distal continuation of this lateral capsular thickening, which extends from the undersurface of the midportion of the lateral meniscus and attaches to the tibia approximately 6 to 7 mm distal to this at the termination of the articular cartilage of the lateral tibial plateau. This portion of the mid-third lateral capsular ligament is commonly found to be injured with posterolateral knee injuries and is well visualized on coronal- or axial-based magnetic resonance imaging (MRI) scans. It can be injured either as a soft tissue avulsion off the tibia, often with the anterior arm of the short head of the biceps femoris and the capsulo-osseous layer of the iliotibial tract, or as a bony fracture, called a Segond fracture.[26]

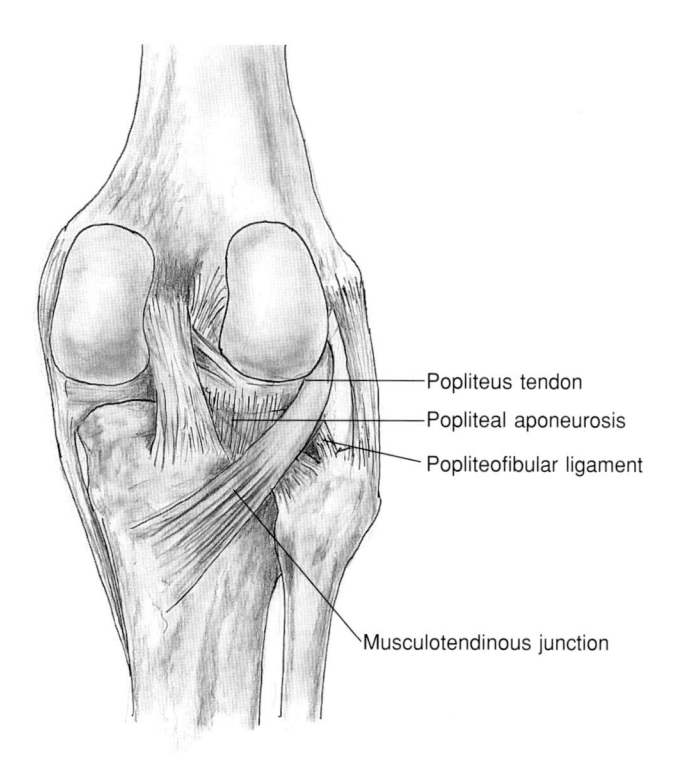

Fig. 2.8. Popliteus complex of the knee. The popliteus tendon, popliteomeniscal fascicles, popliteofibular ligament, aponeurotic attachment to the lateral meniscus, and muscle belly are important components of this complex.

Popliteus Complex of the Knee

The popliteus complex of the knee (Fig. 2.8) is one of the key components to providing static stability for the posterolateral aspect of the knee. It is extremely complex due to its many structural attachments and its varying course along the posterolateral knee. It is important to understand all of the different anatomic portions of the popliteus complex, as injuries to any of a varying number of areas of the popliteus complex can result in functional limitations or a significant amount of external rotation of the knee.

The popliteus tendon originates on the femur at the far anterior and proximal aspect of the popliteal sulcus. While its main tendinous portion is intraarticular in this position, it is also attached intimately to the meniscofemoral portion of the mid-third lateral capsular ligament in this area. In its course distally and posteriorly through the popliteal hiatus, it has three meniscal fascicles that attach to the lateral meniscus in a hoop-like manner. The anteroinferior fascicle, which can be best visualized arthroscopically (Fig. 2.9), forms a broad hoop-like attachment to the lateral meniscus, which extends down to the anterior division of the popliteofibular ligament. There are two posterior popliteomeniscal fascicles: the posterosuperior and posteroinferior popliteomeniscal fascicles. All three of these fascicles allow for the popliteus tendon to have a dynamic interaction with the lateral meniscus. They are frequently found to be torn at the time of arthroscopic evaluation of posterolateral knee injuries.[1,25]

Just distal to the popliteomeniscal fascicles, the popliteofibular ligament passes from the lateral aspect of the popliteus tendon, at its musculotendinous junction, to the fibular styloid. The popliteofibular ligament forms a 45-degree angle at its attachment on the fibular styloid. There are two main divisions of the popliteofibular ligament. The posterior division extends to the posteromedial tip of the fibular styloid. The anterior division is attached anteromedially to this styloid and has a tibial attachment as well. The anterior division of the popliteofibular ligament is that portion

Fig. 2.9. Popliteomeniscal fascicles of a right knee. The popliteus tendon (foreground hemostat) has been taken down off the popliteal sulcus.

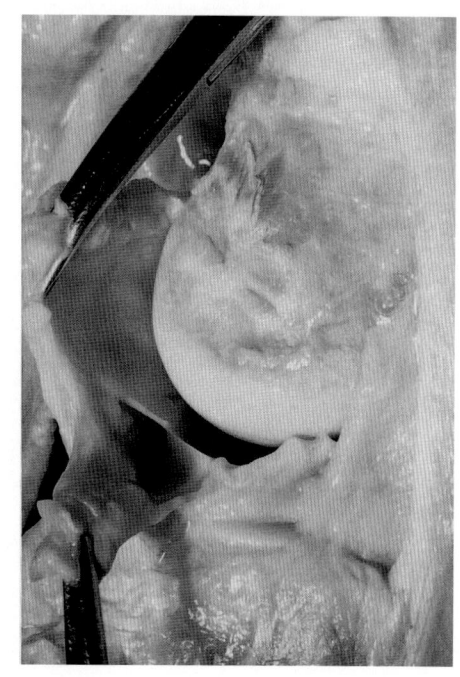

of the ligament that can be best viewed arthroscopically as a distal continuation of the anteroinferior popliteomeniscal fascicle. The anterior division also blends with the proximal anterior tibiofibular ligament joint capsule and its thickenings. The popliteofibular ligament is the main component of the lateral arch of the arcuate (arched) ligament on the posterior aspect of the knee.[5] Just distal to the popliteus musculotendinous junction and extending from the proximal end of the muscular portion of the popliteus tendon is a thick fascial expansion that reinforces the posterior capsule (coronary ligament or meniscotibial portion of the posterior capsule) at the level of the posterior horn of the lateral meniscus.

All of the components of the popliteus complex act in unison, and injuries to its attachment on the femur, the popliteofibular ligament, or its musculotendinous junction can contribute to rotatory instability. Due to its complex nature and varied angles of attachment, the popliteus complex is difficult to visualize with standard MRI imaging techniques. We have found it especially useful to use coronal oblique images along the course of the popliteus tendon to allow for assessment of these main components (Fig. 2.10).

Oblique Popliteal Ligament

The oblique popliteal ligament is a tendinous expansion across the posterior aspect of the knee that forms the medial arch of the arcuate ligament[5] (Fig. 2.11). It is formed from an expansion from the semimembranosus muscle-tendon complex on the posteromedial aspect of the knee and extends over the posterior aspect of the knee to attach to the posterolateral joint capsule in the region of the plantaris muscle and lateral gastrocnemius tendon at the region of the fabella or fabella analogue. It is located just proximal to the posterior cruciate ligament (PCL) attachment on the tibia. It can be well visualized on axial cuts on MRI imaging. No published studies to date have looked at its function biomechanically.

Ligament of Wrisberg

The ligament of Wrisberg is the posterior meniscofemoral ligament of the knee. It extends from its attachment posterior to the PCL on the lateral as-

Fig. 2.10. Coronal oblique magnetic resonance imaging (MRI) technique to visualize the popliteus complex of the knee. The MRI scans are obtained along the intraarticular course of the popliteus tendon.

Fig. 2.11. Oblique popliteal ligament, posterior view of a left knee. It is a thick fascial expansion from the semimembranosus, which courses to the posterolateral joint capsule near the plantaris-fabella region.

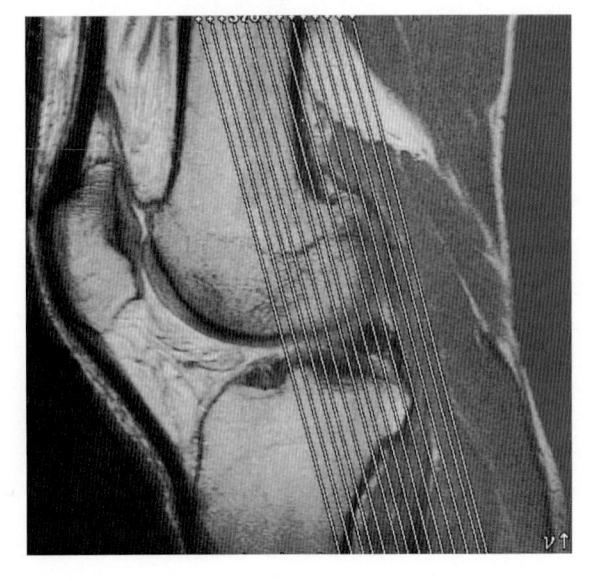

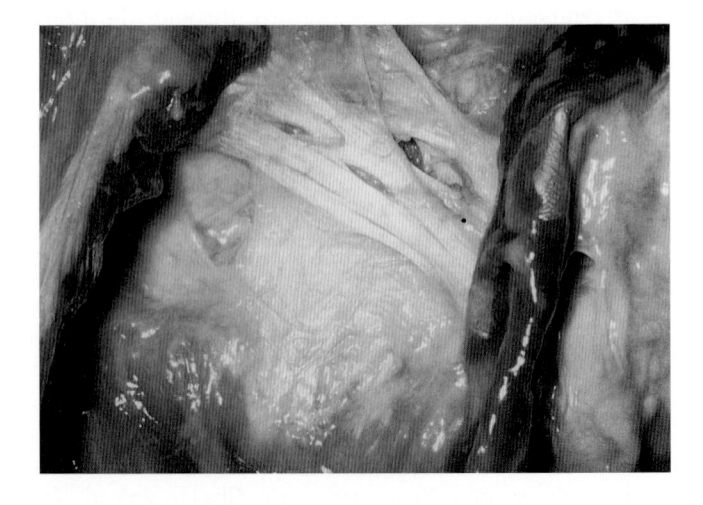

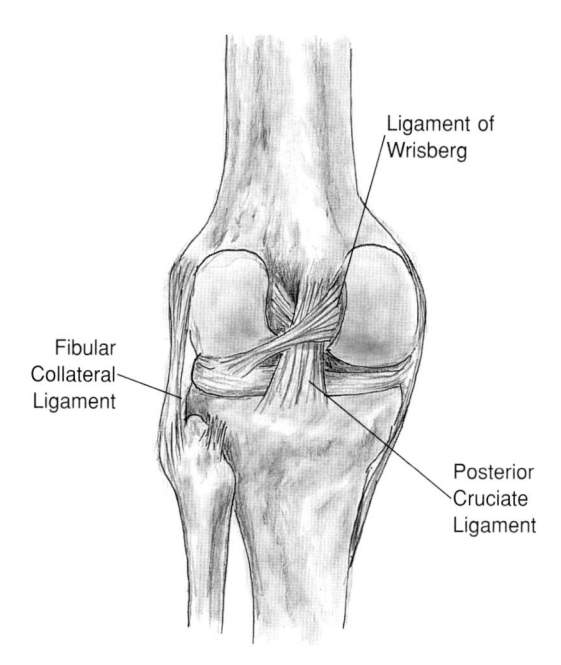

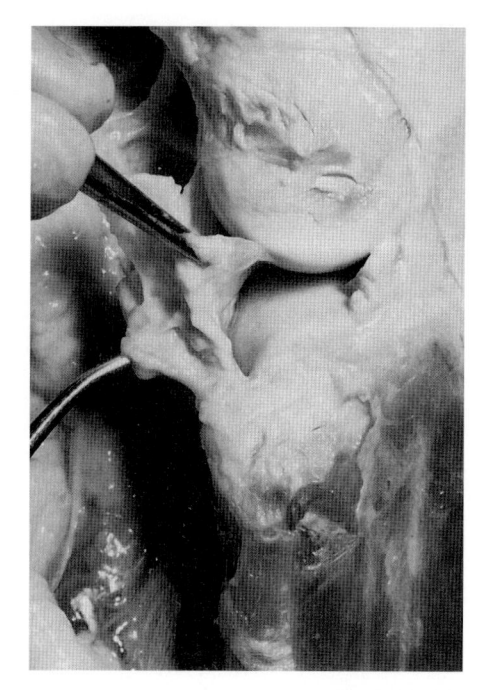

pect of the medial femoral condyle, to the posterior horn of the lateral meniscus (Fig. 2.12). In this area, it is found to intimately blend with the posterior horn of the meniscus. A unique relationship has been found in this regard in that when one pulls on the ligament of Wrisberg, tensioning can be seen to extend across the posterior horn of the lateral meniscus through the musculotendinous junction of the popliteus, extending through the popliteofibular ligament to the fibular styloid. Likewise, tension on the popliteofibular ligament can be seen to exert tension on the ligament of Wrisberg. The exact role of the ligament of Wrisberg in providing posterolateral stability to the knee is not currently known.

Fig. 2.12. Ligament of Wrisberg. The posterior meniscofemoral ligament extends from the posteromedial wall of the intercondylar notch to blend intimately with the posterior horn of the lateral meniscus.

Fig. 2.13. Coronary ligament of the lateral meniscus of a right knee. The coronary ligament is the meniscotibial portion of the posterior capsule.

Coronary Ligament of the Lateral Meniscus

The coronary ligament of the lateral meniscus is that portion of the posterior joint capsule that extends from the meniscus to the tibia (meniscotibial portion of the posterior capsule) (Fig. 2.13). Its boundaries extend from the posterosuperior and posteroinferior popliteomeniscal fascicles laterally to the attachment of the posterior horn of the lateral meniscus medially.[5] It is reinforced posteriorly by a fascial expansion from the proximal border of the popliteus muscle (popliteus aponeurosis). Its exact role in providing static stability to the posterolateral knee is unknown, but it has been found to be injured frequently at the time of arthroscopic evaluation of posterolateral knee complex injuries.[25]

Fabellofibular Ligament and Lateral Gastrocnemius Complex

The fabellofibular ligament is the distal edge of the capsular arm of the short head of the biceps femoris.[5,22] It courses from its attachment on the distolateral aspect of the fabella, or fabella analogue, to attach just lateral to the tip of the fibular styloid (Fig. 2.14). Its exact role in providing static stability to the posterolateral knee is unknown, but it appears to be injured frequently at the time of open surgical reconstruction. It may serve

Fig. 2.14. Fabellofibular ligament and lateral gastrocnemius complex of the right knee. The lateral gastrocnemius (and common peroneal nerve) is under the foreground retractor, while the hemostat is under the fabellofibular ligament.

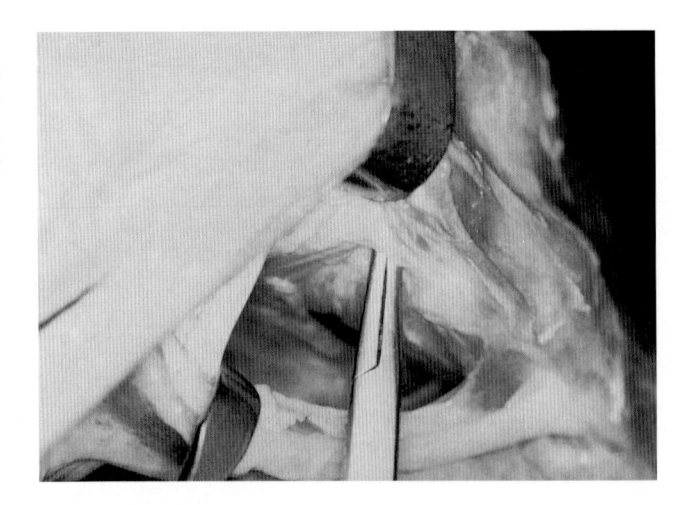

to provide some stability to the knee near extension. This ligament is most readily identified when the knee is near extension, where it is taut, rather than in flexion, where it is lax and appears to be a fine fascial sling.[5,22]

The lateral gastrocnemius tendon is at the far lateral aspect of the lateral gastrocnemius muscle.[5] It attaches to the posterior joint capsule at the region of the fabella and is intimately attached to the posterior joint capsule. It is only a relative thickening of the capsule until it attaches at the supracondylar tubercle region of the distal femur, approximately 4 to 5 mm posterior to the FCL attachment on the femur. The lateral gastrocnemius tendon can be well visualized on sagittal MRI sequences. It serves as a landmark to the surgical interval between the lateral gastrocnemius and soleus muscles in the approach to the posterior portion of the posterolateral knee.

Biomechanics of the Posterolateral Aspect of the Knee

The understanding of the biomechanics of the posterolateral knee has progressed as its complex anatomy has become better understood. Biomechanics has been studied largely through cadaveric sequential sectioning studies. More recently, an emphasis has been placed on understanding the role of stabilizing influence of the posterolateral structures and the consequences of injury to these structures. The clinical significance of a firm understanding of biomechanics of the posterolateral knee has also been shown. Recent studies have demonstrated the important role that intact individual anatomic components of the posterolateral knee play with respect to maintaining normal anterior cruciate ligament (ACL) graft reconstruction forces.[27] It is clear that the recent emphasis on the anatomy, imaging, diagnosis, and treatment of posterolateral knee injuries will only lead to a better understanding of the biomechanics of the posterolateral knee.

Varus Stabilization and the Posterolateral Knee

Many of the biomechanical studies of the posterolateral knee are difficult to compare due to the many aforementioned differences in nomenclature and the fact that many sequential cutting studies have grouped a large number of posterolateral structures together. However, if one constant exists with respect to the stabilization of the posterolateral knee from a biomechanical standpoint, it is that studies have shown that the FCL (lateral collateral ligament, LCL) is the major restraint to primary varus rotation at all positions of knee flexion.[28–30]

Isolated sectioning of the FCL or LCL resulted in at least a small but significant increase in varus rotation at any angle of knee flexion. Some debate remains on the importance of many of the other posterolateral structures with respect to varus stabilization. Gollehon et al.[28] sectioned the popliteal tendon from the posterolateral capsular structures, while Grood et al.[29] cut the popliteus tendon attachment on the femur and the posterolateral part of the capsule, including the arcuate complex (*popliteofibular ligament* and *capsular arm of the short head of the biceps femoris*), along with the capsular attachment of the popliteus muscle (*popliteal aponeurosis*), and any fabellofibular or short lateral ligaments that were present. Both of the aforementioned studies demonstrated that with combined sectioning of the FCL and some variety of sectioning of the deeper ligament complex of the posterolateral knee, varus rotation is increased.

The FCL and other posterolateral structures act to prevent lateral opening of the knee joint, but when they are absent, studies have shown that both cruciate ligaments are then recruited to help resist an applied varus moment. Markoff et al.[30] found that when measuring resultant forces in the cruciate ligaments of cadaveric knees subjected to varus moments, once the FCL was sectioned along with the posterolateral structures, cruciate ligament forces were significantly increased. They noted that when a varus moment was applied, ligamentous section increased the mean force in the ACL at all angles of flexion. The PCL was also noted to have a significant increase in mean force at angles of flexion of 45 degrees or more. Gollehon et al.[28] demonstrated that if the PCL was sectioned, after the FCL and their "deep ligament complex" was sectioned, a large increase in varus rotation occurred (from 14% to 19%) versus just 1% to 4% with isolated sectioning of the FCL.

Thus, with respect to stabilization of the knee with the application of a varus force, clearly an intact FCL is of paramount importance. In addition, the popliteus complex, posterolateral capsule, and PCL play an important role in preventing varus instability.

Anterior-Posterior Stabilization and the Posterolateral Knee

Several studies have demonstrated that sectioning of the posterolateral structures, including the PCL, result in no significant increase in anterior tibial translation.[28,29] The main stabilizer of the knee with respect to posterior tibial translation is an intact PCL. It is known that isolated sectioning of the PCL results in large increases in posterior tibial translation. Indeed, the PCL is the only isolated ligament, when sectioned, capable of providing initial restraint to primary posterior translation at all angles of knee flexion.[28,31,32] The amount of posterior tibial translation that occurred in cadaveric knees with an isolated sectioning of the PCL also increases with an increase in the angle of knee flexion. The greatest amount of posterior translation in a PCL-deficient knee was found to be at 90 degrees of knee flexion (values ranging from 1 to 5 mm at 0 degrees flexion to 11 to 20 mm at 90 degrees of flexion).[28,29]

The posterolateral structures may also play a significant role in restricting posterior tibial translation. Selective cutting studies have demonstrated that section of both the posterolateral structures (dividing the popliteus tendon from the posterolateral capsular structures) and section of the FCL results in slight increases of posterior translation at all angles of flexion.[28,29] It has also been demonstrated that combined section of the PCL,

Fig. 2.15. Role of secondary restraint of the posterolateral structures of the knee in preventing posterior translation. Isolated sectioning of lateral collateral ligament (LCL) or all deep posterolateral structures compared to isolated and combined PCL sectioning. (From Gollehon et al.,[28] with permission.)

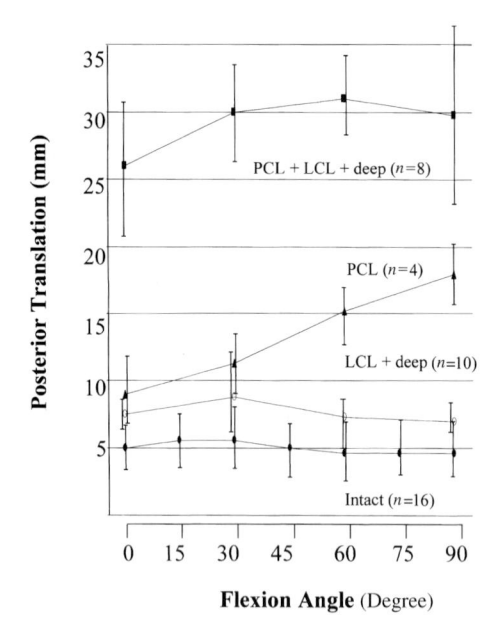

FCL, and posterolateral structures significantly increases posterior translation.[28,29] Interestingly, Gollehon et al.[28] found that at 0 and 30 degrees of knee flexion there was no significant difference in posterior translation between specimens in which a combined sectioning of the FCL and "deep ligament complex" (*fibular collateral ligament, popliteus tendon, mid-third lateral capsular ligament, popliteofibular ligament, capsular arm of the short head of the biceps femoris*), and those that had undergone an isolated PCL sectioning.

In summary, with respect to the role of anterior-posterior stabilization for the knee, it may be stated that across all angles of knee flexion the PCL remains the primary stabilizer. However, studies have shown that the posterolateral structures (popliteus, FCL, etc.) definitely act as secondary restraints to posterior tibial translation.[6,28,29] The action of secondary restraint by the posterolateral structures is most important between 0 and 30 degrees of knee flexion (Fig. 2.15), where the amount of posterior tibial translation measured in motion studies may be as much for a knee with an isolated posterolateral corner injury as for an isolated PCL injury.

Internal and External Tibial Rotation and the Posterolateral Knee

Studies have shown that when an internal tibial torque is placed on the knee, no isolated or combined loss of the FCL, deeper posterolateral structures (popliteus/joint capsule), or the PCL causes a significant increase in internal rotation of the tibia.[29,33] In contrast, external rotation of the tibia is very dependent on the aforementioned structures. Grood et al.[29] found that after sectioning the posterolateral structures (FCL, arcuate complex [*popliteofibular ligament, fabellofibular ligament, capsular arm of the short head of the biceps femoris*], popliteus tendon-femoral attachment), a significant increase in external tibial rotation at both 30 and 90 degrees of knee flexion (13- and 5.3-degree increase, respectively) occurred.

When the PCL was then sectioned after the posterolateral structures had been sectioned, an additional increase in external rotation was noted, but this was found to be significant only at 90 degrees of knee flexion. Gollehon et al.[28] demonstrated the importance of intact posterolateral structures with respect to external tibial rotation. They noted that isolated sectioning of the FCL as well as the popliteus complex brought about a significant increase in external tibial rotation.

Coupled Internal and External Torque on the Tibia with Anterior and Posterior Force and the Posterolateral Knee

In the normal intact knee when an anterior force is applied, the tibia tends to rotate internally. It has also been demonstrated that when a posterior force is applied the tibia rotates externally. When an anterior force is applied to the knee and damage to the posterolateral structures, PCL, or FCL has occurred, there is a small increase in internal tibial rotation.[27,34] However, at least two studies have shown that combined sectioning of the FCL and popliteus cause a marked increase in the coupled external rotation associated with a posterior directed force at all angles of knee flexion (the largest increase noted at 30 degrees).[33,34] The role of the PCL in coupled external rotation is less clear. Some studies have suggested that a loss of the PCL equates to a complete loss of the coupled external rotation when a posterior force is applied.[29] Other studies have not noted this loss.[34] Regardless of the relationship of the PCL to coupled external rotation, it is clear that posterolateral complex injury will result in significant increased coupled posterior translation and external rotation.

Popliteomeniscal Fascicles and Lateral Meniscus Stability

While numerous studies have described disruption of the popliteomeniscal fascicles at the time of surgery,[1,22,25] Simonian et al.[35] were the first to objectively evaluate the stability of the lateral meniscus before and after sequentially sectioning these fascicles. They found a significant increase in lateral meniscus motion with both an isolated anteroinferior fascicle disruption (50%), and a combined anteroinferior and posterosuperior popliteomeniscal fascicle disruption (78%) with application of a simulated 10-N arthroscopically applied load. While the lateral meniscus never became locked in the joint during this study, the authors speculated that meniscal locking could occur with the variable loads encountered in normal daily activities.

Clinical Implications

The effects of injury to the posterolateral corner has become a growing area of concern. Isolated posterolateral corner injuries are not common.[22,25,36] Structures of the posterolateral knee are more commonly injured with concomitant ACL or PCL tears.[22,25,36] LaPrade et al.[27] and Markoff et al.[30] have emphasized the importance of repairing posterolateral injuries prior to or concurrently with ACL graft reconstruction. LaPrade et al.[37] described the strain placed on several structures of the posterolateral knee during clinical limits-of-motion testing. They measured the elongation of specific posterolateral knee structures by placement of differential variable reluctance transducers (DVRTs) in posterolateral knee structures for both the normal state and with sequential cutting of structures in fresh frozen cadaveric knees. Limits-of-motion testing was performed for the relevant posterolateral knee exam: posterolateral tibial rotation at 30 degrees of knee flexion (posterolateral Lachman test), adduction

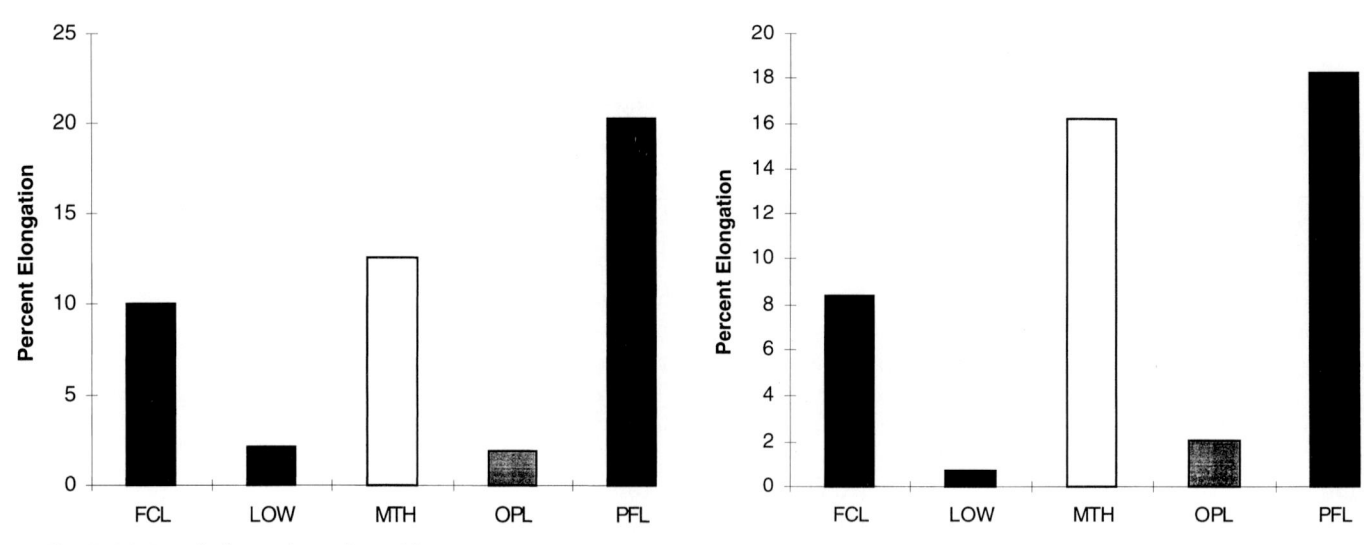

Fig. 2.16. Local elongation of specific posterolateral knee structures with motion testing at 30 degrees for the dial test. FCL, fibular collateral ligament; LOW, ligament of Wrisberg; MTH, mid-third lateral capsular ligament; OPL, oblique popliteal ligament; PFL, popliteofibular ligament.

Fig. 2.17. Local elongation of specific posterolateral knee structures with motion testing at 30 degrees for the posterolateral Lachman test.

and abduction at 0 and 30 degrees, posterolateral drawer, the dial test, and posterior and anterior drawer in neutral and external rotation. Increases in local strain (elongation) were seen in many of the secondary restraints of the posterolateral knee. The mid-third lateral capsular ligament, FCL, and popliteofibular ligament all demonstrated increases in the amounts of strain with sequential cutting studies. The tests demonstrating the most significant increases in local strain were the dial test at 30 degrees of knee flexion and posterolateral tibial rotation at 30 degrees (or the posterolateral Lachman test) (Figs. 2.16 and 2.17).

Markoff et al.[30] noted that the FCL and posterolateral structures act to prevent both cruciate ligaments from being recruited to help resist applied varus moment. This function is protective of an ACL graft reconstruction by reducing potentially injurious forces of a varus applied force. Thus, they recommend concomitant posterolateral repair or reconstruction in knees with multiple ligamentous injuries.

With respect to the effects of posterolateral complex injuries on forces in an ACL reconstruction, LaPrade et al.[27] noted that ACL graft forces are significantly higher with concomitant grade III posterolateral injuries. During joint loading a significant increase in force on an ACL graft occurred during varus and coupled varus–internal rotation moments in knees with grade III posterolateral structure injury compared to the same reconstruction in the posterolaterally intact knee. This study supports the clinical observation that an untreated grade III posterolateral injury contributes to ACL graft failure by developing higher forces on the ACL graft.

Conclusion

With a firm understanding of the anatomy and biomechanics of the posterolateral knee, one may understand the underlying importance of the diagnosis and treatment of a posterolateral knee injury. The FCL, the popliteus/popliteofibular ligament and the associated arcuate ligament complex (fabellofibular ligament, capsular arm of the short head of the biceps femoris, and the oblique popliteal ligament), and the PCL play crucial roles with respect to stabilization of the knee. Further advances in understanding and research regarding the posterolateral knee are dependent on a uniform description of anatomic structures. It is clear that when these structures are not functioning properly, patients may suffer significant disability. Our understanding of both anatomy and biomechanics can lead to a more correct and complete diagnosis and to improved surgical reconstructive techniques that will ultimately benefit our patients.

References

1. Stäubli HU, Birrer S. The popliteus tendon and its fascicles at the popliteus hiatus: gross anatomy and functional evaluation with and without ACL deficiency. Arthroscopy 1990;6:209–220.
2. Stäubli HU, Rauschning W. Popliteus tendon and lateral meniscus. Gross and multiplanar cryosectional anatomy of the knee. Am J Knee Surg 1991;4:110–121.
3. Terry GC, Hughston JC, Norwood LA. The anatomy of the iliopatellar band and iliotibial tract. Am J Sports Med 1986;14:39–45.
4. Terry GC, LaPrade RF. The biceps femoris complex at the knee: its anatomy and injury patterns associated with acute ALRI-AMRI. Am J Sports Med 1996;24:2–8.
5. Terry GC, LaPrade RF. The posterolateral aspect of the knee: anatomy and surgical approach. Am J Sports Med 1996;24:732–739.
6. Veltri DM, Deng X-H, Torzilli PA, et al. The role of the popliteofibular ligament in stability of the human knee: a biomechanical study. Am J Sports Med 1996;24:19–27.
7. Sutton JB. The nature of certain ligaments. J Anat Physio 1884;18:225–238.
8. Last RJ. Some anatomical details of the knee joint. J Bone Joint Surg 1948;30B:683–688.
9. Sneath RS. The insertion of the biceps femoris. J Anat 1955;89:550–553.
10. Kaplan EB. Surgical approach to the lateral (peroneal) side of the knee joint. Surg Gynecol Obstet 1957;104:346–356.
11. Kaplan EB. The iliotibial tract. Clinical and morphological significance. J Bone Joint Surg 1958;40A:817–832.
12. Kaplan EB. The fabellofibular and short lateral ligaments of the knee joint. J Bone Joint Surg 1961;43A:169–179.
13. Kaplan EB. Some aspects of functional anatomy of the human knee joint. Clin Orthop Rel Res 1962;23:18–29.
14. Marshall JL, Girgis FG, Zelko RR. The biceps femoris tendon and its functional significance. J Bone Joint Surg 1972;54A:1444–1450.
15. Reis FP, Ferraz de Carvalho CA. Anatomical study on the proximal muscle. Rev Bras Pesq Med Biol 1975;8:373–380.
16. Cohn AK, Mains DB. Popliteal hiatus of the lateral meniscus. Am J Sports Med 1979;7:221–226.
17. Fabbriciani C, Oransky M, Zoppi U. Il musculo popliteo; studio anatomico. Arch Ital Anat 1982;87:203–217.
18. Seebacher JR, Inglis AE, Marshall JL, et al. The structure of the posterolateral aspect of the knee. J Bone Joint Surg 1982;64A:536–541.
19. Lobenhoffer P, Posel P, Witt S, et al. Distal femoral fixation of the iliotibial tract. Arch Orthop Trauma Surg 1987;106:285–290.
20. Oransky M, Canero G, Maiotti M. Embryonic development of the posterolateral structures of the knee. Anat Rec 1989;225:347–354.
21. Watanabe Y, Moriya H, Takahashi K, et al. Functional anatomy of the posterolateral structures of the knee. Arthroscopy 1993;9:57–62.
22. LaPrade RF, Hamilton CD, Engebretsen L. Treatment of acute and chronic combined ACL and posterolateral knee ligament injuries. Sports Med Arthrosc Rev 1997;5:91–99.
23. LaPrade RF, Hamilton CD. The anatomy of the fibular collateral ligament–biceps femoris bursa. An anatomic study. Am J Sports Med 1997;25:439–443.
24. Hughston JC, Andrews JR, Cross MJ, et al. Classification of knee ligament instabilities. Part II. The lateral compartment. J Bone Joint Surg 1976;58A:173–179.
25. LaPrade RF. Arthroscopic evaluation of the lateral compartment of knees with grade 3 posterolateral knee complex injuries. Am J Sports Med 1997;25:596–602.

26. Segond P. Recherches Cliniques et Experimentales sur les Epanchemants Sanguins du Genou oar Entorse. Progres Med (Paris) 1879;7:1–84.
27. LaPrade RF, Resig S, Wentorf FA, et al. The effects of grade III posterolateral knee complex injuries on force in an anterior cruciate ligament reconstruction graft: a biomechanical analysis. Am J Sports Med 1999;27:469–475.
28. Gollehon DL, Torzilli PA, Warren RF. The role of the posterolateral and cruciate ligament in the stability of the human knee. J Bone Joint Surg 1987;69A:233–242.
29. Grood ES, Stowers SF, Noyes FR. Limits of movement in the human knee. Effect of sectioning the posterior cruciate ligament and posterolateral structures. J Bone Joint Surg 1988;70A:88–97.
30. Markoff, KL, Burchfield DM, Shapiro MM, Cha CW, Finerman GA, Slauterbeck JL. Biomechanical consequences of replacement of the anterior cruciate ligament with a patellar ligament allograft. Part II: forces in the graft compared with forces in the intact ligament. J Bone Joint Surg 1996;78A:1728–1734.
31. Brantigan OC, Voshell AF. The mechanics of the ligaments and menisci of the knee joint. J Bone Joint Surg 1941;23A:44–66.
32. Fukubayashi T, Torzilli PA, Sherman MF, et al. An in vitro biomechanical evaluation of anterior-posterior motion of the knee: tibial displacement, rotation, and torque. J Bone Joint Surg 1982;64A:258–264.
33. Noyes FR, Stowers SF, Grood ES, et al. Posterior subluxations of the medial and lateral tibiofemoral compartments. Am J Sports Med 1993;21:407–414.
34. Veltri DM, Deng X-H, Torzilli PA, et al. The role of the cruciate and posterolateral ligaments in stability of the knee. Am J Sports Med 1995;23:436–443.
35. Simonian PT, Sussman PS, van Trommel M, et al. Popliteomeniscal fasciculi and lateral meniscal stability. Am J Sports Med 1997;25:849–853.
36. LaPrade RF, Terry GC. Injuries to the posterolateral aspect of the knee: association of anatomic injury patterns with clinical instability. Am J Sports Med 1997;25:433–438.
37. LaPrade RF, Hamilton CD, Wentorf F, et al. The measurement of elongation in specific posterolateral knee structures during clinical limits-of-motion testing. Paper presented at the International Arthroscopy Association/International Society of the Knee Meeting, Buenos Aires, May, 1997.

Chapter Three

The Anatomy and Biomechanics of the Posteromedial Aspect of the Knee

Fred Flandry and Christian C. Perry

This chapter addresses the anatomy and the biomechanical aspects of the anatomy of the medial side of the knee within the narrow context of medial knee ligament injuries associated with posterior cruciate ligament (PCL) sprains. As such, it is not an all-encompassing discourse, but rather is intended to provide the clinician with the salient background requisite for managing this specific injury entity.

Anatomy

The capsular ligaments of the knee may be thought of as an aponeurosis of the more proximal thigh and, to a lesser extent, the more distal tibial musculature. The meniscus is the terminal element of this musculotendinous unit (Fig. 3.1). Through this arrangement, the muscles, particularly those of the thigh, are able to not only activate motion of the joint, but also to impart stability by modulating ligament tension and, as a result of this, by increasing joint contact forces. Thus, the "static" ligaments are not really static at all, but rather are dynamic structures. Failure to appreciate

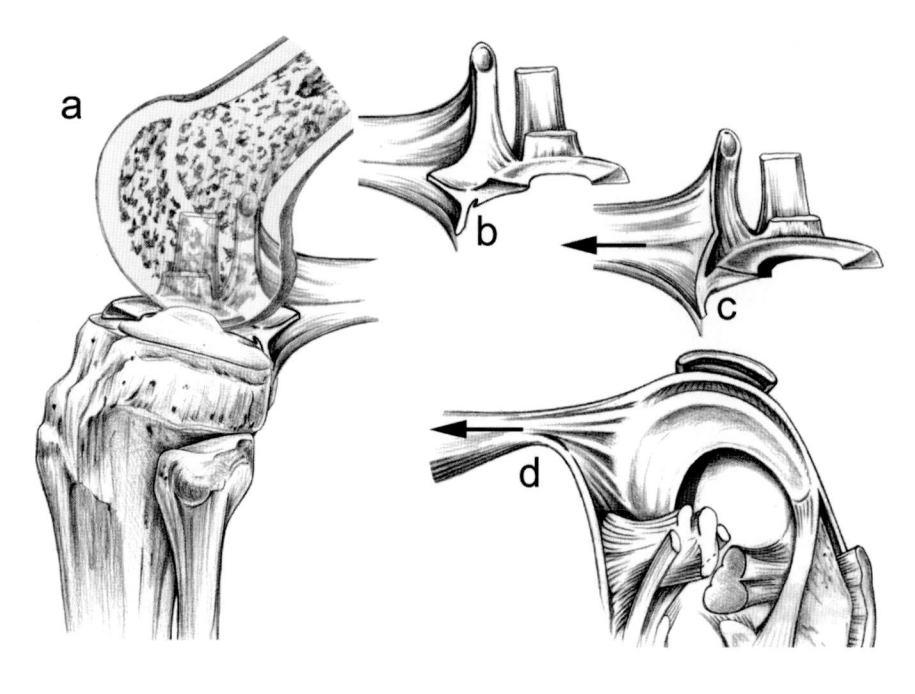

Fig. 3.1. (a) The posteromedial corner can better be appreciated as an aponeurotic continuum of the semimembranosus. (b) As such, the semimembranosus and the deeper portions of the capsular ligaments have direct and strong attachments to the periphery of the medial meniscus. (c) When the semimembranosus is contracting, the capsular ligaments bowstring, which increases joint stability by decreasing compliance of the ligamentous tissue and increasing contact pressure of the articular surfaces. (d) The meniscus posterior horn, as well, is pulled posteriorly with hamstring contraction, which may protect it from becoming impinged between the femoral condyle and tibial plateau and possibly torn. (Copyright © 2000 by the Hughston Sports Medicine Foundation, Inc.)

these subtle but essential attributes may, in part, explain why less attention is given to restoring anatomic integrity to the capsular ligament injuries in knee sprains.

Layers Versus Ligaments

The "layer" concept[1] has been advocated as an approach to describing the medial anatomy of the knee. While true that the capsular ligaments are to some extent (with the exception of the tibial collateral ligament) continuous with regional thickenings and fiber orientations, this approach places the clinician at a disadvantage because it de-emphasizes the discrete descriptions and biomechanical functions of the individual ligaments. Layer I consists merely of the sartorius fascia, and as such, is not actually a ligament. Further, the concept of three layers holds only in the middle third, where the tibial collateral ligament composes layer II. The anatomic approach that describes ligaments rather than layers provides the tools to appreciate biomechanical function, identify injury pathology, and formulate repair and reconstruction strategies.

Ligaments

Blueprint of the Medial Capsular Ligaments

Before discussing specific structures, it is worthwhile to develop some generalized concepts regarding capsular ligament structure and function. The capsular ligaments are a continuous sheet of ligamentous connective tissue with varying attachments, varying regional thickenings and fiber orientations, and varying muscular attachments. The primary structures medially are the medial retinacular ligament, medial mid-third capsular ligament, posterior oblique ligament, oblique popliteal ligament, tibial collateral ligament, and semimembranosus capsular insertion.

In cross section, the capsular ligaments may be thought of as having anterior, middle, and posterior thirds on both the medial and lateral sides (Fig. 3.2). The anterior third consists of the medial and lateral retinacular

Fig. 3.2. Blueprint for the capsular ligaments. In this cross section, or bird's-eye view, of the tibial articular surface, the capsular ligaments can be divided into thirds. Note that the tibial collateral ligament is separated from the medial mid-third capsular ligament by a deep bursa. (Copyright © 2000 by the Hughston Sports Medicine Foundation, Inc.)

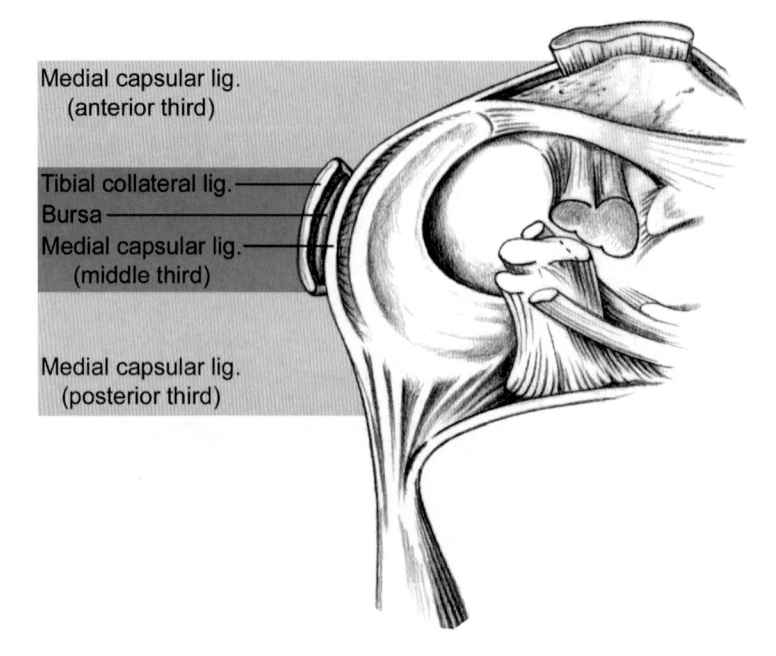

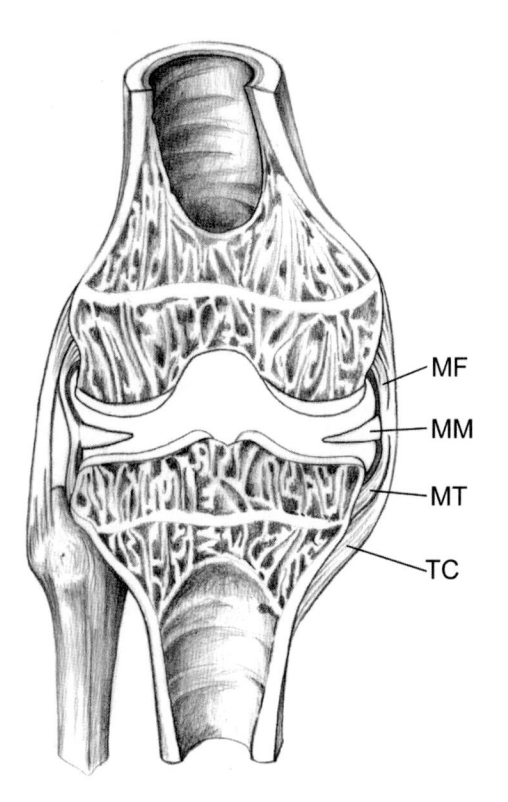

Fig. 3.3. Illustrated in a midcoronal section, the tibial collateral (TC) ligament, medial mid-third (MM) capsular ligaments, and medial meniscus. The medial mid-third capsular ligament can be subdivided into meniscofemoral (MF) and meniscotibial (MT) segments. (Copyright © 2000 by the Hughston Sports Medicine Foundation, Inc.)

ligaments of the extensor aponeurosis. This segment of the capsule has tibial and meniscal attachments, but no femoral bony attachment. Medially, the mid-third capsule contains the deep medial mid-third capsular ligament and superficial to this, the tibial collateral ligament. The posterior third contains the posterior oblique ligament and the origin of the oblique popliteal ligament. The posterior third ligaments, in particular, are the direct termination of the capsular arm of the semimembranosus tendon.

In the coronal plane, each capsular ligament can be functionally divided into a meniscofemoral and a meniscotibial segment (Fig. 3.3). The meniscofemoral and meniscotibial segments, sometimes referred to as ligaments, are really portions of the described ligaments. Medially, the meniscofemoral segments span a relatively longer distance (from the medial epicondyle/adductor tubercle to the meniscus) than the meniscotibial segments, which anchor the meniscus to the tibia.

Anterior Third Capsular Ligament (Medial Retinaculum)

This ligament is functionally a component of the extensor mechanism, and, as such, is biomechanically more significant to the patellofemoral rather than the tibiofemoral joint. It is, however, important to realize the association of medial extensor mechanism injury with medial ligament injury[2] in approaching repairs to these ligaments.

Deep Mid-Third Capsular Ligament

The mid-third capsular ligament arises from the medial epicondyle. Its meniscofemoral segment is relatively longer and thinner than its meniscotibial segment. The meniscotibial segment spans the meniscal peripheral rim to insert on the tibia approximately 5 mm from the articular margin. It is separated from the overlying tibial collateral ligament by a confluence of bursae,[3] which allow the tibial collateral ligament to glide freely over deeper structures. As implied by its name, the mid-third medial capsular

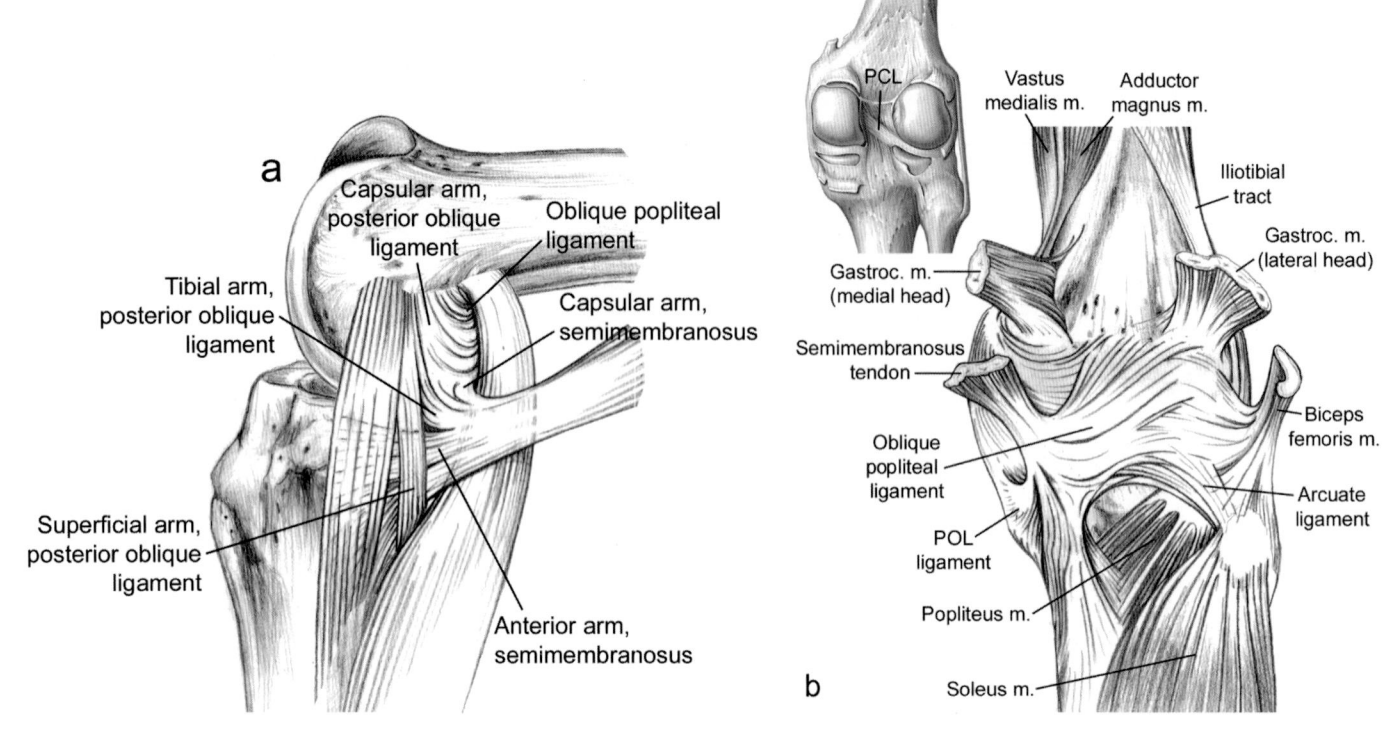

Fig. 3.4. (a) The superficial arm of the posterior oblique ligament (POL) courses distally and superficially, while the tibial arm of the POL courses distally and deep to the anterior arm of the semimembranosus. The capsular arm inserts posteriorly, becoming confluent with (b) the oblique popliteal ligament (OPL) and receiving the insertion of the capsular arm of the semimembranosus. (Copyright © 2000 by the Hughston Sports Medicine Foundation, Inc.)

ligament provides the attachment point for the middle third of the medial meniscus.

Posterior Oblique Ligament and Oblique Popliteal Ligament

This ligament complex takes its origin from the adductor tubercle just posterior and slightly proximal to the medial epicondyle. The combined structures, the medial epicondyle and the adductor tubercle, have been referred to as the "saddle" because their shape resembles the seat of a western horse saddle. The saddle is an important palpable bony landmark for surgical anatomic orientation.

The posterior oblique ligament (POL)[4] has three distinct arms: superficial, tibial, and capsular (Fig. 3.4). The superficial arm is the thin, fibrous, and more anterior portion of the ligament. It passes superficial to the anterior arm of the semimembranosus as it courses distally to become confluent with the fascia of the pes anserinus. The tibial arm is deep to the superficial arm and passes deep to the anterior arm of the semimembranosus as it courses distally. The deeper portion of the POL in its segment between the adductor tubercle and the tibia forms a strong attachment to the posterior horn of the medial meniscus. The capsular arm also originates at the adductor tubercle, but courses more posteriorly, becoming confluent with the oblique popliteal ligament (OPL) and inserting on the posterior intercondylar rim of the lateral femoral condyle. The POL capsular arm–OPL complex lies deep to the medial head of the gastrocnemius muscle and is separated from the medial head by a gastrocnemius (popliteal) bursa. The capsular arm is the conduit for transmission of dynamic input from the semimembranosus to the POL and the OPL. The capsular arm also provides the direct connection that allows the semimembranosus to retract the medial meniscus posteriorly, protecting it from encroachment by the femoral and tibial condyles as the knee flexes.

A thin, synovial fascial space separates the anterior border of the tibial

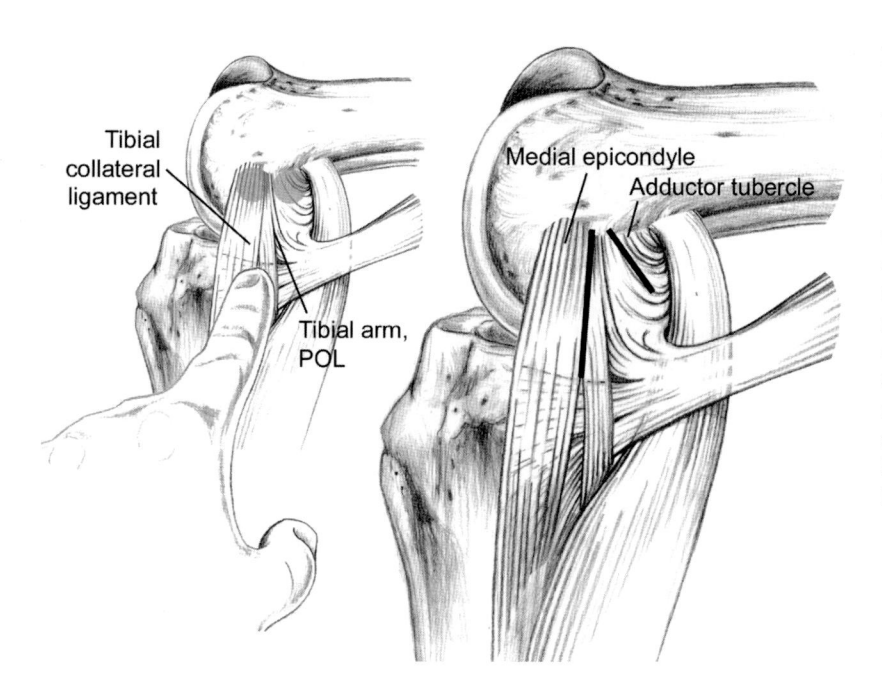

Fig. 3.5. The interval between the posterior border of the tibial collateral ligament and the anterior border of the tibial arm of the POL can be palpated as a soft spot on a line distal to the medial epicondyle/adductor tubercle complex (saddle) when the knee is flexed to 90 degrees. This interval is the location of the standard posteromedial arthrotomy incision. A more distal arthrotomy may be made in line with the fibers of the capsular arm of the POL, if needed, to gain access to the posterocentral attachment of the medial meniscus or the posterior cruciate ligament. (Copyright © 2000 by the Hughston Sports Medicine Foundation, Inc.)

arm of the POL from the posterior border of the tibial collateral ligament. This thin interval extends as a line distally from the saddle and parallel to the anterior tibial crest; it can be palpated at surgery unless injury or swelling have obscured normal anatomy (Fig. 3.5). This fascial space is the interval in which a posteromedial arthrotomy is placed. The OPL technically is a component of the posterior capsule, but its origin is from the posteromedial corner, and it thus functions in concert with the POL and the semimembranosus. The OPL takes its origin from the capsular arm of the POL and the aponeurotic insertion of the semimembranosus into the capsular arm of the POL (Fig. 3.4b). It continues as the posterior capsule and composes the meniscotibial segment of the capsular ligament for the posterior horn of the medial meniscus, inserting just below the tibial articular rim of the posterior tibial plateau. It passes laterally and superficial to the PCL as the posterior capsule of the intercondylar notch, attaching to the notch margin of the posterior lateral femoral condyle as well as merging into the posteromedially oriented fibers of the arcuate ligament. More proximally, the OPL lies deep to and forms the deep surface of the medial gastrocnemius tendon bursa.

Tibial Collateral Ligament

The tibial collateral ligament (TCL) originates from the medial epicondyle and lies superficial to the medial mid-third capsular ligament. It inserts on the medial face of the tibia, usually about 4.6 cm distal to the tibial articular rim and just posterior to the pes anserinus insertion.[3] It has no attachment to the tibia proximal to this insertion point, and in fact is separated from the tibia by a series of deep bursae.[3] The inferior medial geniculate artery, as it courses anteriorly, passes between the TCL and the tibia in this bursal plane. More proximally, the TCL is similarly separated from the medial mid-third capsular ligament by more bursae and areolar tissue. The ligament has no direct fibrous connection to the medial meniscus.[3,5] The separation of the TCL from the deeper structures by bursae and areolar tissue allows it to glide posteriorly during knee flexion and anteriorly during extension. The more anterior fibers of the ligament parallel the line from origin to insertion. The fibers in the posterior portion of the ligament fol-

Fig. 3.6. (a) The posteromedial meniscocapsular musculotendinous unit as viewed from inside the joint with the femur and tibia removed. Note how the semimembranosus blends into the POL and OPL and sends direct fibers through the capsule into the posterior horn of the medial meniscus. (b) When the semimembranosus contracts, the posterior horn of the meniscus is pulled posteriorly, protecting it from entrapment by the condyles and possible injury. (Copyright © 2000 by the Hughston Sports Medicine Foundation, Inc.)

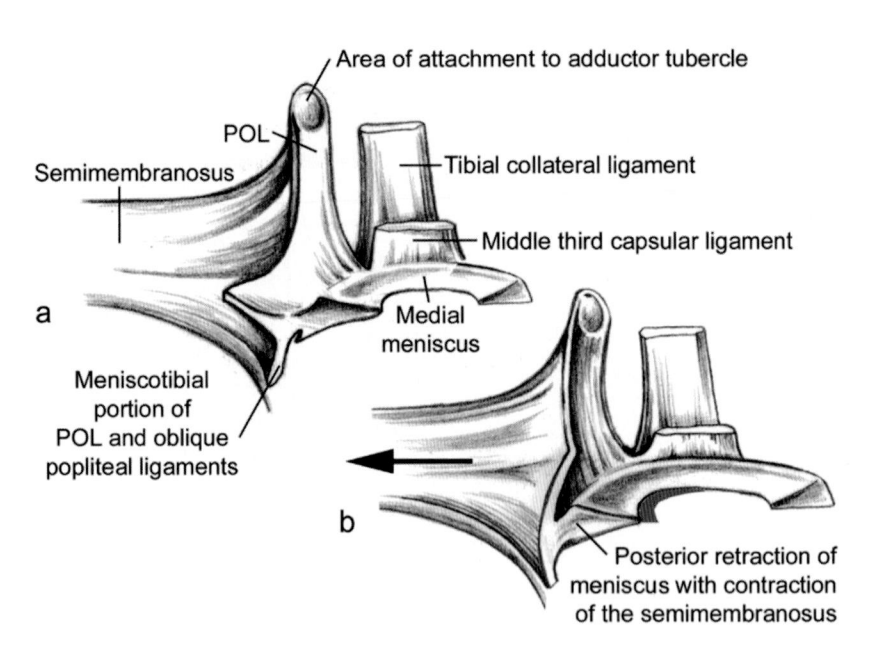

low an oblique course, giving rise to the ligament sometimes being described as having a parallel (anterior) and an oblique (posterior) portion.[3]

Meniscus

The medial meniscus provides a cushioning effect in hyperextension and full flexion. The peripheral outline of the medial meniscus is much larger in diameter than its lateral analogue; however, it is thinner at its periphery and narrower, and thus more of the medial tibial articular surface is exposed. The topography of the medial tibial plateau is more concave than the lateral plateau; therefore, the meniscus has less of a role in "cupping" the medial femoral condyle than the lateral meniscus does in cupping the lateral condyle.

Unlike the lateral meniscus, the medial meniscus has no direct attachment to the cruciate ligaments; its central anterior and posterior attachments terminate in bone. The periphery is confluent with the deep capsular ligaments. The meniscotibial segments of these ligaments closely associate the meniscus with the tibial articular margin. The femoral surface of the meniscus is concave or wedge shaped in cross section, while the tibial or undersurface of the meniscus is relatively flat. This wedge contour aids in providing resistance to anterior tibial translation if the meniscotibial attachments remain competent.

Muscular Input

The semimembranosus muscle serves two functions: as a direct actuator of tibial motion through direct insertions to bone and as a dynamic stabilizer of the joint through aponeurotic insertions into posteromedial corner capsular structures (Fig. 3.6). The complex insertion of the semimembranosus consists of five major arms. The tibial arm, or anterior arm, inserts on the anteromedial tibial condyle deep to and near the insertion of the TCL. The inferior arm inserts broadly on the posteromedial ridge of the medial tibial condyle. As its name implies, its fibers course distally. The direct arm inserts into the posterior tibial tubercle. The remaining two arms form an aponeurosis merging into capsular structures. The capsular arm (some-

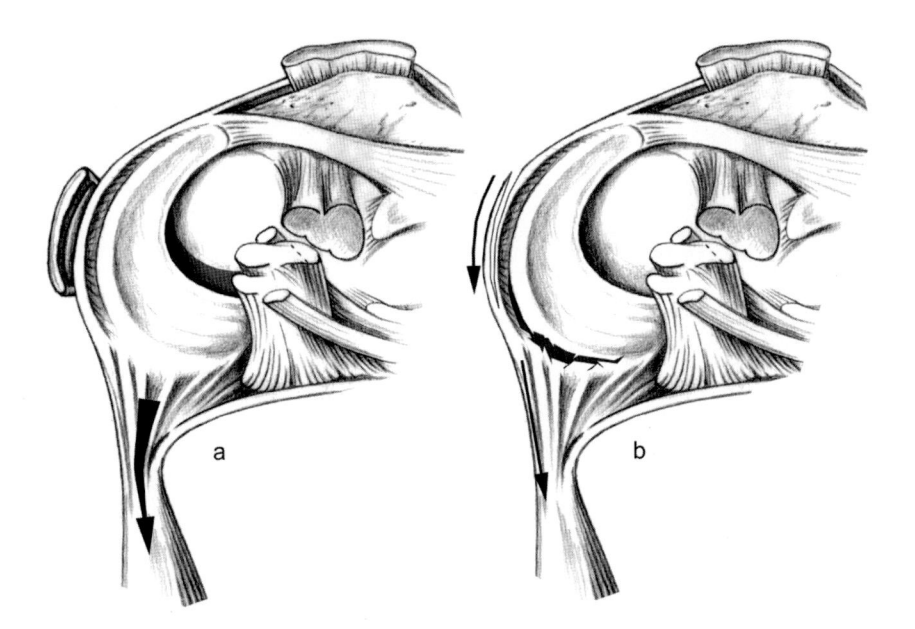

3. The Anatomy and Biomechanics of the Posteromedial Aspect of the Knee

Fig. 3.7. Normal semimembranosus–capsular ligament–meniscus function. The importance of associated ligament injury in the presence of a peripheral tear of the medial meniscus can be appreciated. The meniscus should be reattached with the ligament adequately retensioned. Without dynamic input into the meniscus, the semimembranosus cannot pull the posterior horn of the meniscus to protect it from impingement by the posterior medial femoral condyle as the knee flexes. (Copyright © 2000 by the Hughston Sports Medicine Foundation, Inc.)

times used to refer to both aponeurotic arms) blends with the capsular arm of the POL. The lateral arm courses posteriorly to contribute to the OPL.

Musculotendinous Meniscocapsular Aponeurosis

By now, it has become clear that the concept of the meniscus or ligament as an isolated structure oversimplifies the complexity of the posteromedial corner. The semimembranosus is the dynamic motor of the posteromedial corner. As its tendon approaches the joint capsule, it ramifies, and like the tentacles of an octopus (or in this case "pentopus"), it reaches in all directions around the posteromedial corner, developing a broad grasp on both bone and capsule. Its dynamic aponeurosis blends into the POL and the OPL, and through these and the more anterior capsular ligaments, terminates in the medial meniscus. When viewed graphically (Fig. 3.6), this image of a unified structure becomes more comprehensible.

Once this concept is grasped, injury to this area is no longer thought of as a simple torn ligament. Repair becomes more than simply suturing the torn ends of a ligament back together. Repair must be undertaken so that the integrity of the entire complex, not just the ligament, is restored for normal function to be reestablished. One can imagine in the case of a severe tear how much a semimembranosus in spasm will displace the torn ends to which it remains intact. The notion that they will simply "heal back" with normal becomes unfathomable.

As the semimembranosus contracts, the knee flexes. By its pull through the aponeurotic segments, the meniscus moves posteriorly, providing protection from impingement between the femoral and tibial condyles and possibly tearing from its periphery (Fig. 3.7). Additionally, the pull on the capsular ligaments causes these ligaments to bowstring, and thus contributes to stability in two ways. First, the bowstringing itself negates mild laxity in the ligament and provides a mechanism for central nervous system modulation of tension within the ligament. The concept of mechanoreceptors in knee ligaments is now well accepted,[6] and the role for these receptors may be to function in a feedback loop arrangement. Second, increasing tension in the ligament increases joint contact forces, which by itself is a method of increasing stability.[7]

Fig. 3.8. As the knee progressively flexes, tension diminishes on the POL. At 60 degrees of flexion, the distance from origin to insertion is 5 mm shorter than at 30 degrees of flexion. This information may be relevant in avoiding undue strain on a healing ligament. (Copyright © 2000 by the Hughston Sports Medicine Foundation, Inc.)

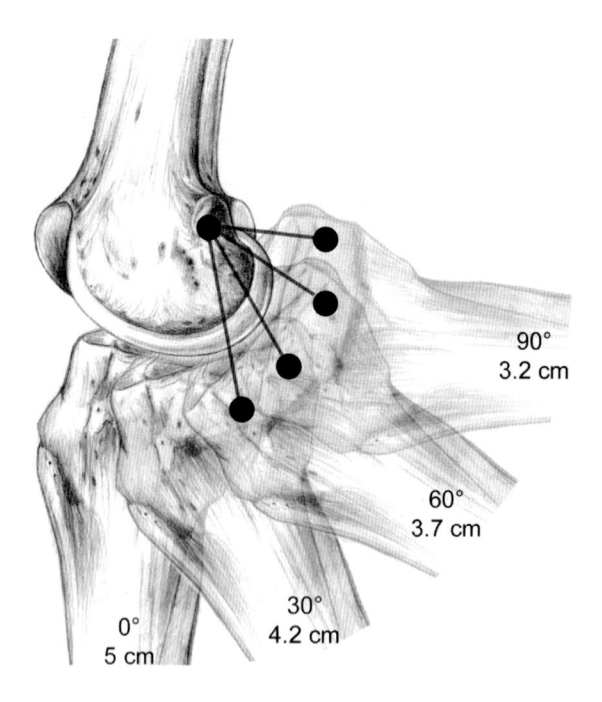

Biomechanics

Many biomechanical functions of the posteromedial corner have already been addressed:

- Capsular structures of the posteromedial corner resist external rotation torque, valgus moment, and anterior translation (tibia externally rotated or semimembranosus contraction) when the knee is flexed beyond 30 degrees. In extension, the PCL primarily resists these motions.
- Fibers of the TCL remain in constant tension over the flexion range of 0 to 90 degrees of flexion. The TCL glides over the deeper capsular structures posteriorly as the knee flexes.
- The distance from the origin to the insertion of the POL fibers shortens with progressive knee flexion and these fibers relax. Semimembranosus contraction retensions these ligaments (Fig. 3.8).
- The meniscus with intact meniscotibial segments resists anterior translation of the tibia when the knee is flexed to 90 degrees and the tibia is externally rotated.
- The meniscus deforms with flexion-extension arcs of the knee. The posterior horn is pulled posteriorly with knee flexion by the semimembranosus. The anterior horn glides anteriorly during knee extension.
- Semimembranosus contraction, in addition to retensioning the POL, results in increased joint contact forces, and thereby increases stability.
- Through its connection to the OPL, the semimembranosus may contribute to lateral capsular stability as well.

Biomechanical functions beneficial to the knee are attributable to structures in the posteromedial corner. Shoemaker and Markolf[8] have been able to demonstrate a protective function of the semimembranosus. With maximal muscle contraction, the rotational torque required to produce ligament failure is doubled. The posteromedial capsular ligaments resist external torque, valgus stress, and anterior translation (tibia externally rotated).

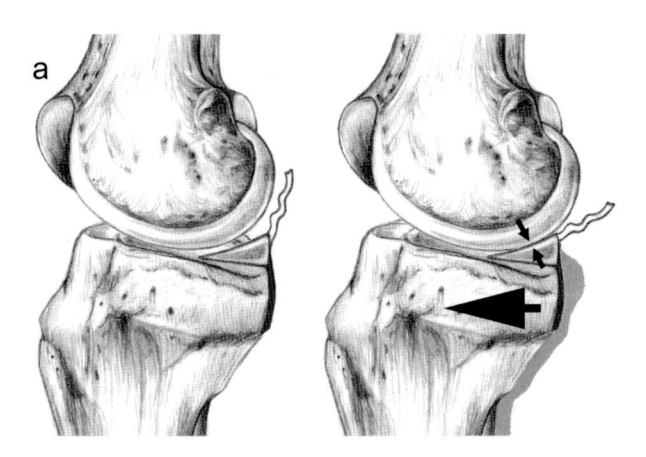

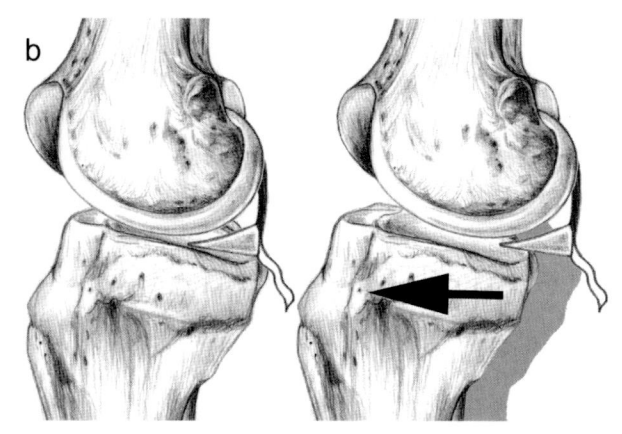

Fig. 3.9. When tears occur in the meniscofemoral segment of a ligament and the meniscus remains anchored to the tibia, the wedge shape of the meniscus body acts as a bushing, diminishing the amount of anterior translation. Because the undersurface of the meniscus is relatively flat, this effect does not occur when tears affect the meniscotibial segment. The resulting anterior translation is of a greater magnitude. PCL, posterior cruciate ligament; POL, posterior oblique ligament. (Copyright © 2000 by the Hughston Sports Medicine Foundation, Inc.)

Strain gauge analysis of the anterior cruciate ligament (ACL) showed increased strain when these ligaments were sectioned and those functions negated. This implies that an injury that goes unrepaired, resulting in a nonfunctional posteromedial corner, may place the knee at increased risk for subsequent ACL injury.[9]

The anatomic organization of the meniscofemoral and meniscotibial segments of the capsular ligament has several implications for function. As Last[10] observed, flexion and extension occur proximal to the meniscus, where the longer lengths of the meniscofemoral segments allow the femoral condyles to rotate between the axes of the epicondyles and the articular surface. Normal axial rotation of the tibia conversely occurs submeniscally and only to a minor extent, due to the shorter lengths of the meniscotibial segments.

Injuries restricted to the meniscofemoral segments may result in lesser demonstrable instability. If the meniscus remains anchored to the tibia, it may act as a "bushing" to buffer the magnitude of, for example, a 90-degree drawer test. If, however, the meniscotibial segment is disrupted, the tibia readily glides beneath the flat undersurface of the meniscus, resulting in significant instability, which is appreciated in standard clinical tests (Fig. 3.9). It follows that meniscotibial segment injuries may be expected in more severe instability patterns. The surgeon should be knowledgeable about the normal relative lengths of these segments, and both segments of each ligament should be explored if capsular ligament repair or reconstruction is undertaken.

Classification of Medial Knee Ligament Instability

The classification scheme reported by Hughston and colleagues[11] remains one of the few published classifications of tibiofemoral instability. No scheme published since has been based on an extensive database of documented examination compared with injury pathology found at operation. With the current minimally invasive procedures and no imaging modality as reliable as operative dissection, it is unlikely that such a study could be undertaken again in the foreseeable future. In this scheme, examination predicts injury pathology, which guides treatment decisions. Injury pathologies can be subgrouped both by the pathology present and by the response to a given treatment approach. These approaches can then be validated by long-term prospective studies.

The complete classification is found in the original article and subsequent reviews,[11–13] but several points can be made regarding medial capsular ligament injuries coupled with PCL injuries. Injury to the PCL falls in the class of straight instabilities. There are four broad types of straight instabilities that are named for the direction of their major instability pattern. Straight medial instability involves tears of the PCL, medial capsular structures, and likely the ACL as well. Medial capsular structures can be torn to varying degrees in straight anterior and straight posterior instability. Finally, the terms *posteromedial instability* and *posteromedial rotatory instability* pervade the literature. They are imprecise and incorrect terms if the above classification is accepted. If a straight instability had both medial and posterior components, it would be classified by the major component. If components of this instability were felt to be equivalent, it would be classified as a combined instability (i.e., straight medial, straight posterior instability). Posteromedial rotatory instability cannot exist since it would involve a posteromedial subluxation of the tibia when it was internally rotated. Recalling that internal rotation of the tibia tensions the PCL, instability in this position mandates incompetency of the PCL. A straight, rather than a rotatory, instability would then be present.

Implications for Examination

The strategy for ligament examination of the knee is to position the joint so as to isolate the individual ligaments or ligamentous complexes that will primarily resist a stress and then apply that stress. If no abnormal motion is measured, those isolated ligaments are presumed to be functionally intact. If abnormal motion is measured, the converse is true. By systematically isolating all the ligament complexes, a total picture of the joint injury is developed. To do this effectively, a sound basis in the anatomy of individual ligaments as described above is required.

An isolated ligament or ligament complex's abnormal motion can be individually graded. Most investigators still use the conventions established by the American Medical Association (AMA),[14] which provide a clinical gradation of 1+ (mild), 2+ (moderate), or 3+ (severe) with an implied measured, albeit subjective, amount of abnormal motion or laxity corresponding to each gradation. The instability of a ligament or ligament complex should not be confused with a clinical instability pattern of the joint. To do so leads to confusion regarding treatment patterns.[15,16] For example, it is possible to have a complete (grade III) tear of the TCL without an associated straight medial or anteromedial rotatory instability of the joint. While the latter two would require surgery, the isolated TCL injury can be managed nonoperatively.

In full extension or hyperextension, the PCL is the primary restraint to coronal plane varus or valgus rotation, posterior tibial translation, and (at least with internal tibial rotation) anterior tibial translation. When the knee is internally rotated, the intact cruciate ligaments "wrap" on each other, increasing their stiffness, and the bundles of the PCL further wrap on themselves. Thus, stresses placed on a knee in full extension, hyperextension, or internal tibial rotation primarily test the competency of the PCL and secondarily test the competency of the capsular structures involved with that abnormal motion. For example, medial opening (a positive abduction stress in extension) will not occur with an intact PCL, even if all the medial capsular ligaments are torn. If the PCL is torn, however, the magnitude of the medial opening will increase with the severity of injury to the medial capsular structures.

To begin to isolate and evaluate the capsular ligaments, the knee must be positioned such that the capsular ligaments are relatively more taut than the PCL (and thus, the first structure to resist the abnormal motion). These

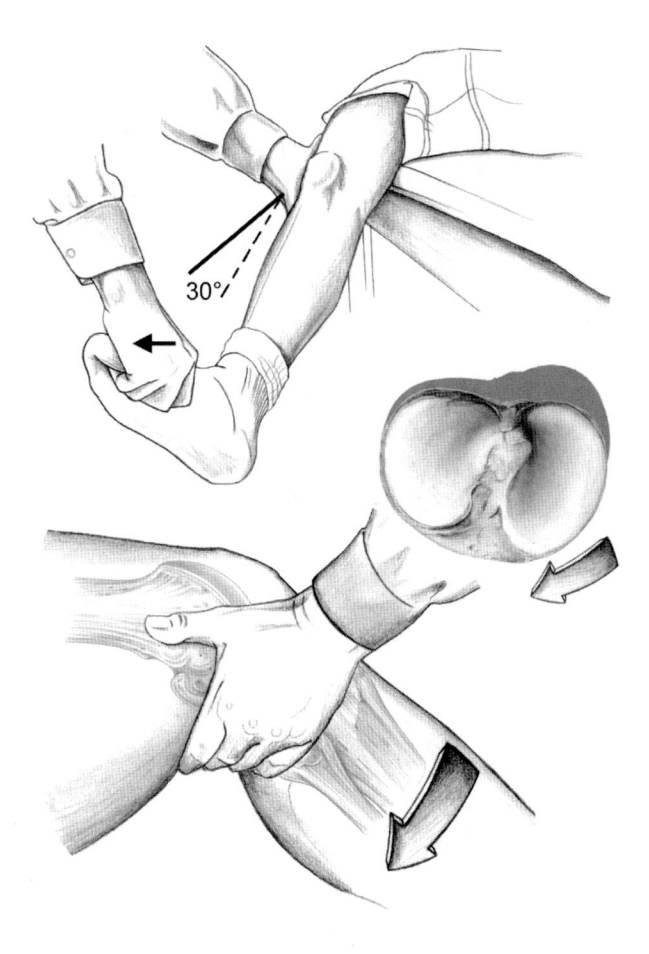

3. The Anatomy and Biomechanics of the Posteromedial Aspect of the Knee

Fig. 3.10. The abduction stress or valgus stress test is performed by grasping the forefoot and not the ankle or leg, which would constrain tibial axial rotation and possibly result in a false-negative interpretation of the test. (Copyright © 2000 by the Hughston Sports Medicine Foundation, Inc.)

positions have been well documented.[11,17] Much attention has been directed to a concept of primary and secondary restraints in the past, those restraints deriving their definition from an ultimate tensile strength.[18] Again, this leads to confusion, for it is the slope of a stress-strain curve rather than which bundle of connective tissue is strongest that carries the greatest clinical import. The most severe clinical instability is an abnormal motion that occurs with minimal rather than maximal stress.

Therefore, rather than being concerned with which ligament absorbs the greatest load to strain in a cadaver knee mounted on a materials testing machine with no dynamic input affecting ligament tension, no axial joint load, and tibial rotation constrained, the examiner should focus on which ligament more than any other is resisting the applied stress. In the case of the medial capsular ligaments, two tests are helpful to evaluate the ligaments in the medial and posteromedial corners: the abduction stress test at 30 degrees of flexion (ABD 30) and the anterior drawer with the tibia externally rotated at 90 degrees of flexion (ADER 90).

In the ABD 30, the patient is supine and the knee is positioned in 30 degrees of flexion (Fig. 3.10). It is important in any test to have as near-total relaxation of the patient as possible to prevent muscular guarding by the patient. If guarding occurs, the magnitude of abnormal motion may be diminished and the actual severity of injury missed. The test is thus best performed on an examination table or bench. In this way, the thigh can rest on the table in slight abduction of the hip, which aids in relaxation. The knee can then be flexed over the side of the table. In contrast, examining a patient on the ground requires flexing the hip and holding the leg in the air. Control of the thigh in this position is transmitted from the leg through an injured knee; consequently, significant guarding will usually

Fig. 3.11. The anterior drawer test, performed in 70 to 90 degrees of knee flexion with the tibia externally rotated, provides information regarding injury and functional integrity of capsular ligaments in the posteromedial corner (see text for details). (a) Patient and examiner position and hand placement to facilitate hamstring relaxation. (b) The index fingers relax the hamstring tendons, and the thumbs palpate the femoral condyles and tibial rims to detect anterior subluxation. Abnormal anteromedial subluxation is indicated by the overlay shadow and signifies damage to posteromedial structures. (Copyright © 2000 by the Hughston Sports Medicine Foundation, Inc.)

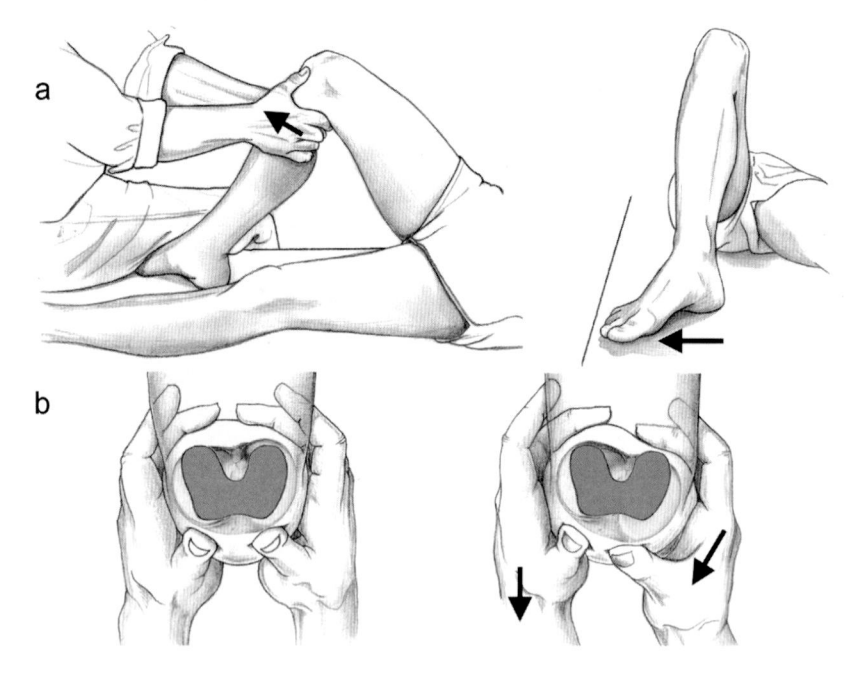

occur as the patient subconsciously attempts to posture the thigh and avoid pain from the sprained ligaments. Further, placing the hip in extension promotes relaxation of the hamstring muscles.

As has been demonstrated in many other joints, what may appear as a pure single-plane rotation is in fact coupled to a rotation or translation in another plane.[7,19] If the coupled motion is constrained, the observed motion will be of less magnitude. It is, therefore, important to appreciate the coupled motions inherent in each test and perform that test so that the coupled motion is unconstrained. In the ABD 30, valgus opening is coupled to external tibial rotation. The tibia, therefore, must rotate freely during performance of the test, or the perceived laxity will be less than the actual laxity. If the examiner holds the extremity by the leg or by the ankle, tibial rotation will be constrained. Applying the valgus stress by grasping the forefoot allows both coupled motions to occur unconstrained (Fig. 3.10).

The specifics of the test beyond this point are well described elsewhere and will not be reiterated.[11] A positive ABD 30 indicates damage to the TCL, mid-third capsular ligament, and POL.[11] Determining which individual ligaments or segments are torn is further assisted by palpation performed after all other ligament tests are concluded (again, to prevent guarding).

The ADER 90 is one of a series of drawer tests performed with the patient in this position. Our remarks will concern the ADER 90 only. The ADER 90 is similarly performed with the patient supine and resting comfortably. The hip is flexed to 45 degrees and the knee to 90 degrees, with the foot resting on the examining surface, which promotes relaxation (Fig. 3.11). The leg is rotated externally, which tensions the medial and posteromedial capsular ligaments, and the examiner sits on the forefoot to maintain this position, but again not to fully constrain the tibia. Translation in this case is coupled to tibial axial rotation. Significant anterior translation elicited by this test in this position indicates injury to the mid-third capsular ligament and the POL. A positive result will be further accentuated by a torn ACL. The severity of the drawer as well may provide insight into which segment of the ligament is torn. Tears primarily involving the meniscofemoral segment of a ligament leave the meniscus well anchored to the tibia. As the tibia translates anteriorly, the wedge shape of the meniscus abuts the posterior femoral condyle, limiting the magnitude of the translation. If, however, the meniscotibial segment is torn, the tibia can more readily slip forward, as the wedge effect of the meniscus is defeated by the torn segment (see Fig. 3.9).

The salient features of examination for medial capsular ligament injury are thus as follows:

- Abduction opening in extension or a significant drawer test in 90 degrees of flexion with internal rotation of the tibia indicates a torn PCL and thus straight instability.
- The more severe laxity obtained with these tests suggests associated capsular injury.
- The contribution of the PCL to medial laxity can be eliminated by performing the ABD 30 and ADER 90 tests, which relax the PCL relative to the mid-third capsular ligament and POL.
- Proper technique is important in performing these tests so as not to constrain coupled motions or elicit guarding, both of which may result in false-negative tests results.
- The positive ADER 90 may be accentuated if the capsular ligament is torn in the meniscotibial segment or if there is a concomitant tear of the ACL.
- Palpation of the origin, insertion, and segments of each described ligament should follow standard ligament tests to verify what is torn and where.

By following these principles, the clinician can determine if surgery is indicated, and if so, formulate an operative strategy based on the anatomic derangement rather than rote.

Implications for Surgical Repair

To assume that a completely torn ligament will reform with proper orientation, tension, and dynamic function, unless guided to do so by suture and retensioning, is as naive as to believe a broken bone will heal with proper alignment and length without fixation or appropriate splinting. Incomplete tears can be expected to merge and contract; complete tears will fill a void with scar or, worse, areolar tissue, and remain lax and nonfunctional.

When the anatomic and biomechanical relationships of the semimembranosus, capsular ligaments, and medial meniscus are understood, the importance of associated ligament injury in the presence of a peripheral tear of the medial meniscus can be appreciated. Because the peripheral meniscus tear cannot occur without a degree of capsular ligament sprain, it is not sufficient to simply sew the meniscus back to the capsule. Care must be exercised that the meniscus is reattached with the ligament adequately retensioned. Failure to do so will leave the semimembranosus with no dynamic input into the meniscus. Without this dynamic input, the posterior horn of the meniscus will not be pulled posteriorly with semimembranosus contraction and will, thus, be vulnerable to entrapment and injury (see Fig. 3.6).

Studies of length versus flexion angle have served as guides for surgeons performing purely extraarticular ligament reconstructions. It is important to immobilize the knee in 60 degrees of flexion postoperatively whenever the medial side is repaired or reconstructed. The extra 5 mm of relaxation gained between 30 and 60 degrees of flexion results in less tension on the repair and a more stable knee once healed. When an intraarticular ligament reconstruction is performed, however, the knee must be splinted in exten-

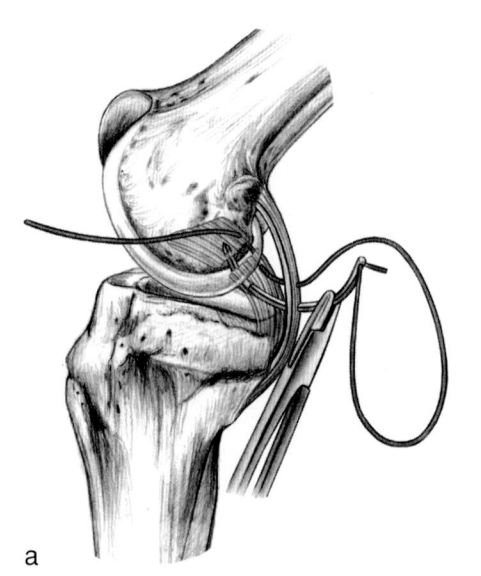

Fig. 3.12. The posterior capsule (OPL) is used to augment repair of a midsubstance tear of the PCL. (a) The suture is advanced from anterior to posterior. (b) The suture is advanced from posterior to anterior. (c) Augmentation is complete. (Copyright © 2000 by the Hughston Sports Medicine Foundation, Inc.)

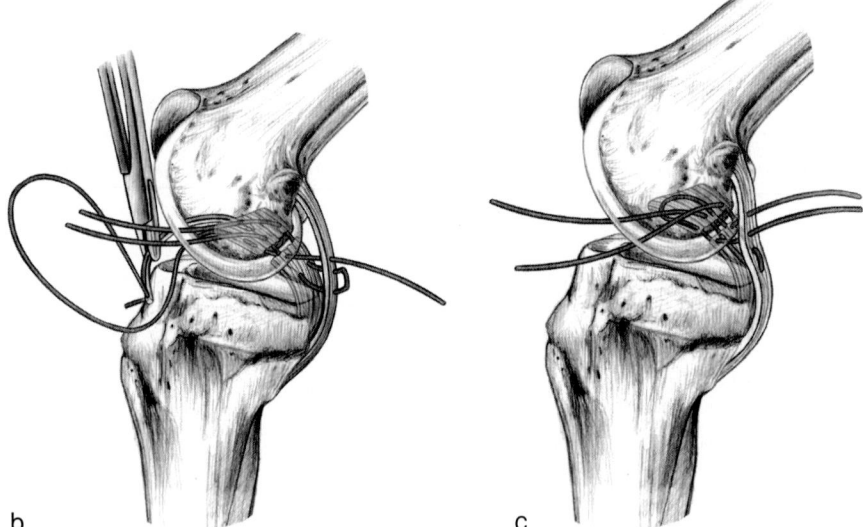

sion and moved early, or flexion contracture and ankylosis will result. The effect of this postoperative protocol on the medial repair, when simultaneous capsular and intraarticular surgery is performed, is unknown. The additional immediate stability gained by the cruciate reconstruction may spare the capsular ligaments additional strain, which would occur as a result of the absence of a functional cruciate ligament. This, in turn, may compensate for the strain imposed by immobilization in extension.

Knowledge of how the semimembranosus functions as a unit with the posteromedial capsule provides an indication to repair the semimembranosus tendon to the POL and OPL if it has been avulsed. Access to the PCL for open procedures is safer from a medial approach (to avoid the neurovascular structures). An interval through the gastrocnemius bursa may be developed in which the PCL lies just lateral to this bursa and deep to the OPL. The OPL may be recruited to augment a PCL repair by suturing the OPL to the PCL repair (Fig. 3.12).

Sound knowledge of the anatomy and function of the capsular structures in the posteromedial knee forms the basis for diagnosis, treatment, and surgical repair or reconstruction of injuries to these structures. Appreciating the many important roles of these structures precludes their being "left to heal closed" if the injury is severe.

Acknowledgment

The true credit for this work belongs to Jack C. Hughston, M.D., an inspiring teacher, a colleague, a friend, a student of anatomy. Most of the concepts presented here spring from his life's quest to improve the care of athletic injuries through academic and applied anatomy.

References

1. Warren LF, Marshall JL. The supporting structures and layers on the medial side of the knee: an anatomical analysis. J Bone Joint Surg 1979;61A:56–62.
2. Hunter SC, Marascalco R, Hughston JC. Disruption of the vastus medialis obliquus with medial knee ligament injuries. Am J Sports Med 1983;11:427–431.

3. Brantigan OC, Voshell AF. The tibial collateral ligament: its function, its bursae, and its relation to the medial meniscus. J Bone Joint Surg 1943;25A:121–131.
4. Hughston JC, Eilers AF. The role of the posterior oblique ligament in repairs of acute medial (collateral) ligament tears of the knee. J Bone Joint Surg 1973;55A:923–940.
5. Brantigan OC, Voshell AF. The mechanics of the ligaments and menisci of the knee joint. J Bone Joint Surg 1941;23A:44–66.
6. Schultz RA, Miller DC, Kerr CS, Micheli L. Mechanoreceptors in human cruciate ligaments. J Bone Joint Surg 1984;66A:1072–1076.
7. Hsieh HH, Walker PS. Stabilizing mechanisms of the loaded and unloaded knee joint. J Bone Joint Surg 1977;58A:87–93.
8. Shoemaker BS, Markolf KL. In vivo rotatory knee stability. J Bone Joint Surg 1982;64A:208–216.
9. Shapiro MS, Markolf KL, Finerman GAM, Mitchell PW. The effect of section of the medial collateral ligament on force generated in the anterior cruciate ligament. J Bone Joint Surg 1991;73A:248–256.
10. Last RJ. The popliteus muscle and the lateral meniscus. J Bone Joint Surg 1950;32B:93–99.
11. Hughston JC, Andrews JR, Cross MJ, Moschi A. Classification of knee ligament instabilities. Part I: the medial compartment and cruciate ligaments. J Bone Joint Surg 1976;58A:159–172.
12. Hughston JC, Andrews JR, Cross MJ, Moschi A. Classification of knee ligament instabilities. Part II: the lateral compartment. J Bone Joint Surg 1976;58A:173–179.
13. Flandry F. A classification of knee ligament instability. In: Baker CL, Flandry F, Henderson JM, ed. The Hughston Clinic Sports Medicine Book. Baltimore: Williams & Wilkins, 1995:481–493.
14. Committee on the Medical Aspects of Sports. Standard Nomenclature of Athletic Injuries. Chicago: AMA, 1968:99–101.
15. Indelicato PA. Non-operative treatment of complete tears of the medial collateral ligament of the knee. J Bone Joint Surg 1983;65A:323–329.
16. Hughston JC. The importance of the posterior oblique ligament in repairs of acute tears of the medial ligaments in knees with and without an associated rupture of the anterior cruciate ligament: results of long-term follow-up. J Bone Joint Surg 1994;76A:1328–1344.
17. Kennedy JC, Fowler PJ. Medial and anterior instability of the knee. J Bone Joint Surg 1971;53A:1257–1270.
18. Butler D, Noyes FR, Grood ES. Ligamentous restraints to anterior-posterior drawer in the human knee: a biomechanical study. J Bone Joint Surg 1980;62A:259–270.
19. White AA, Panjabi MM. The basic kinematics of the human spine. A review of past and current knowledge. Spine 1978;3:12–20.

II

Diagnosis and Evaluation

Chapter Four

Clinical Examination of the Posterior Cruciate Ligament–Deficient Knee

Darren L. Johnson

Injury to the posterior cruciate ligament (PCL) may represent up to 20% of surgically treated knee ligament injuries.[1-3] This may represent a low estimate in that there are a significant amount of true isolated injuries that go undiagnosed or even misdiagnosed because of their benign functional problems. While 75% of anterior cruciate ligament (ACL) injuries are isolated, only 50% of PCL injuries are.[4] Although the natural history of the PCL-deficient knee is still being defined with ongoing research in the basic science and clinical arenas, treatment decisions are based on accurate clinical diagnosis. Over the last 10 years, significant new information has become available on the anatomy and biomechanics of the PCL that has allowed us to be more precise in our diagnostic skills, which have an ultimate effect on treatment decisions.

The primary function of the PCL is to prevent posterior tibial translation in the anterior-posterior plane,[5,6] and its secondary function is to prevent external rotation.[7] It is the coupled movement of posterior displacement and external rotation that is responsible for the frequent association of posterior cruciate and posterolateral ligament injury.[8-12] Accurate diagnosis of the PCL-injured knee requires a complete history, meticulous physical examination, and ancillary studies as dictated by the patient's clinical presentation.

History

An accurate diagnosis requires obtaining as much information as possible about the injury and the associated problems the knee is experiencing. Knowledge of the mechanism of injury can be beneficial in helping to determine the location of the injury within the length of the ligament. Different mechanisms have been found to tear the PCL in different anatomic locations. PCL injuries occur as a result of athletics (40%), motor vehicle accidents (MVAs) (50%), and industrial accidents (10%). The incidence found within athletics is often sports specific. There are more PCL injuries in American football than in basketball.[4] It is also apparent from the literature that isolated PCL injuries appear to be more frequent in athletics.[13]

A PCL injury may be classified with respect to timing (acute verus chronic) and severity (isolated/unidirectional or combined/multidirectional). Both variables directly affect treatment and prognosis. Another way to consider injury mechanisms is by the amount of energy absorbed by the knee; athletic injuries entail low energy, and motor vehicle and industrial accidents entail high energy. The energy imparted to the knee affects the extent of soft tissue injury, and has an impact on surgical decision making and prognosis.

Athletic injuries to the PCL are often caused by a contact mechanism. The knee is often in a relatively flexed position, and it sustains a posterior directed force on the proximal tibia, injuring the ligament. This may occur as the athlete falls to the ground on a flexed knee with the foot in maximal plantar flexion, with the ground acting as the force producer (Fig. 4.1). Or another athlete may act as the force producer, such as a football player's helmet or a soccer player's leg. These types of mechanisms usually result in an isolated midsubstance tear of the PCL. The amount of rotation of the tibia on the femur at that time of impact and the exact vector line of force determine the extent of damage to the posteromedial or posterolateral structures. For example, a PCL/medial collateral ligament (MCL) injury may occur during a valgus-directed force on a knee about which there is external rotation torque as a result of a fixed but unloaded foot.

Noncontact athletic injuries of the PCL are very uncommon. Hyperflexion alone, without a posteriorly directed force from the ground, may result in PCL injury. Importantly, this mechanism frequently results in proximal avulsion of the PCL from the femur with adjacent perichondrium or periosteum. This type of PCL injury often can be repaired with a reasonable expectation of a stable knee. An unusual mechanism is one in which the athlete changes direction on a planted but unloaded foot, and the femur anteriorly translates and internally rotates, creating a relative posterior external rotation force at the knee that results in an isolated injury to the PCL with little or no posterior capsular injury. In these instances the PCL may be in continuity but is functionally lax. There is little swelling of the knee and the patient may not be aware of the significance of the injury. Another mechanism of injury is an anteriorly directed force applied to a fully hyperextended knee with the foot fixed on the ground. Hyperextension of the knee results in injury first to the ACL, then the posterior capsule and the PCL at 30 degrees of hyperextension, and finally to the popliteal artery at 50 degrees of hyperextension.[14] This mechanism, therefore, often results in a combined or multiligament injury.

High-energy mechanisms, such as an MVA, motorcycle accident, or a fall from a height of over 20 feet, often cause additional ligamentous damage to the PCL-injured knee. The typical automobile dashboard impact on the

Fig. 4.1. Pretibial trauma as a cause of posterior cruciate ligament (PCL) disruption in athletes. A posterior directed force is delivered through the tibial tubercle as the athlete falls on the flexed knee with the foot in plantar flexion.

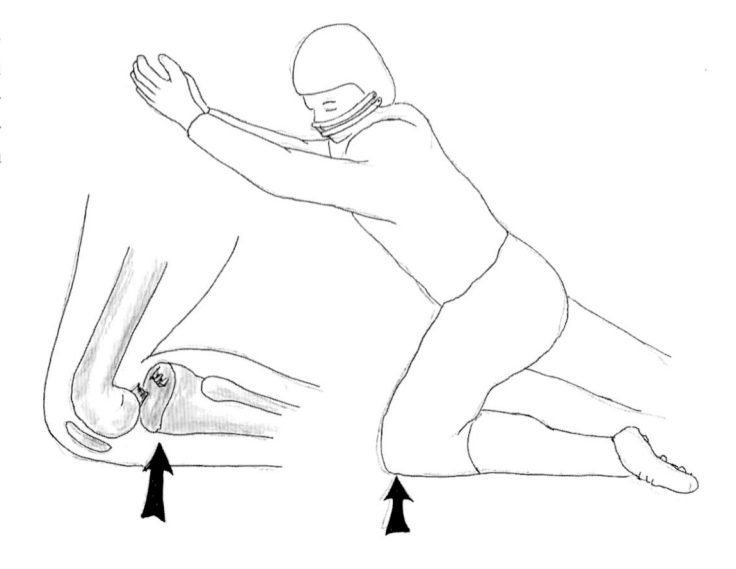

knee can cause significant ligamentous injury, and Fanelli et al.[23] noted that 95% of PCL injuries seen in the emergency department were combined with other ligament damage. Combined PCL and posterolateral injuries are the most common. The exact position of the tibia on the femur and the vector line of force determine specific collateral ligament damage. A significant percentage of PCL injuries in combination with injuries to other capsuloligamentous structures are avulsions from the bony insertion, most commonly from the femur but occasionally from the tibia. In these combined injuries, the PCL avulsions may be amenable to primary repair. An often discussed but uncommonly seen injury is an avulsion of the bony insertion of the cruciate ligament that generally involves the tibial origin of the PCL. This injury may be repaired primarily, with the expectation of good stability.

Patients with an acute/chronic and isolated/combined injury of the PCL have different functional complaints and presentations. Athletes with an acute/isolated injury to the PCL give a different history from those with an ACL injury or combined ligament injury. Unlike an individual with an ACL injury, PCL-injured athletes deny feeling or hearing a "pop" or buckling episode. They often continue playing following their injury, and it is not until the next day, when a very mild effusion or an ache in the posterior aspect of the knee develops, that they notify the team physician that there has been an injury. Those patients with an acute/combined ligament injury often suspect they have sustained a significant major injury to their knee and seek medical attention immediately.

Patients with a chronic PCL tear may present to the clinician with different complaints depending on the time since the injury and the status of additional structures injured within the knee. Those with an isolated complete tear often present with pain as their chief complaint. The pain, often retropatellar, is not present initially, but develops later. This occurs in very active individuals who participate in level 1 sports (defined as sports that usually involve contact and starting, stopping, or change in direction, such as football, soccer, or basketball) as well as in those involved in extensor mechanism–dependent activities, such as ascending or descending stairs or a slope, initiating a run, or picking up a load. Basic science studies have shown increased articulator cartilage contact forces within the patellofemoral joint and medial compartment in those knees with isolated PCL deficiency.[15] During ambulation the proximal tibia subluxates posteriorly as the knee is flexed. This alters the vector forces of the extensor mechanism and results in increased compression of the articular surfaces within the medial and patellofemoral compartments.

The chronicity of the injury and the activity level of the individual dictate the exact level of symptoms and the physical examination/radiographic findings. Patients with a chronic combined ligament injury may complain that the knee is "not right" but are often not able to elaborate specifically what is wrong. They may give a history of a sensation of instability, particularly on uneven ground. Those with combined PCL/posterolateral corner injury complain that the knee tends to give way posterolaterally and that pain is felt on the medial aspect and at the posterolateral corner. The medial pain is believed to be secondary to medial compartment overload due to a varus thrust in gait. The posterolateral discomfort is due to stretching of the posterolateral soft tissues resulting from the varus thrust. These patients tend to walk with the foot in external rotation and the tibia held in the subluxed position. Athletes may report a decreased ability to rapidly change directions. They may state that the knee gives way but often not to the degree of an ACL deficiency.

Physical Examination

A thorough knee examination is essential for making the correct diagnosis as well as for implementing a treatment program. The injury mechanism and degree of acuity dictate the tests to be performed (Table 4.1). For example, a patient in a high-energy MVA that injures multiple ligaments and an athlete who complains of isolated medial compartment or retropatellar pain for the last two years in level 1 sports secondary to chronic PCL insufficiency require different tests.

Examination of the high-energy acutely injured knee begins with detailed documentation of the neurovascular status of the injured limb followed by inspection of the overlying skin and soft tissues in the pretibial and popliteal area for ecchymosis or abrasions. Absence of a large knee effusion does not eliminate the possibility of a spontaneously reduced knee dislocation with multiple knee ligaments injured. In these serious knee injuries, the capsule of the knee is often torn, and blood escapes from the knee cavity into the thigh or calf. Significant pain or tenderness with associated hematoma overlying the posterolateral or medial structures should alert the examiner to possible injury of those structures. The clinician must assume a dislocated or multiple ligament–injured knee until proven otherwise. Wascher et al.[16] found that in a population of patients with knee dislocations, regardless of whether the patients had gross or reduced dislocations when first seen, the incidence of vascular injury was 14%. Any abnormality of vascular status must be appropriately worked up and managed urgently. Patients with PCL and lateral collateral injuries have an incidence of peroneal nerve injury that approaches 30%.

Physical exam begins with the knee in full extension; a gently applied varus or valgus load will indicate a possible knee dislocation. If the knee opens up in full extension, a knee dislocation is assumed with injury of three of the four major ligaments, and further workup may be instituted using magnetic resonance imaging (MRI) as an adjunct to accurately define the extent of the injury. The amount of discomfort the knee is experiencing determines how aggressive the examiner's diagnostic maneuvers can be. If the knee can be placed in 90 degrees of flexion without pain or quadriceps and hamstring spasm, then accurate assessment of the PCL ligament can be done.

While the patient is resting comfortably supine on the table with the knee flexed 90 degrees and the foot resting on the table in neutral rotation,

Table 4.1. Clinical tests of the posterior cruciate ligament (PCL)-deficient knee

Clinical test	PCL	PCL/posterolateral	PCL/posteromedial
Posterior drawer	++++	+++	+++
Posterolateral drawer	−	+++	+++
Posterior sag/Godrey	+++	+++	+++
Quadriceps active	++++	+++	+++
Posteromedial drawer	−	−	+++
External rotation 30 degrees	++	+++	−
Varus stress 30 degrees	−	+++	−
Valgus stress 30 degrees	−	−	+++
Reverse pivot-shift test	−/+	++	−
Posteromedial pivot-shift test	−/+	−	++

Symbols represent grading scale for usefulness in detecting type of injury ranging from − (not useful) to ++++ (most useful).

4. Clinical Examination of the Posterior Cruciate Ligament–Deficient Knee

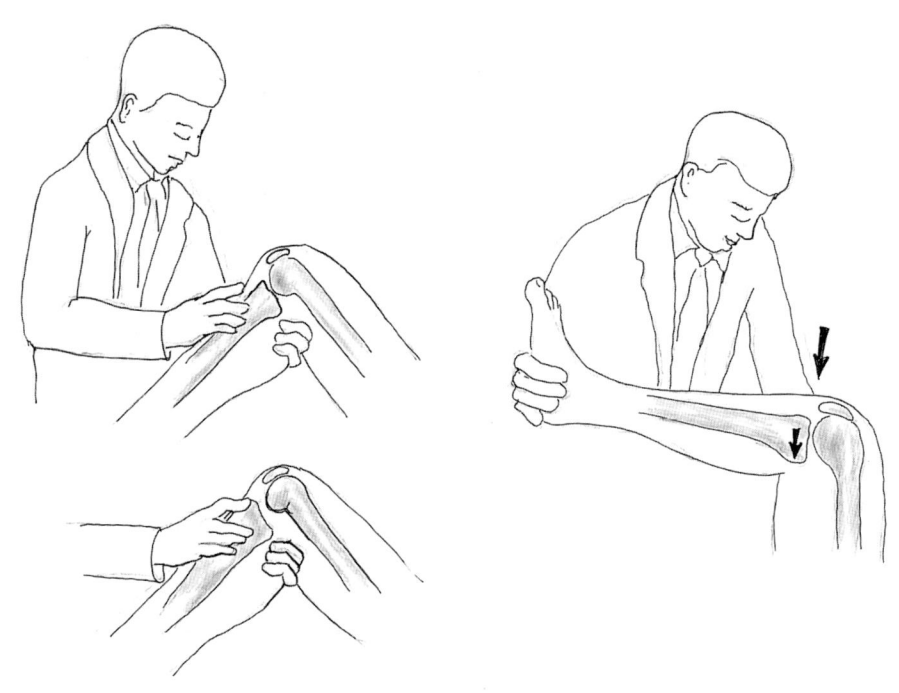

Fig. 4.2. The unsupported proximal tibia is displaced by gravity, giving a concave anterior contour, tilting the patella downward, and displacing the patellar tendon/tibial tubercle posteriorly as compared with the uninvolved opposite knee. Using the index finger over the anteromedial joint line adjacent to the patellar tibial tendon, the examiner approximates the step-off.

the examiner can observe the relative position of the tibia on the femur compared to that in the uninvolved, noninjured knee. A positive *posterior sag test*[1] (complete PCL tear) is present with the observation of an asymmetric concave anterior contour of the proximal tibia/patellar tendon and the decreased prominence of the tibial tubercle resulting from posterior subluxation of the tibia (Fig. 4.2). Additionally, palpation of the normal tibial step-off will alert the examiner to a PCL injury (Fig. 4.2). With the patient's knees flexed to 90 degrees, the hips flexed 45 degrees, and the feet planted on the examination surface, the examiner places his thumbs just medial to the patellar tendon on the joint line and palpates the normal tibial step-off. There is normally a 1-cm step-off from the tibia condyle to the femoral condyle. Any decrease should alert the examiner to an injury to the PCL.

The most sensitive and specific as well as reproducible test to detect injury to the PCL is the *posterior drawer test* (Fig. 4.3). Rubinstein et al.,[17] in a prospective randomized study, showed that the posterior drawer test was the best of eight tests assessed to detect PCL injuries, with a 90% sensitivity and 96% overall accuracy. This test is performed with the patient's knee in the same position as the previous two tests. Accuracy of the test depends on ensuring the comfort of the patient, including the absence of spasm and contraction of the thigh musculature. It is most important to assess the accurate starting point (zero position) by using the *tibial step-off test*. A gentle posterior force is applied to the proximal tibia, and the extent of translation and quality of the "end point" (i.e., hard or soft) is documented.

Treatment decisions are based on the degree of PCL laxity or injury; therefore, the grade of injury needs to be determined. In the grade I posterior drawer sign, the tibia can be moved posteriorly 0 to 5 mm on the injured side. Grade II shows 5 to 10 mm of excessive posterior translation, which corresponds to posterior displacement of the tibial condyles until they are flush with the femoral condyles. In grade III, the tibia can be displaced more than 10 mm posteriorly, which corresponds to a displacement of the tibial condyles posterior to the femoral condyles. In a grade III knee

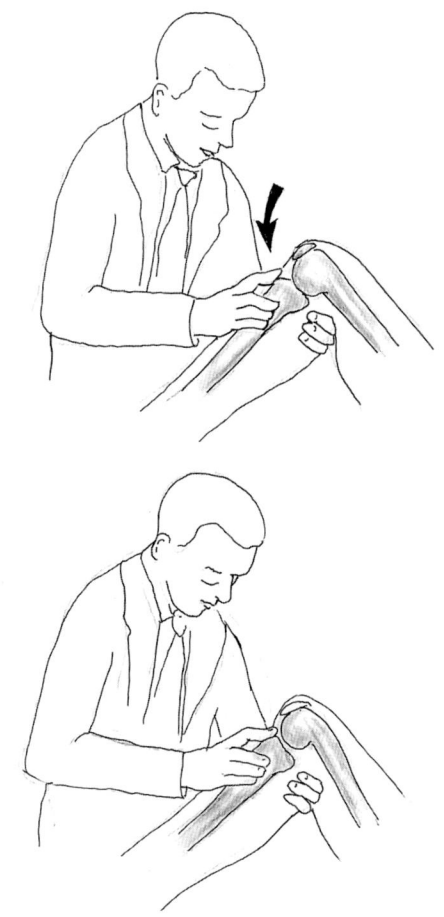

Fig. 4.3. With the knee flexed to 90 degrees and neutral rotation, the posterior drawer is performed after ensuring the "zero" starting position of the tibia on the femur with internal and external rotation of the tibia on the femur. The posteromedial and posterolateral drawer are also performed. The rotation drawer tests are considered positive when the respective tibial plateau moves posteriorly on the femoral condyle during the push phase of the test.

Fig. 4.4. Application of a posterior force to the tibia, with the patient prone and the knee flexed 90 degrees, results in posterior subluxation and foot deviation toward the side of an associated capsule injury.

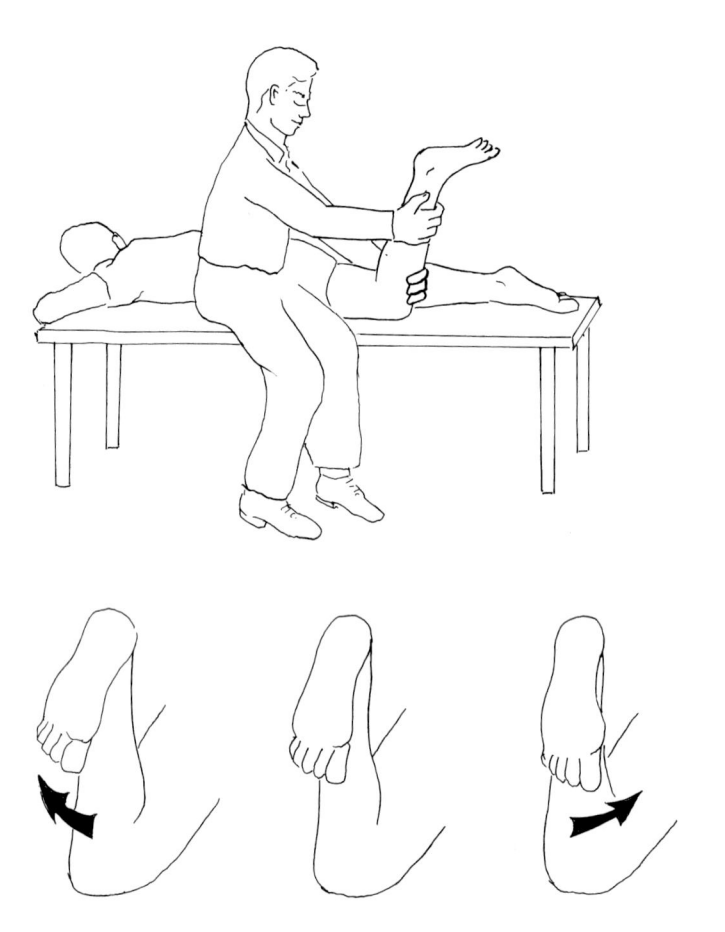

there is often concomitant injury to other ligamentous structures, so careful examination of collateral ligamentous structures is required.[18] An isolated PCL injury with a finding of Grade III is very uncommon. In the patient with pain and the inability to relax the musculature of the extremity, the *prone posterior drawer test* may be helpful (Fig. 4.4).[19] As the patient lies prone on the examination table, the examiner flexes the knee to 90 degrees, supports the tibia without influencing rotation, and applies a posterior force to the proximal tibia. In a PCL-deficient knee, the tibial subluxes posteriorly, and the foot deviates toward the side of an associated posterolateral or posteromedial capsular injury.

The examiner must also assess the quality of the end point of the posterior drawer test. Most acutely injured PCL-deficient knees have an altered end point. However, the posterior end point may return to normal in the chronically deficient PCL knee. In this situation, the posterior drawer test end point is less sensitive than the end point in a Lachman test done for an ACL injury. Examination of an injured knee always includes a Lachman test. In a PCL-deficient knee, the tibia is starting in a subluxed position posteriorly; therefore, the Lachman test may appear to be positive (false-positive Lachman). There is increased excursion in the anterior-posterior plane. What the examiner is doing is effectively reducing the knee from a subluxed position. The most common reason for a false-positive Lachman exam is a PCL tear.

The posterior drawer test should also be performed with the foot in internal (*posteromedial drawer test*) and external (*posterolateral drawer test*) rotation to assess the integrity of the posteromedial and posterolateral corners, respectively (Fig. 4.5). Often patients with a positive posterior drawer

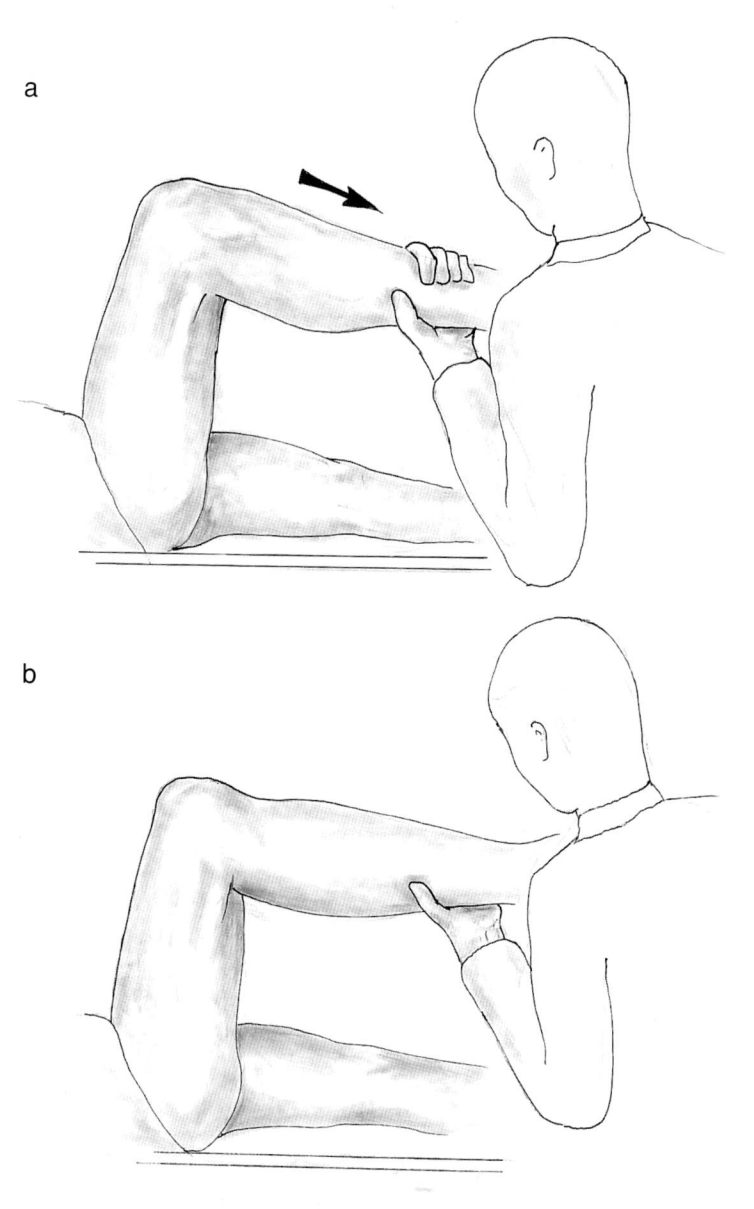

Fig. 4.5. With the patient's hips and knees flexed to 90 degrees, the examiner stands on the side of the injured extremity to observe the contour of the proximal tibia, tibial tubercle, and patellar tendon compared to the opposite knee. A positive test is an indication of a complete rupture as well as associated ligamentous damage.

sign in neutral rotation have a decreased excursion when the drawer test is performed in internal rotation. Debate continues surrounding the primary anatomic structure responsible for this finding—intact meniscofemoral ligaments or the superficial medial collateral ligament.[20] This finding may also indicate maintenance of the integrity of the posterolateral corner, which provides the secondary restraint to posterior displacement. The findings with these tests, as with all tests, must be compared with those in the intact uninjured knee.

Other passive tests of PCL insufficiency include Godfrey's test and the gravity sign near extension test.[21] *Godfrey's test* is performed with the patient resting comfortably supine and both hips and knees flexed to 90 degrees. The examiner observes the asymmetric concave anterior contour of the patellar tendon and proximal tibia with associated decreased prominence of the tibial tubercle resulting from posterior subluxation of the tibia. This accentuates what is observed in the posterior sag test. The *gravity sign near extension test* begins with the patient in the supine position and the distal femur supported by a 15-cm bolster, which flexes the knee 10 to 15 degrees. The examiner observes the asymmetric concave anterior

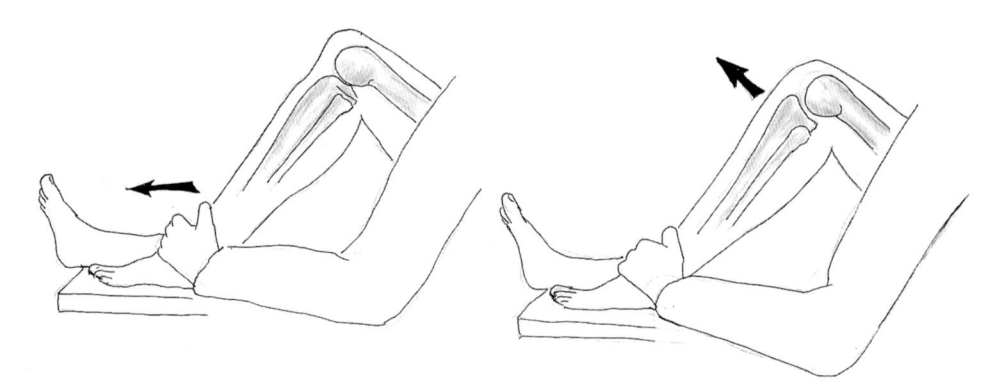

Fig. 4.6. In the PCL-deficient knee, the tibia is posteriorly subluxated. A quadriceps contraction causes proximal tibial reduction, which is visible when the examiner is observing the tibial movement from the affected side.

contour of the proximal tibia, the downward tilt of the patella, and the decreased prominence of the tibial tubercle resulting from posterior subluxation of the tibia by gravity.

Dynamic tests for isolated PCL insufficiency include the quadriceps active drawer test and the dynamic posterior shift test. Daniel et al.[22] were the first to describe the *quadriceps active drawer test* (Fig. 4.6), which is based on the concept that between 60 and 90 degrees of knee flexion, there exists a specific angle of flexion in which the shear component of patellar tendon vector forces is eliminated during a quadriceps contraction. At this specific angle (quadriceps neutral angle), no anterior or posterior shift of the tibia is detected with an isometric quadriceps contraction. In the presence of a PCL injury, if the knee is flexed to the patient's quadriceps-neutral angle (which can be determined by the normal knee) or to 90 degrees, an active contraction of the quadriceps muscle without extending the knee will cause a visible and palpable anterior tibial shift because of an abnormal anterior orientation of the patellar tendon. The examiner must adequately support the patient's thigh so that the muscles are completely relaxed for successful performance of this test. The *dynamic posterior shift test* as advocated by Shelbourne et al.[23] is performed with the patient's hip and knees flexed to 90 degrees; then the examiner slowly extends the patient's knee. Hamstring tension is maintained to provide axial loading and assist gravity in posterior subluxation of the tibia. In the posteriorly unstable knee, as the knee approaches full extension, the subluxed tibia reduces, and a palpable or visible clunk occurs, at which time the patient may state, "My knee doesn't feel right."

A key point that must be emphasized is that the involvement of other capsuloligamentous structures (posterolateral or posteromedial) must be ruled out when the posterior tibial translation is greater than 10 mm or when there is elimination of the tibial step-off (grade III posterior drawer). I, as well as others, have found very few cases that exhibit a grade III posterior drawer sign (greater than 10 mm of posterior translation) that do not have significant involvement of additional structures. In the knee with a posterolateral corner injury, posterior tibial subluxation may diminish or even negate posterolateral corner laxity. If the surgeon only reconstructs the PCL in the presence of an involvement of the posterolateral corner, the PCL graft will experience nonphysiologic forces, with subsequent failure.

Posterolateral instability, which may often be associated with PCL injuries, is assessed by varus laxity in extension and 25 degrees of flexion, the posterolateral drawer test, the external rotation recurvatum test,[24] the reverse pivot shift test,[25] the passive tibial external rotation test or thigh-foot angle test, the voluntary evoked posterolateral drawer sign, and observation of a lateral or varus thrust during the stance phase of gait.

Performance of the *external rotation recurvatum test* requires the examiner to lift the supine patient's extended legs off the table by grasping the great toes (Fig. 4.7). A positive test finding shows varus angulation, hyperextension, and tibial external rotation compared to the noninjured extremity. The *reverse pivot shift test of Jakob* is performed by the examiner's applying a valgus stress to the patient's flexed knee with the foot externally rotated (Fig. 4.8). As the examiner slowly extends the patient's knee, palpable reduction of the posteriorly subluxed tibial plateau at 20 to 30 degrees of flexion is considered a positive test finding.

This test may be positive in up to 35% of patients under anesthesia, and does not represent an abnormality unless the test result is negative in the contralateral knee. With the patient resting comfortably supine on the examination table, the *tibial external rotation test* or *thigh-foot angle test* is performed with the knee at both 30 and 90 degrees of flexion (Fig. 4.9). The examiner rotates the tibia using the medial border of the foot or the tibial tubercle as a reference point. A difference of greater than 15 degrees when compared with the contralateral side is a positive test finding. Increased external rotation at 30 degrees of flexion is consistent with damage to the arcuate ligament complex, lateral collateral ligament (LCL), and the popliteus.

The *voluntary evoked posterolateral drawer sign*[26] may be present in up to 60% of patients with posterolateral instability. These patients can actively sublux and reduce the tibial plateau by active muscle contraction. Electromyographic analysis suggests that the biceps femoris muscle causes posterior subluxation of the lateral plateau, and the popliteus muscle causes the reduction.

Evaluation of limb alignment and gait is critical in the clinical diagnosis and treatment of the knee with PCL injury and is even more important in the multiple ligament–injured knee. Chronic injury of the PCL and posterolateral corner may lead to a double varus knee. Double varus results from loss of the medial compartment articular cartilage and a shift of the weight-bearing line into the medial compartment, combined with stretching of the lateral structures or primary injury to the PCL and lateral ligament structures. When the weight-bearing axis shifts medially, the lateral muscular forces are not strong enough to maintain lateral tibiofemoral joint compression during ambulation. Separation of the lateral tibiofemoral compartment occurs with weight bearing, producing lateral condylar liftoff. When this happens, the patient clinically has a visible varus thrust during gait. This has been termed a double varus knee. More severe involvement of the lateral and posterolateral structures may lead to a triple varus knee, which has, in addition to geometric varus and varus thrust with walking, varus recurvatum. These patients also complain of medial joint pain in sports, but they also have posterolateral knee pain with weight bearing.

Posteromedial instability, which is less common, is diagnosed by valgus stress in full extension and 25 degrees of flexion, the *posteromedial drawer test*, and internal rotation of the foot with prone drawer testing. When PCL injury is combined with injury to the MCL and posteromedial capsular structures, the increased valgus laxity in full extension is the key to recognition of the PCL injury. Additionally, the physician may perform the *posteromedial pivot-shift test*,[27] which signifies injury of the PCL, MCL, and posterior oblique ligament (Fig. 4.10). If one of these three ligaments is intact, it will not occur. The physician performs the posteromedial pivot-shift test by first producing a posterior subluxation of the medial tibial plateau on the medial femoral condyle, and by flexing the patient's knee

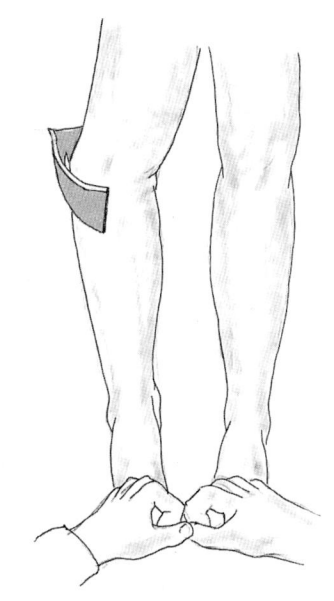

Fig. 4.7. The external rotation recurvatum is considered positive if the affected knee goes into hyperextension, the tibia externally rotates, and the knee shows an apparent tibia vara when standing at the foot of the bed.

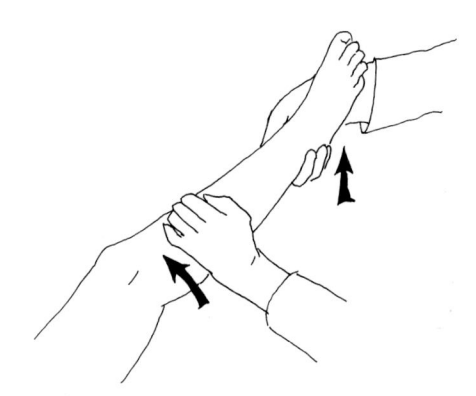

Fig. 4.8. Knee in 45 degrees of flexion and external rotation with a valgus stress. Moving the knee slowly externally results in reduction with a palpable clunk, confirming the posterolateral corner injury.

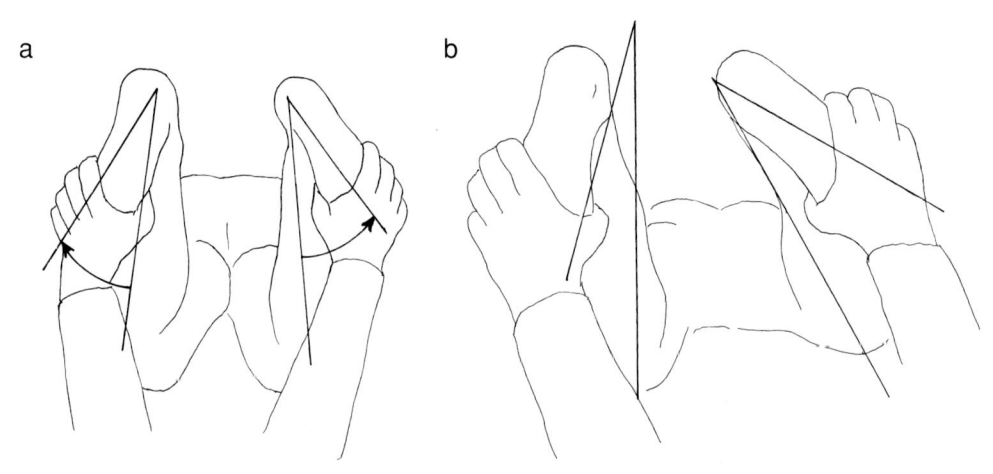

Fig. 4.9. (a) With the knee flexed 30 degrees and the patient lying prone, the feet are externally rotated by the examiner. External rotation of the affected foot relative to the thigh is compared to the normal side. The test is positive if external rotation on the affected side is 10 degrees or more greater than that achieved on the normal side. (b) The test may also be performed with the patient's knee flexed 90 degrees.

Fig. 4.10. (a) The physician produces posterior subluxation of the medial tibial plateau on the medial femoral condyle by flexing the knee more than 45 degrees while applying varus stress, compression, and internal rotation. (b) As the knee is brought into extension, the tibia suddenly shifts back into a reduced position, 20 to 40 degrees short of full extension.

more than 45 degrees while applying varus stress, compression, and internal rotation. As the knee is brought into extension, the tibia suddenly shifts back into a reduced position, 20 to 40 degrees short of full extension. As with all tests, it is helpful to compare the behavior of the involved knee with that of the contralateral, normal knee, since extension of even a normal knee seems to produce some internal rotation of the femur.

Stress Radiography

Stress radiographs can be used in the workup of the PCL-deficient knee. Hewett et al.[28] showed that stress radiographs were superior to both KT-1000 arthrometer and clinical posterior drawer testing for determining PCL stability in 21 patients (10 complete tears, 11 partial tears). Eight millimeters or more of increased posterior translation on stress radiographs was indicative of complete PCL rupture.

Stress radiography overcomes some of the limitations seen with the clinical examination and arthrometer testing. Unfortunately, only a limited number of studies have examined the use of stress radiography in the PCL-deficient knee. The methods have differed in amount of load, application of load, knee flexion angle, and tibial rotation and measurement techniques. A lateral stress radiograph with the knee between 70 and 90 de-

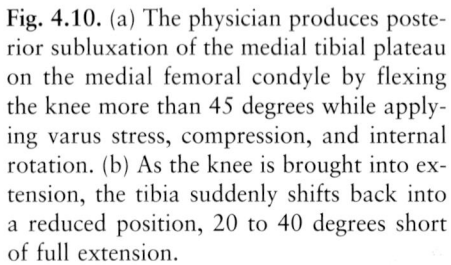

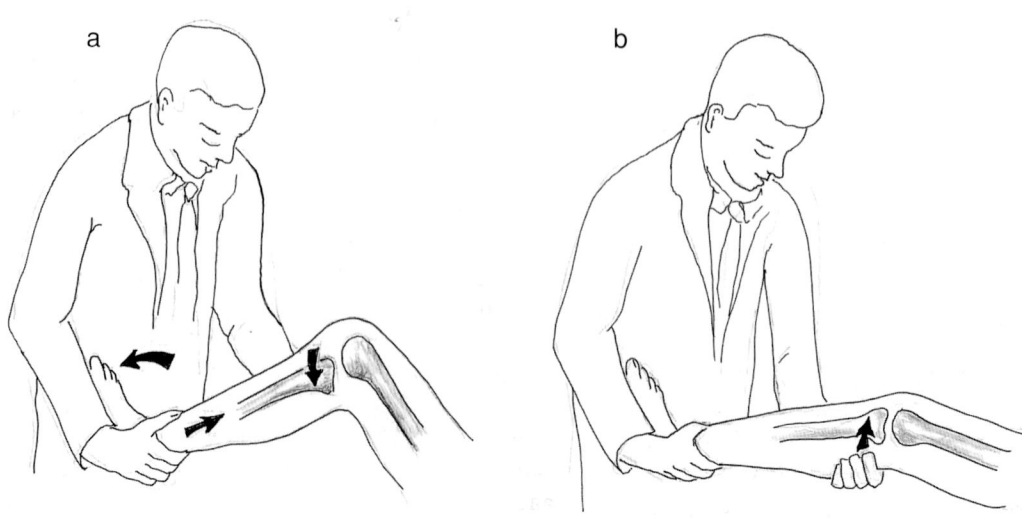

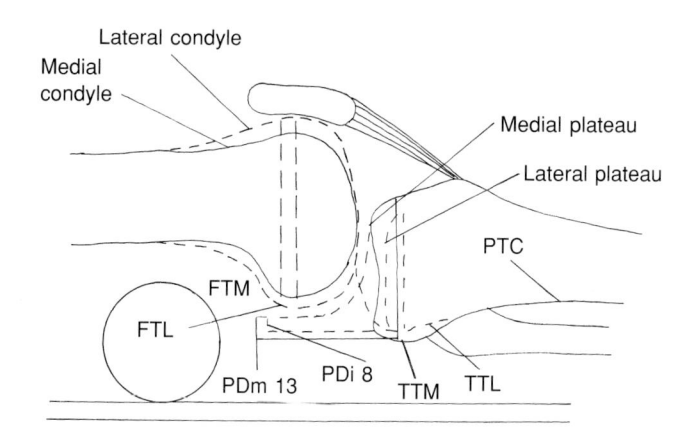

Fig. 4.11. Posterior stress radiography documents posterior translation of the tibia on femur. Posteromedial and posterolateral displacement should be measured accurately. Lines parallel to the posterior tibial cortex (PTC) are drawn tangential to the most posterior contour of medial (FTM) and lateral (FTL) femoral condyles and the medial (TTM) and lateral (TTL) tibial plateaus. In this knee the posterior displacement of the lateral tibial plateau is 8 mm and the medial tibial plateau is 13 mm.

grees of flexion should be taken. The most common technique involves separately measuring the sagittal displacement of the medial and lateral compartments[21] (Fig. 4.11). The posterior aspects of the tibial plateaus are identified and the amount of translation that occurs in relation to their respective posterior femoral condyles is determined.

Stress radiography has many advantages. It provides a noninvasive, objective, reproducible method for recording the amount of sagittal translation of the intact and injured knee. This is important preoperatively as well as for comparing different surgical techniques and long-term results following surgical reconstruction. Additionally, more information may be obtained with natural history studies if the absolute amount of pathologic translation is documented and prospectively recorded. Disadvantages of stress radiograph techniques include a learning curve in filming and measuring stress radiographs, its inappropriateness for combined ligament injury with tibial rotation, and patient variables such as pain, muscle activity, and possible healing of chronic injuries.

References

1. Clancy WG Jr, Shelbourne KD, Zoellner GB, et al. Treatment of knee joint instability secondary to rupture of the posterior cruciate ligament. Report of a new procedure. J Bone Joint Surg 1983;65A:310–322.
2. Fanelli GC. Posterior cruciate ligament injuries in trauma patients. Arthroscopy 1993;9:291–294.
3. Fanelli GC, Edson CJ. Posterior cruciate ligament injuries in trauma patients. Part II. Arthroscopy 1995;11:526–529.
4. Bergfeld JA. Diagnosis and nonoperative treatment of acute posterior cruciate ligament injury. Instr Course Lect 1990;208.
5. Butler DL, Noyes FR, Grood ES. Ligamentous restraints to anterior posterior drawer in the human knee. A biomechanical study. J Bone Joint Surg 1980;62A:259–270.
6. Grood ES, Stowers SF, Noyes FR. Limits of movement in the human knee. Effect of sectioning the posterior cruciate ligament and posterolateral structures. J Bone Joint Surg 1988;70A:88–97.
7. Gollehon DL, Torzill PA, Warren RF. The role of the posterolateral and cruciate ligaments in the stability of the human knee. A biomechanical study. J Bone Joint Surg 1987;69A:233–242.
8. Harner CD, Höher J, Bogrin TM, et al. The effect of sectioning of the posterolateral structures on in situ forces in the human posterior cruciate ligament. Trans Orthop Res Soc 1998;23:47.
9. Harner CD, Bogrin TM, Höher J, et al. The effects of loading the popliteus muscle on the intact and PCL deficient knee. Trans Orthop Res Soc 1997;43:863.

10. Veltri DM, Deng X-H, Torzilli PA, et al. The role of the popliteofibular ligament in stability of the human knee. A biomechanical study. Am J Sports Med 1996;24:19–27.
11. Veltri DM, Deng X-H, Torzilli PA, et al. The role of the cruciate and posterolateral ligaments in stability of the knee. A biomechanical study. Am J Sports Med 1995;23:436–443.
12. Veltri DM, Warren RF. Posterolateral instability of the knee. Instr Course Lect 1994;44:441–453.
13. Fowler PJ, Messieh SS. Isolated posterior cruciate ligament injuries in athletes. Am J Sports Med 1987;15:553–557.
14. Kennedy JC, Hawkins RJ, Willis RB, Danylchuk KD. Tension studies of human ligaments. Yield point, ultimate failure and disruption of the cruciate and tibial collateral ligaments. J Bone Joint Surg 1976;58A:350–355.
15. Skyhar MF, Warren RF, Ortiz GJ, et al. The effects of sectioning of the posterior cruciate ligament and the posterolateral complex on the articular contact pressures within the knee. J Bone Joint Surg 1993;75A:694–699.
16. Wascher DC, Dvirnak PC, DeCoster TA. Knee dislocation: initial assessment and implications for treatment. J Orthop Trauma 1997;11:525–529.
17. Rubinstein RA Jr, Shelbourne KD, McCarroll JR, et al. The accuracy of the clinical examination in the setting of posterior cruciate ligament injuries. Am J Sports Med 1994;22:550–557.
18. Fukubayashi T, Torzilli PA, Sherman MF, et al. An in vitro biomechanical evaluation of anterior-posterior motion of the knee. J Bone Joint Surg 1982;64:258–264.
19. Whipple TL, Ellis FD. Posterior cruciate ligament injuries. Clin Sports Med 1991;10:515–527.
20. Ritchie JR, Bergfeld JA, Kambic H, Manning T. Isolated sectioning of the medial and posteromedial capsular ligaments in the posterior cruciate ligament-deficient knee. Am J Sports Med 1998;26:389–394.
21. Staubli HU, Jakob RP. Posterior instability of the knee near extension. J Bone Joint Surg 1990;72B:225–230.
22. Daniel DM, Stone ML, Barnett P, Sachs R. Use of the quadriceps active test to diagnose posterior cruciate-ligament disruption and measure posterior laxity of the knee. J Bone Joint Surg 1988;70A:386–391.
23. Shelbourne KD, Benedict F, McCarroll JR, Rettig AC. Dynamic posterior shift test: an adjuvant in evaluation of posterior tibial subluxation. Am J Sports Med 1989;17:275–277.
24. Hughston JC, Norwood LA. The posterolateral drawer test and external rotation recurvatum test for posterolateral rotary instability of the knee. Clin Orthop 1980;147:82–87.
25. Jakob RP, Hassler H, Staubli H-U. The reverse pivot shift sign—a new diagnostic aid for posterolateral rotary instability of the knee: II. Acta Orthop Scand 1981;52:18–29.
26. Shino K, Horible S, Ono K. The voluntarily evoked posterolateral drawer sign in the knee with posterolateral instability. Clin Orthop 1987;215:179–186.
27. Owens TC. Posteromedial pivot shift of the knee: a new test for rupture of the posterior cruciate ligament. A demonstration in six patients and a study of anatomical specimens. J Bone Joint Surg 1994;76A:532–539.
28. Hewett TE, Noyes FR, Lee MD. Diagnosis of complete and partial posterior cruciate ligament ruptures. Am J Sports Med 1997;25:648.

Chapter Five

Imaging of the Posterior Cruciate Ligament

D.C. Peterson, L.M.F. Thain, and P.J. Fowler

Images of any structure, either in its normal state or when it has been subjected to a pathologic process, add objectivity to treatment decisions and to assessment of outcome. In the posterior cruciate ligament (PCL)-injured knee, utilization of new advances in this discipline together with existing techniques continue to improve our diagnostic and therapeutic acumen.

The major stabilizing function of the PCL from both a biomechanical and neurosensory perspective is well recognized. Most often (but not always) injury to this ligament is clinically evident. The PCL has an intricate relationship with its neighboring anatomic structures. Consequently, PCL injuries are frequently not isolated. Conversely, a damaged PCL may be a component of other bone and soft tissue injury patterns.

Besides identifying associated injury, imaging may provide information about the extent and location of a PCL tear—information that is important in planning treatment and predicting outcome. Secondary degenerative change is a recognized sequela in the chronic PCL-deficient knee. Identification of degenerative processes earlier rather than later should be advantageous as one follows these patients.

Roentgenographic Imaging

Routine Approach

At our center the initial x-ray series for an acute knee injury includes an anteroposterior (AP), cross table lateral, tunnel view, and a skyline patella view. Two oblique views of the knee can be added if additional information is sought. These are assessed for bony avulsion of the PCL (Fig. 5.1). Fracture of the tibial plateau (more than just an avulsed fragment) may coexist with a significant ligamentous disruption.[1] A fibular head fracture may alert one to a posterolateral injury. As reported by Hall and Hochman,[2] a medial capsular avulsion fracture (reverse Segond-type fracture) may occasionally be seen in PCL tears. They described one case of a cortical avulsion off the medial rim of the medial tibial plateau associated with a tear of both the PCL and the medial meniscus.

The soft tissues must be assessed specifically for the presence of joint fluid such as hemarthrosis or a lipohemarthrosis (an indicator of an intraarticular fracture). The cross-table lateral film allows identification of a fat-fluid level in the suprapatellar recess when there is a lipohemarthrosis. This may be the only plain film view that provides a clue to an occult intraarticular fracture. Capsular rupture should be ruled out in the patient

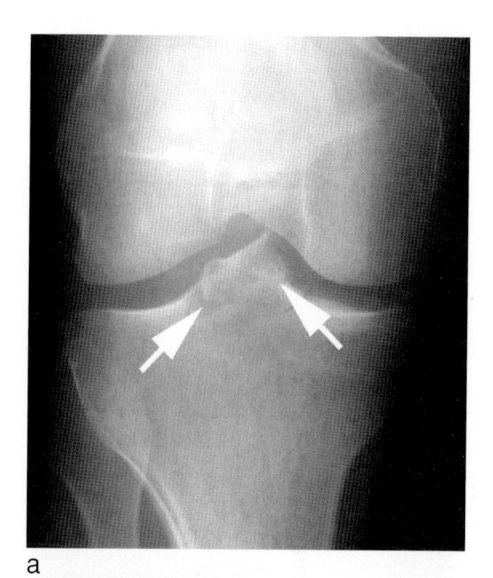

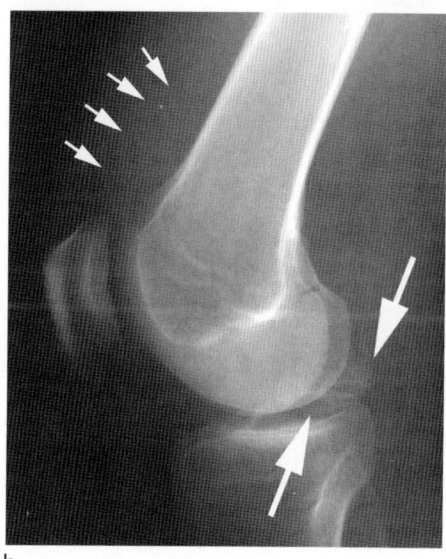

Fig. 5.1. Posterior cruciate ligament (PCL) avulsion fracture. (a) Anteroposterior (AP) x-ray. Arrows point to avulsed fragment. (b) Lateral x-ray. Large arrows point to avulsed fragment. Small arrows point to joint effusion.

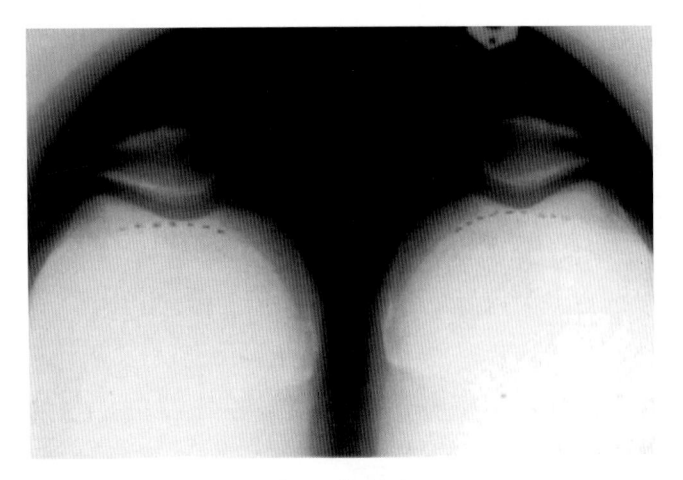

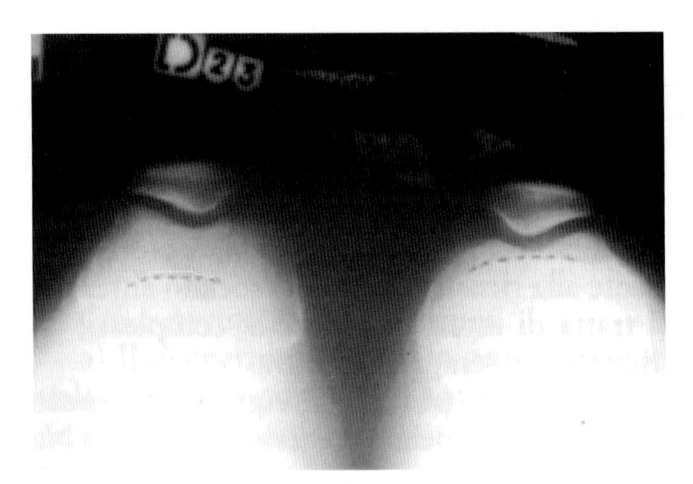

Fig. 5.2. Axial radiographs with the knee at 70 degrees of flexion. Left image demonstrates a normal knee. Right image demonstrates posterior translation of the tibia in the knee on the left as you view the figure, evidence of a PCL lesion. (From Bonnani et al.[3])

who has an acute PCL tear on clinical examination and the absence of joint fluid on plain film. In this situation a dislocation of the knee should be suspected and the limb carefully assessed for a neurovascular injury. A capsular rupture may also lead to compartment syndrome as intraarticular fluid flows into the calf. Capsular rupture is not a contraindication to arthroscopy, but one needs to proceed very cautiously and be prepared to abort the procedure if significant leakage occurs.

A subtly posteriorly subluxated tibia may be the only radiographic clue to a PCL rupture. An axial view of both knees in the 70-degree flexed position may enhance one's ability to visualize posterior subluxation of the tibia on the affected side (Fig. 5.2).[3]

Roentgenograms may also be useful in the assessment of the chronic PCL-deficient knee. Besides posterior subluxation, one should look for degenerative changes, especially in the medial tibiofemoral and the patellofemoral compartments.[4]

Pitfalls

A reduced knee does not exclude a dislocation injury.

Additional Techniques

Tomography (Plain or Computed)

We prefer computed tomography (CT). The availability of helical CT scanners has allowed very fast (less than 1 minute and often less than 30 seconds) thin-section scanning with excellent quality reformatted images in any plane as well as the ability to perform three-dimensional (3D) reconstructions. Tomography is useful to fully evaluate the extent of a fracture and the position of fracture fragments that may be associated with a PCL injury.

Some authors have used CT to assess the PCL itself.[5,6] Passariello et al.[6] compared their CT findings with surgical findings in 45 patients suspected of having capsuloligamentous injuries. They found five true positive and 40 true negative PCL CT examinations, reporting a sensitivity and specificity of 100%. We do not routinely use CT to image the cruciate ligaments specifically. One should look for this information, however, when CT scans have been used to assess other injuries such as tibial plateau fractures.

Stress Radiography

Hewett et al.[7] compared stress radiography with KT-1000 arthrometer and posterior drawer testing. Ten patients with complete tears and 11 with partial tears of the PCL were examined. They found that stress radiography distinguished complete from partial tears, and the other two methods did not. Furthermore, when they compared stress radiographic measurements in patients with complete PCL tears to electrogoniometric measurements in cadaveric PCL-sectioned knees, the results were similar. Although many feel that this technique is rarely necessary in the initial assessment of the acutely injured knee, it may provide important documentation for clinical trials and may be useful in difficult diagnostic cases.

Arthrography

Imaging of the intact anterior cruciate ligament (ACL) is a requisite for adequate arthrographic assessment of a PCL tear.[8] We feel there is little or no role for this modality here. It is an invasive test and its accuracy in diagnosing PCL injuries has, to our knowledge, yet to be reported.

Nuclear Medicine

Routine Approach

A three-phase bone scan with blood flow, blood pool, and delayed images of the knee may be helpful for assessment of acute PCL injuries. Bone scan will identify areas of increased bone turnover. In the acute knee injury, bone bruises (subcortical infractions) and transcortical fractures can be identified.[9] Serial bone scans may be useful to help determine a safer time to return to sporting activity in nonoperatively treated PCL-injured athletes.

Bone scans may also play a role in the chronic PCL-injured patient to monitor for early signs of degenerative change.[4] It has not been determined if reconstruction of the PCL in these patients will alter the progress of degeneration.

Pitfalls

Bone scintigraphy is nonspecific and does not distinguish between a subcortical infraction and an undisplaced transcortical fracture. Other atraumatic causes of increased bone turnover will also "light up," giving a positive scan. The spatial resolution of bone scan is poor compared to all other imaging modalities. There is radiation exposure not only to the knee but to the entire skeleton. Of particular concern in this predominantly younger population is radiation exposure to the gonads as scatter from the bladder, since the isotope is excreted by the kidneys.

Additional Technique: Single Photon Emission Computed Tomography (SPECT)

SPECT provides slices in three planes, allowing improved localization of abnormalities. It is not usually necessary in the knee.

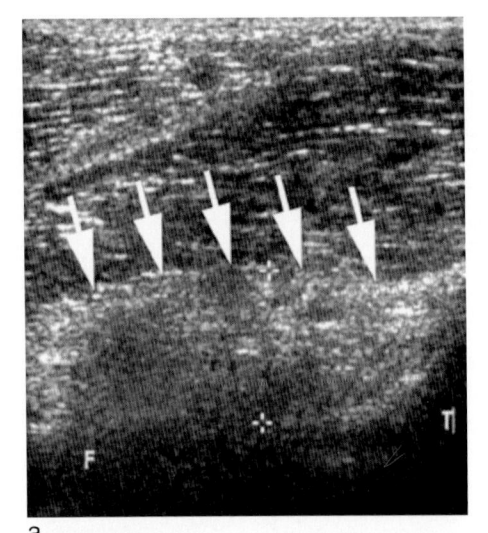

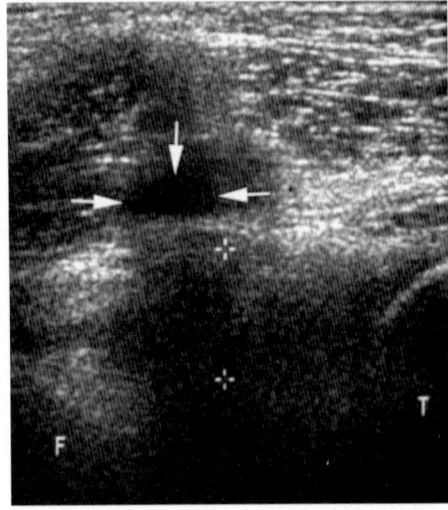

Ultrasound

Routine Approach

The knee is scanned in the longitudinal plane posteriorly along the expected course of the PCL with the patient lying prone and the knee in extension. One can easily compare the abnormal to the normal side. A normal PCL is slightly triangular in shape, narrower proximally and wider distally. It dives deep proximally toward its origin on the lateral aspect of the medial femoral condyle. The echogenicity may vary depending on the angle of the ligament with the transducer (anisotropy artifact), but a structure with linear echoes should be identified and both the origin and insertion visualized. A torn PCL will result in nonvisualization or a redundant or discontinuous ligament (Fig. 5.3). Ultrasound can also assess the status of the posterolateral structures such as the popliteus tendon, fibular collateral ligament, and biceps femoris. As well, it will demonstrate an avulsion fracture of the tibial insertion of the PCL. As an additional benefit, ultrasound provides a fast, inexpensive, and noninvasive evaluation of the popliteal vessels.

Ultrasound has not received a great deal of attention in the area of PCL injury. Suzuki et al.[10] studied 100 normal knees and five knees with rupture of the PCL. They noted that with the posterior approach the image was very clear, and they found ultrasound to be effective for the diagnosis of cruciate ligament rupture. Although more research is needed in this area, the cost, safety, and relative availability of ultrasound compared to magnetic resonance imaging (MRI), for instance, makes ultrasound an attractive prospect.

Pitfalls

Ultrasound provides limited assessment of the lateral meniscus and does not demonstrate bone bruises. A partial in-substance tear or a femoral peel-off type of PCL tear may also be missed. The ACL cannot be assessed well by ultrasound in an acute injury as the patient is required to fully flex the knee. Finally, the benefits of ultrasound imaging are highly operator dependent. An accurate assessment of the PCL relies on sufficient experience and expertise.

Magnetic Resonance Imaging

Routine Approach

A combination of T1- and T2-weighted images should be obtained in at least two planes. Usually sagittal and coronal images are obtained with optional oblique sagittal images, along the length of the PCL. Our routine assessment of a patient with a suspected PCL injury includes sagittal T1 spin echo, sagittal 3D volume Gradient Refocused Acquisition in the Steady State (GRASS), coronal T1 spin echo, coronal fast spin echo dual echo (proton density and T2) with fat saturation, and oblique sagittal fast spin echo T2 along the length of the PCL and along the length of the ACL. If there is concern about the posterolateral structures, a 45-degree oblique (between coronal and sagittal) fast spin echo proton density sequence with fat saturation is also obtained. The normal PCL is homogeneously low in signal intensity on all sequences (Fig. 5.4). It is gently curved, convex pos-

Fig. 5.3. Sagittal ultrasound images of PCL. (a) Long axis posterior view of normal PCL. Large arrows bound superficial margin and + cursor marks deep margin. The normal PCL is homogeneously echogenic with a fibrillar pattern in its long axis. F, femur; T, tibia. Since all sound bounces off of bone (none penetrates it), the cortical margin of the bone is seen as a bright interface and the medullary space cannot be assessed. (b) Long axis posterior view of completely torn PCL. The + cursors mark where the PCL should be but is replaced by a hypoechoic area without a fibrillar pattern. Small arrows bound the adjacent hematoma.

5. Imaging of the Posterior Cruciate Ligament

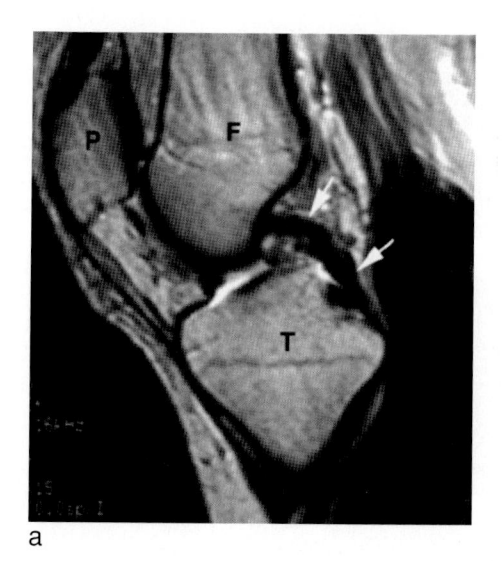

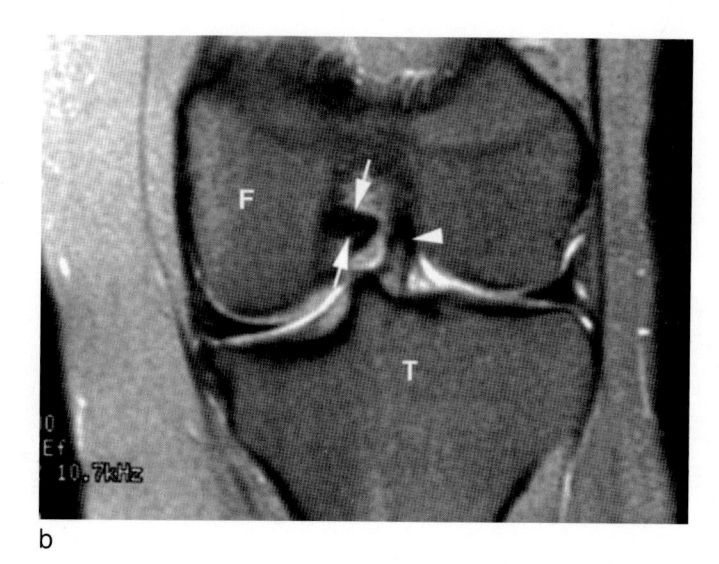

teriorly, in its proximal portion with the knee in extension The meniscofemoral ligaments of Humphry (anterior) and/or Wrisberg (posterior) may be anatomically absent.[11] However, when present they can be identified on MRI and assessed for injury. The integrity of the PCL can be assessed directly. Table 5.1 outlines a number of studies that have addressed MRI of the PCL. The ability to detect injury to this ligament is impressive, with some studies having a specificity, sensitivity, and an accuracy of 100%.[13,14] Partial (Fig. 5.5) and complete (Fig. 5.6) tears can also be distinguished. The location of the tear (femoral, midsubstance, or tibial) and the presence of an avulsed fragment can be identified (Figs. 5.7 and 5.8). Associated injuries to the bone,[15] menisci, other ligaments, tendons, articular cartilage, and capsule can all be assessed. Magnetic resonance angiography of the popliteal vessels can be performed using a 2D time of flight technique, if there is clinical suspicion of a vascular injury.

There are also some indirect signs on MRI that may raise the suspicion of a PCL injury. Lee et al.[16] showed in a cadaver experiment that at maximum joint distention fluid almost completely surrounds the cruciate ligaments, except for a triangular space between the ligaments seen on the midsagittal image. This space corresponds to an extrasynovial space within which the cruciates reside, and the authors felt that fluid seen here on MRI may have resulted from injury to these ligaments. Bone contusions in both the anterior tibial plateau and posterior femoral condyles may be consistent with a forced posterior displacement of the tibia in a flexed knee.[17]

Fig. 5.4. Magnetic resonance imaging (MRI) of normal PCL. (a) Sagittal oblique (parallel to PCL) fast spin echo T2-weighted MR image of a normal PCL (arrows). Note its homogeneously low signal and gentle curvature. T, tibia; F, femur; P, patella. (b) Coronal fast spin echo proton density weighted fat saturated MR image of a normal PCL (arrows) near its origin from medial femoral condyle (F). Note its ovoid shape and homogeneously low signal. Arrowhead, anterior cruciate ligament; T, tibia.

Table 5.1. Studies of magnetic resonance imaging of the posterior cruciate ligament (PCL)

References	Number of knees studied	Specificity	Sensitivity	Accuracy
Kinnunen et al.[29]	33	97	*	
Gross et al.[14]	203	100	100	100
Fischer et al.[30]	1,014	99	†	99
Polly et al.[12]	50	100	*	100
Chen et al.[13]	50	100	100	100
Bui-Mansfield et al.[31]	50	84	100	84

*No PCL tears.
†Too few torn to calculate with assurance.

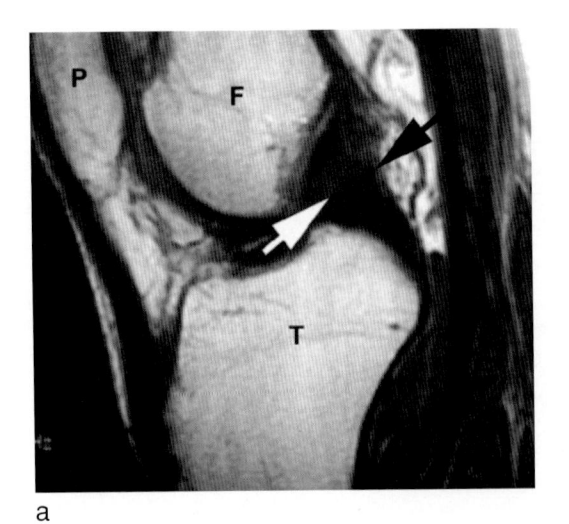

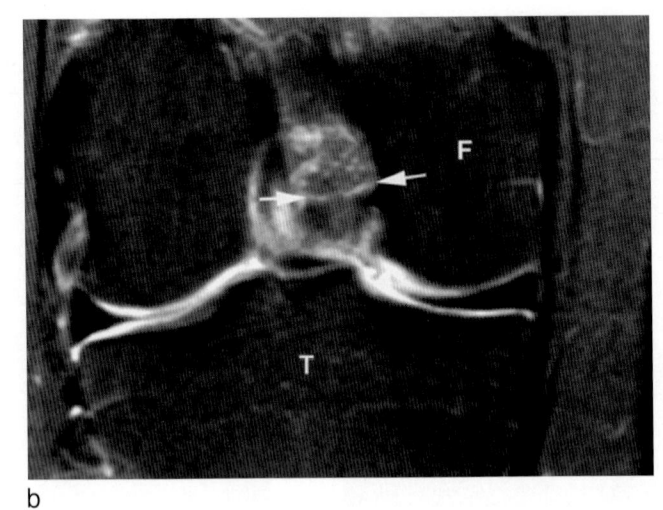

Fig. 5.5. MRI of partial midsubstance PCL tear, confirmed at arthroscopy. Sagittal spin echo T1-weighted (a) and coronal fast spin echo proton density weighted fat saturated (b) MR images demonstrate the increased signal and width of the midsubstance of the ligament (large arrows) as well as the split through the midsubstance of the ligament, seen in cross section (small arrows). F, femur; T, tibia; P, patella.

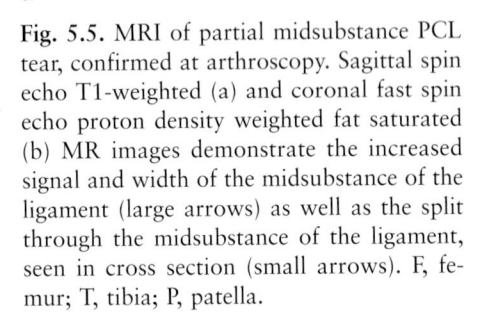

On the other hand, contusions in the anterior tibial articular surface and the anterior aspect of the femoral condyle may be associated with hyperextension injuries.[17] These patterns on an MRI should lead to a very thorough examination of the PCL. With respect to the MRI findings in patients with PCL injuries compared to ACL injuries, meniscal tears are less frequently observed with PCL injuries, and bone contusions tend to be located anteriorly on the tibia with PCL injuries rather than posteriorly on the tibia with most ACL injuries.[18,19]

There is very little information available about the normal appearance of the reconstructed PCL on MRI (Fig. 5.9). Presumably this is similar to that of the reconstructed ACL where the neoligament tends to initially demonstrate a low signal intensity in both T1-weighted and T2-weighted sequences that variably and gradually increases with time, with a few fibers of low signal intensity usually persisting.[20]

Pitfalls

The normal PCL may have intermediate signal intensity proximally on T1-weighted images. This is caused by the magic angle effect and will disappear with T2-weighting. The magic angle effect occurs when a highly

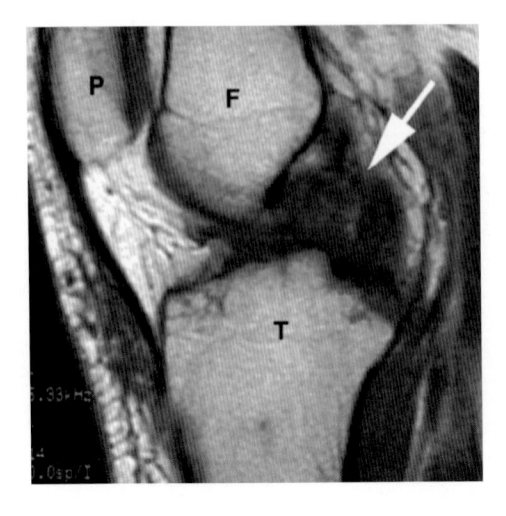

Fig. 5.6. MRI of a complete PCL tear confirmed at arthroscopy. Sagittal oblique (parallel to PCL) fast spin echo T2-weighted MR image of a complete midsubstance PCL tear. Note the high signal intensity and loss of linear shape of the midportion of the ligament (arrow). F, femur; T, tibia; P, patella.

ordered linear structure passes through an angle of 55 degrees to the main magnetic field, as does the PCL as it curves from the femur toward the tibia.

Weiss et al.[21] described a low-signal band anterior and parallel to the PCL on sagittal MR images. This was caused by a displaced bucket-handle tear of the medial meniscus. This "double PCL" should not be mistaken for a PCL injury (Fig. 5.10). Eosinophilic or mucoid degeneration of the PCL produces high signal intensity simulating injury on the MRI.[22] Smith et al.[23] noted that signal intensity in the cruciate ligaments changed with knee position, and that the change from extension to flexion caused a decrease in the signal intensity in five of six PCLs. This may have been due to magic angle artifact since the ligaments were imaged with a T1-weighted spin echo pulse sequence. However, this study also found signal intensity change when tension was applied to the anterior cruciate ligament. The authors concluded that the normal differences in signal intensity observed between the ACL and PCL were in part due to the position of the joint and consequent differences in tension on the two ligaments. Whether the difference in signal intensity of the PCL with changing knee position in this study was due to magic angle artifact or tension remains to be further delineated.

Partial tears or degenerative change to the PCL usually involves the central fibers of the ligament. These can be missed on sagittal images and are best seen on coronal images, which demonstrate the PCL in cross section.

In a complete midsubstance tear the PCL may maintain its normal shape. However, the signal intensity of the ligament will be diffusely increased.

The status of the PCL in the chronic situation can be confusing. Tewes et al.[24] reviewed 13 patients (follow-up ranging from 5 months to 4 years) following a complete PCL tear initially documented within 10 weeks of their injury on MRI. They noted that 77% of these ligaments were continuous from tibia to femur on follow-up MRI. It should be noted that 9 of the 10 continuous ligaments were U shaped or hyperbuckled.

Although buckling of the PCL (also called J-shaped PCL) may be seen when the ligament has been injured, changes in curvature can also be noted when the PCL is normal[25–27] but the patient has an ACL injury. Furthermore, an increase in the curvature of the PCL may be perceived if the knee is hyperextended during the MRI.[28]

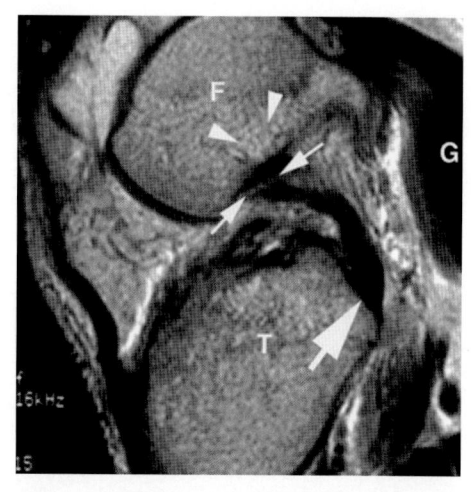

Fig. 5.7. MRI of a femoral peel-off injury of PCL. Sagittal oblique (parallel to PCL) fast spin echo T2-weighted MR image showing normal PCL signal (large arrow) distally and a normal tibial (T) attachment. The PCL is peeled off of its femoral (F) attachment (small arrows), confirmed at arthroscopy. The arrowheads point to bone bruises at the femoral attachment. Note the high signal fluid extending through the disrupted capsule and deep to medial head of gastrocnemius (G) muscle.

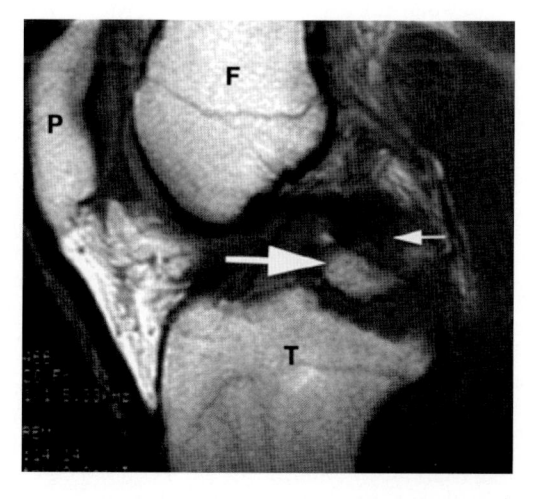

Fig. 5.8. MRI of PCL avulsion fracture. Sagittal T1-weighted MR image demonstrates a large avulsed fragment (large arrow) with the insertion of the PCL attached. Increased signal intensity at the insertion of the ligament (small arrow) may indicate a partial tear; however, the loss of tension on the ligament can sometimes give abnormal signal without a tear. F, femur; T, tibia; P, patella.

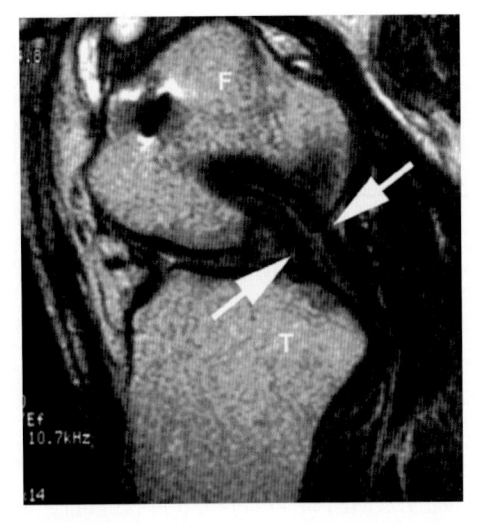

Fig. 5.9. MRI of PCL reconstruction. Sagittal oblique (parallel to PCL) fast spin echo T2-weighted MR image of a double loop PCL reconstruction with a Kennedy Ligament Augmentation Device (large arrows). Note the intermediate signal that can often be seen within normal grafts. A knee scope a few years later noted some laxity in the PCL, but it was intact and well synovialized. F, femur; T, tibia.

Fig. 5.10. MRI of "double PCL" pitfall. Sagittal spin echo T2-weighted MR image of a "double PCL" sign. The large arrow points to the PCL and the smaller arrow to the bucket handle fragment of the torn medial meniscus, confirmed at arthroscopy. F, femur; T, tibia; P, patella.

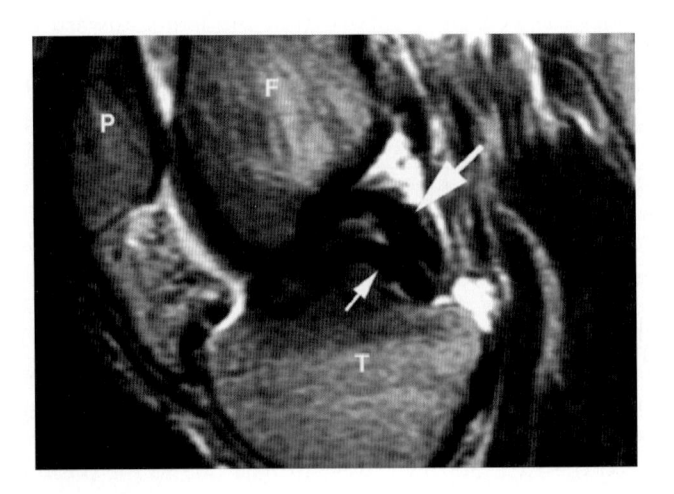

Conclusion

Plain film should be used as the first imaging modality. This allows identification of most fractures, either directly or by visualization of a lipohemarthrosis. It also allows a preliminary assessment of capsular integrity in the acutely injured patient. MRI is the next imaging modality of choice because of its superb demonstration of PCL pathology and all associated injuries. As well, vascular injuries can be assessed by MRI. If MRI is not available, ultrasound in experienced hands can well assess the PCL, in addition to many of the potentially co-injured structures, including the vessels. In the isolated PCL injury that is being treated nonoperatively, bone scan may be helpful in the identification of associated bone bruises to predict duration of pain and to assess early degenerative change.

One should always remember that a thorough history and physical examination must guide the choice of imaging, and not the other way around.

References

1. Harner CD, Höher J. Evaluation and treatment of posterior cruciate ligament injuries. Am J Sports Med 1998;26:471–482.
2. Hall FM, Hochman MG. Medial Segond-type fracture: cortical avulsion off the medial tibial plateau associated with tears of the posterior cruciate ligament and medial meniscus. Skeletal Radiol 1997;26:553–555.
3. Bonnani G, Cacchio A, Candella V, et al. Patologia del Legamento Crociato Posteriore. In: De Paulis F, Puddu G, eds. Ginocchio. Diagnostica per Immagini e Inqadramento Clinico, 1st ed. Naples: Casa Editrice Idelson, 1996:143–159.
4. Andrews JR, Edwards JC, Satterwhite YE. Isolated posterior cruciate ligament injuries. Clin Sport Med 1994;13:519–530.
5. Golimbu C, Firooznia H, Rafii M, Beranbaum E. Computerized tomography of the posterior cruciate ligament. Comput Radiol 1982;6:233–238.
6. Passariello R, Trecco F, DePaulis F, Masciocchi C, Bonanni G, Zobel BB. CT demonstration of capsuloligamentous lesions of the knee joint. J Comput Assist Tomogr 1986;10:450–456.
7. Hewett TE, Noyes FR, Lee MD. Diagnosis of complete and partial posterior cruciate ligament ruptures. Am J Sports Med 1997;25:648–655.
8. Pavlov H, Schneider R. Extrameniscal abnormalities as diagnosed by knee arthrography. Radiol Clin North Am 1981;19:287–304.
9. Marks PH, Goldenberg JA, Vezina WC, Chamberlain MJ, Vellet AD, Fowler PJ. Subchondral bone infractions in acute ligamentous knee injuries demon-

strated on bone scintigraphy and magnetic resonance imaging. J Nucl Med 1992;33:516–520.
10. Suzuki S, Kasahara K, Futami T, Iwasaki R, Ueo T, Yamamuro T. Ultrasound diagnosis of pathology of the anterior and posterior cruciate ligaments of the knee joint. Arch Orthop Trauma Surg 1991;110:200–203.
11. Heller L, Langman J. The menisco-femoral ligaments of the human knee. J Bone Joint Surg 1964;46B:307–313.
12. Polly DW, Callaghan JJ, Sikes RA, McCabe JM, McMahon K, Savory CG. The accuracy of selective magnetic resonance imaging compared with the findings of arthroscopy of the knee. J Bone Joint Surg 1988;70A:192–198.
13. Chen M-C, Shih TT-F, Jiang C-C, Su C-T, Huang K-M. MRI of meniscus and cruciate ligament tears correlated with arthroscopy. J Formos Med Assoc 1995;94:605–611.
14. Gross ML, Grover JS, Bassett LW, Seeger LL, Finerman GA. Magnetic resonance imaging of the posterior cruciate ligament. Am J Sports Med 1992;20:732–737.
15. Vellet AD, Marks PH, Fowler PJ, Munro TG. Occult posttraumatic osteochondral lesions of the knee: prevalence, classification and short term sequelae evaluated with MR imaging. Radiology 1991;178:271–276.
16. Lee SH, Petersilge CA, Trudell DJ, Haghighi P, Resnick DL. Extrasynovial spaces of the cruciate ligaments: anatomy, MR imaging, and diagnostic implications. AJR 1996;166:1433–1437.
17. Sonin AH, Fitzgerald SW, Friedman H, et al. Posterior cruciate ligament injury: MR imaging diagnosis and patterns of injury. Radiology 1994;190:455–458.
18. Sonin AH, Fitzgerald SW, Hoff FL, Friedman H, Bresler ME. MR imaging of the posterior cruciate ligament: normal, abnormal and associated injury patterns. RadioGraphics 1995;15:551–561.
19. Remer EM, Fitzgerald SW, Friedman H, Rogers LF, Hendrix RW, Schafer MF. Anterior cruciate ligament injury: MR imaging diagnosis and patterns of injury. RadioGraphics 1992;12:901–915.
20. Munk PL, Vellet AD, Fowler PJ, Miniaci A, Crues JV. Magnetic resonance imaging of reconstructed knee ligaments. Can Assoc Radiol J 1992;43:411–419.
21. Weiss KL, Morehouse HT, Levy IM. Sagittal MR images of the knee: a low-signal band parallel to the posterior cruciate ligament caused by a displaced bucket-handle tear. AJR 1991;156:117–119.
22. Hodler J, Haghighi P, Trudell D, Resnick D. The cruciate ligaments of the knee: correlation between MR appearance and gross and histologic findings in cadaveric specimens. AJR 1992;159:357–360.
23. Smith KL, Daniels JL, Arnoczky SP, Dodds JA, Cooper TG, Gottschalk A, Shaw DA. Effect of joint position and ligament tension on the MR signal intensity of the cruciate ligaments of the knee. J Magn Reson Imaging 1994;4:819–822.
24. Tewes DP, Fritts HM, Fields RD, Quick DC, Buss DD. Chronically injured posterior cruciate ligament. Clin Orthop Rel Res 1997;335:224–232.
25. Tung GA, Davis LM, Wiggins ME, Fadale PD. Tears of the anterior cruciate ligament: primary and secondary signs at MR imaging. Radiology 1993;188:661–667.
26. McCauley TR, Moses M, Kier R, Lynch JK, Barton JW, Jokl P. MR diagnosis of tears of anterior cruciate ligament of the knee: importance of ancillary findings. AJR 1994;162:115–119.
27. Robertson PL, Schweitzer MR, Bartolozzi AR, Ugoni A. Anterior cruciate ligament tears: evaluation of multiple signs with MR imaging. Radiology 1994;193:829–834.
28. Crotty JM, Monu JUV, Pope TL. Magnetic resonance imaging of the musculoskeletal system. Clin Orthop Rel Res 1996;330:288–303.
29. Kinnunen J, Bonestam S, Kivioja A, et al. Diagnostic performance of low field MRI in acute knee injuries. Magn Reson Imaging 1994;12:1155–1160.
30. Fischer SP, Fox JM, Del Pizzo W, Friedman MJ, Snyder SJ, Ferkel RD. Accuracy of diagnosis from magnetic resonance imaging of the knee. J Bone Joint Surg 1991;73A:2–10.
31. Bui-Mansfield LT, Youngberg RA, Warme W, Pitcher JD, Nguyen P-LL. Potential cost savings of MR imaging obtained before arthroscopy of the knee: evaluation of 50 consecutive patients. AJR 1997;168:913–918.

Chapter Six

Measurement of the Posterior Cruciate Ligament and Posterolateral Corner

Don Johnson

The proper evaluation of the posterior displacement of the knee involves the clinical examination, instrumented measuring devices, and stress x-rays. It is important to measure the amount of posterior displacement of the tibia relative to the femur for the following reasons:

- To differentiate a partial from a complete posterior cruciate ligament (PCL) injury.
- To compare results of treatment of PCL injuries.
- To compare the change over time in laxity after PCL injury or surgical treatment.

The assessment of the results of surgical treatment of the anterior cruciate ligament (ACL) only became precise with the development of an accurate device, the KT-1000 (Medmetric, San Diego, CA).

There are several available methods to measure the posterior displacement of the tibia:

- Clinical measurement
- KT-1000
- Knee laxity tester
- Telos stress radiography
- LARS rotational laxiometer

Clinical Measurement

Posterior Drawer Test

The posterior drawer test is performed with the patient relaxed and lying supine on the examining table (Fig. 6.1). The knee is flexed to 90 degrees. A hand is used to steady the foot, and the other hand is used to push the proximal tibia posterior. The degree of posterior displacement of the tibia relative to the femur is measured. When the tibia is forward, it is referred to as the tibial step. The position of the tibial step is measured relative to the femur. This test is graded from 1 to 3:

- Grade I—0–5 mm (the tibial step is palpable).
- Grade II—6–10 mm (the tibia is equal to the femoral condyle).
- Grade III—>10 mm (the tibia is behind the condyle).

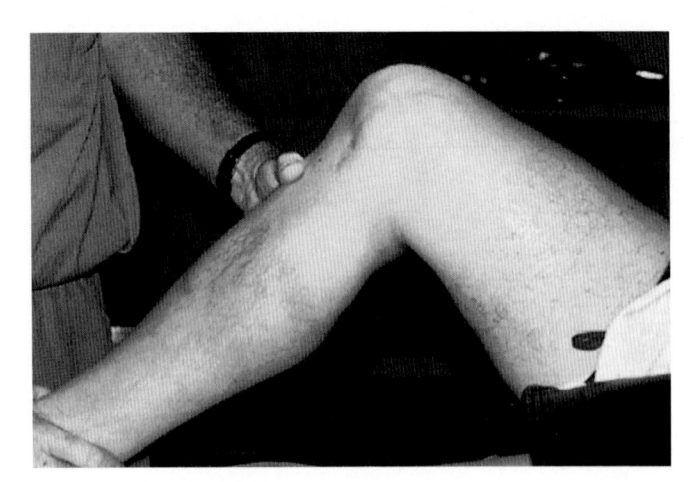

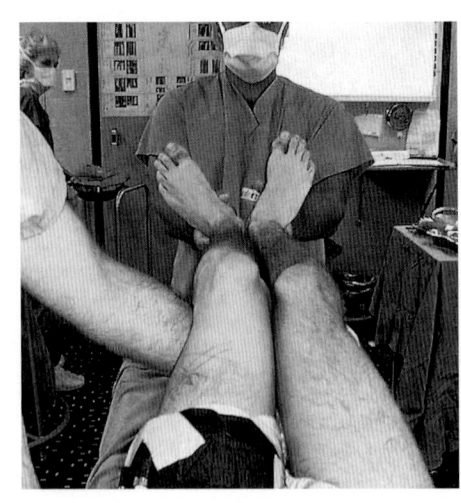

Fig. 6.1. The posterior drawer test. When the tibia is pushed backward, it falls behind the femoral condyle. This is a grade III posterior drawer.

Fig. 6.2. The dial test measures the amount of external tibial rotation. The injured side, the left, demonstrates increased tibial external rotation. This amount of increased external rotation is 15 degrees.

Dial Test

The dial test is performed with the patient relaxed and lying supine on the examining table (Fig. 6.2). The knees are flexed to 30 degrees, while an assistant suports the thighs. The foot is turned outward, and the amount of external rotation is measured in degrees as a dial on a circular instrument. The test is repeated at 90 degrees of knee flexion. If the test is positive only at 30 degrees, this is an isolated injury to the posterolateral corner. If it is positive at 90 degrees, it is associated with a PCL injury. The dial test can also be performed with the patient in the prone position. The external rotation of the tibia is noted at 30 and 90 degrees. The advantage of the prone position is that a second examiner is not required to support the thighs of the patient.

Advantages

- Most clinicians are familiar with the techniques of clinical examination.
- No special equipment is required.

Disadvantage

- There is considerable variation between novice and experienced observers on the degree of displacement.

KT-1000

The KT-1000 has become the standard for the measurement of ACL laxity (Fig. 6.3). However, it has not achieved the same level of acceptance for the PCL. Daniel et al.[1] first described the method of measuring the posterior laxity by determining the quadriceps' neutral point. The principle of the measurement, as described by Daniel et al., is to determine the four levels of anterior to posterior motion:

- Anterior—passive anterior displacement with 30 pounds of force from the quadriceps' neutral position.
- Quadriceps neutral angle—the patient contracts the quadriceps at 71 degrees to bring the tibia forward to the quadriceps neutral position.
- Posterior sag—the gravity posterior displacement.
- Posterior displacement—passive displacement with 30 pounds of force.

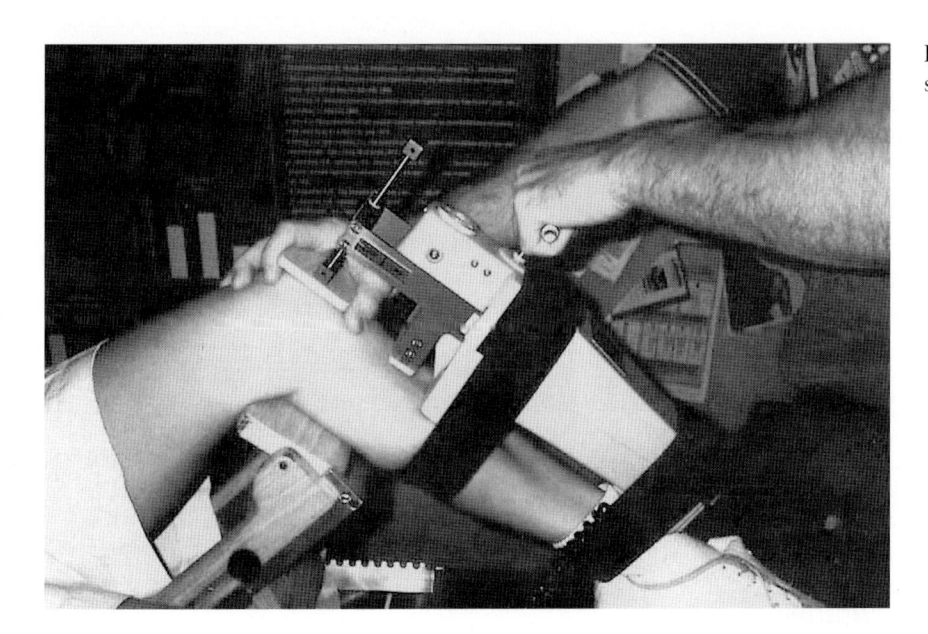

Fig. 6.3. A PCL-deficient knee being measured with the KT-1000 device.

Steps in the Measurement

The patient is measured in a relaxed, supine position. A support for the posterior thigh that positions the knee at 70 degrees is useful. The foot is steadied by the examiner sitting on the toes. The KT-1000 device is applied to the anterior tibia. The patient is asked to contract the quadriceps by sliding the foot down the table. The quadriceps' neutral angle is determined when the tibia comes forward to its limit. The dial is set at 0. The tibia is allowed to sublux posteriorly, and the posterior sag is measured. The tibia is pushed posterior to the beep (usually 30 pounds of force), and the posterior displacement is measured. The posterior motion from the quadriceps' neutral angle to the posterior sag, and then the posterior displacement with 30 pounds of posterior force is totaled. The posterior laxity is determined when these two are added.

In my experience it is difficult to get the patient to contract his quadriceps to bring the tibia fully forward to the neutral position. This amount of forward displacement is often underestimated. I presented the results of a study to the PCL study group in 1995, comparing the KT value against the stress x-ray. When there was >10 mm of posterior displacement, the KT, expressed as a percentage of the Telos, was 65%; when there was <10 mm of posterior displacement, the KT was 72% of the Telos. Thus, the KT measurement is less than that which is measured with the stress x-ray, and this difference is more marked when the displacement is greater than 10 mm.

In summary, the PCL-deficient knee should be measured with the stress x-ray. The underestimation of displacement by the KT-1000 was also confirmed by Frank Noyes' group[2] in a study reported in the *American Journal of Sports Medicine*. In this study, the average displacements were as follows:

- COMPLETE PCL TEARS

Clinical posterior drawer	9.2 mm
KT-1000	7.6 mm
Stress x-ray	12.2 mm

- PARTIAL PCL TEARS

  Clinical posterior drawer 7.4 mm

  KT-1000 6.7 mm

  Stress x-ray 5.6 mm

This study confirms that the measurement of the posterior displacement is more accurate with stress x-ray, especially in those cases that are greater than 10 mm. Another study, by Harner's group,[3] compared a novice and an experienced user of the KT-1000 device and found that the device was a moderately reliable tool to evaluate PCL laxity. This was a small group of patients, most of whom had less than 10 mm of laxity.

Advantages

- KT-1000 is a widely used and accepted method of measurement of the ACL-deficient knee.
- KT-1000 is available in most sports medicine clinics.

Disadvantage

- The technique of the quadriceps active test underestimates the posterior displacement, especially when measuring more than 10 mm.

Telos Stress Radiography

Staubli[4] has demonstrated that stress radiography is an accurate method to measure anterior-posterior motion of the knee. The method of measurement, using the Telos device to perform a stress x-ray, requires the patient to lie in the lateral position with the injured side down (Fig. 6.4). The knee is positioned at 90 degrees, and a 15-kg (32-pound) force is exerted on the anterior tibial tubercle. Then the x-ray is taken. The knee must be positioned in a true lateral position, which lateral is evident when the femoral condyles are exactly overlapped. The common difficulty is in obtaining a true lateral. This is made easier if the technician places a piece of soft foam under the patient's ankle and foot to elevate the foot. The tibia must be parallel to the table top to get a lateral view of the knee.

It is possible to perform the examination without the Telos device, and achieve similar results, by using the technique of maximum hamstring contraction. In Fig. 6.5, the patient is performing an active maximal hamstring contraction against resistance in the lateral position. The x-ray is done during the maximal contraction. The tibia must be parallel to the table to get

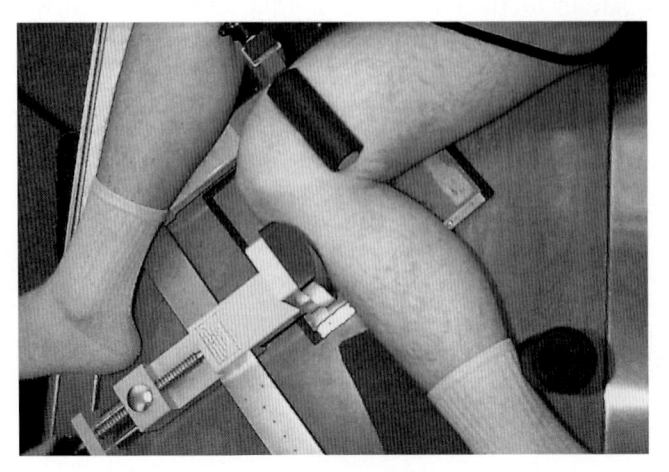

Fig. 6.4. The stress x-ray examination of the PCL deficient knee with the Telos (Austin and Associates, Fallston, PA) device.

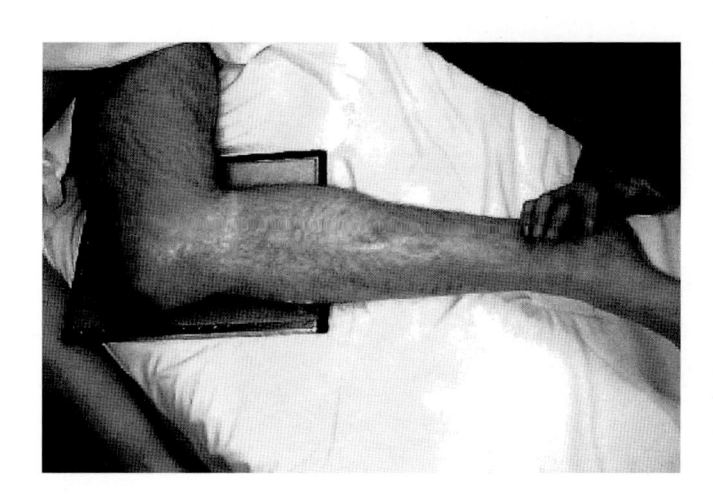

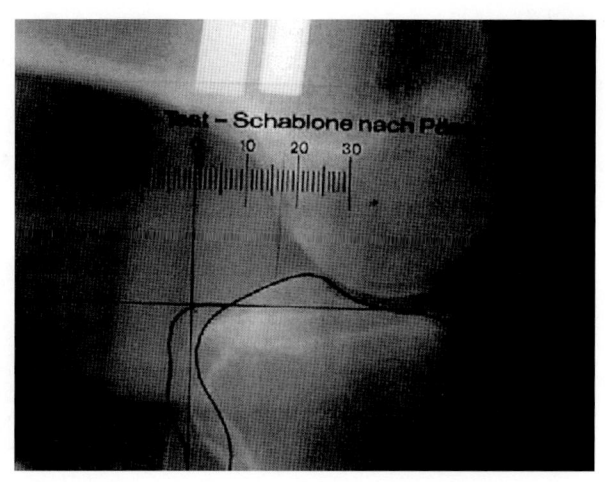

the true lateral. This has been described and validated by Professor Pascal Christal of Paris, France.[5]

The degree of posterior displacement is measured with a template on the lateral stress x-ray. In the example shown in Fig. 6.6, the posterior displacement is 17 mm. It is important to note that the femoral condyles are overlapping, indicating a true lateral x-ray. This figure also demonstrates the correct position of the template for measuring the posterior displacement. A study at our clinic was undertaken to validate the use of the template. Two observers independently reviewed 80 patients with PCL laxity. The conclusion of the study was that the measurement could be reliably reproduced.

Fig. 6.5. The stress x-ray performed with active resisted hamstring contraction.

Fig. 6.6. The stress x-ray with the measuring template.

Advantages

- This method allows accurate measurement of the posterior displacement with a template.
- Reproducible and accurate force is applied to the anterior tibia.

Disadvantages

- It is essential to have a true lateral x-ray with the femoral condyles overlapping, as in Fig. 6.6.
- The template must be accurately positioned to reproduce the same measurement each time.
- The use of x-rays entails exposure for the patient and cost in terms of materials and labor.
- The radiologic technician must be trained in the correct use of the device.
- The Telos device is expensive.

Knee Laxity Tester

The use of the knee laxity tester (KLT) arthrometer (Orthopedic Systems, Hayward, CA) has been described by Cannon and Vittori.[6] The technique is similar to the KT-1000. The patient is positioned sitting with the knee flexed to 90 degrees over the end of a table. The patient actively contracts the quadriceps. At this quadriceps neutral point, the instrument is set to 0. The tibia is then displaced posteriorly with a 20- and 40-pound force. The displacements are recorded. Eakin and Cannon[7] found that the arthrometric measurements correlated well with the clinical examination. The arthrometer was also able to detect subtle grade I injuries.

Fig. 6.7. The rotational laxiometer used to measure the external rotation of the tibia at 30 to 90 degrees of knee flexion.

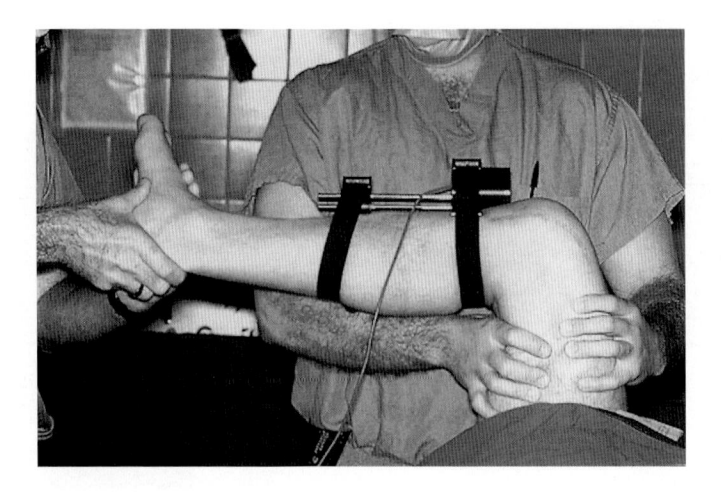

Advantage
- The knee is held in the 90-degree position and it may be easier for the patient to perform the quadriceps active test.

Disadvantages
- The instrument is not widely available.
- The 71-degree position was determined by Daniel to be the optimum knee flexion angle to measure the quadriceps' active position.

Rotational Laxiometer

The rotational laxiometer (LARS, Dijon, France) was developed by Dr. Jonathon Beacon to measure the degree of rotation of the tibia. This device is essentially an electrical goniometer to measure the internal and external rotation of the tibia. The patient is positioned supine and relaxed on the examining table. The device is applied to the anterior tibia (Fig. 6.7). The assistant flexes the knee to 30 degrees, and steadies the distal femur. The foot is held in the neutral position, and the device set to 0. The foot is then maximally externally and internally rotated and the degree of rotation measured. This procedure is repeated with the knee at 90 degrees and both internal and external rotation measured and recorded. The opposite knee is also measured in the same manner.

This device has been validated to measure the normal variation of tibial rotation by Fannelli's group[8] and reported in *Arthroscopy*; the objective of this study was to establish a baseline of normal measurement of the degree of normal external rotation of the tibia at 30 and 90 degrees. Three authors each examined 30 asymptomatic patients to determine the side-to-side difference. At 90 degrees the side-to-side difference was 4.4 degrees (range 3.7–5.1) and at 30 degrees the difference was 5.5 degrees (range 4.7–6.3). This means that any measurement above this number is abnormal. This gives us a baseline to determine which patients need to have a posterolateral reconstruction. It also gives us a measurement device to assess reconstructed knees postoperatively.

Advantages
- There is an accurate method to measure in degrees the amount of internal and external rotation of the tibia. Standard deviations of normal have also been recorded.

- There is no other device available on the market to measure external and internal tibial rotation.

Disadvantages
- The device requires two people to operate.
- The device is expensive and not widely available.

Conclusion

The clinical examination of the PCL still remains the cornerstone of measurement. There are available sophisticated instruments to more accurately measure the posterior displacement of the tibia relative to the femur. At the present time these are expensive and difficult to use, and they entail a specific learning curve. The main value is to those who are performing posterior cruciate reconstructive surgery and need to accurately measure and report their results.

References

1. Daniel DM, Stone ML, Barnett P, et al. Use of the quadriceps active test to diagnose posterior cruciate ligament disruption and measure posterior laxity of the knee. J Bone Joint Surg 1988;70A:386–391.
2. Hewett TE, Noyes FR, Lee MD. Diagnosis of complete and partial posterior cruciate ligament ruptures: stress radiography compared with KT1000 arthrometer and posterior drawer testing. Am J Sports Med 1997;25(5):648–655.
3. Huber FE, Irragang JJ, Harner C, Lephart S. Intratester and Intertester reliability of the KT-1000 arthrometer in the assessment of posterior laxity of the knee. Am J Sports Med 1997;25(4):479–485.
4. Staubli H-U. Stress radiography. Mesurements of knee motion limits. In: Daniel D, ed. Knee Ligaments. Structure, Function, Injury, and Repair. New York: Raven Press, 1990:449–459.
5. Chassaing V, Deltour F, Touzard R, Ceccaldi JP, Miremad C, Lemaire M. Etude radiologique du LCP a 90 degrees de flexion. Rev Chir Orthop 1995;suppl 2:65–88.
6. Cannon WD, Vittori JM. The use of instrumented testing to measure knee displacement. In: Aichroth P, Cannon WD, eds. Knee Surgery: Current Practice. London: Martin Dunitz, 1992:206–216.
7. Eakin CL, Cannon WD. Arthrometric evaluation of posterior cruciate ligament injuries. Am J Sports Med 1998;26(1):96–102.
8. Bleday RM, Fanelli GC, Giannotti BF, Edson CJ, Barrett TA. Instrumented measurement of the posterolateral corner. Arthroscopy 1998;14(5):489.

Chapter Seven

Arthroscopic Evaluation of the Posterior Cruciate Ligament and Posterolateral Corner

Gregory C. Fanelli

The incidence of posterior cruciate ligament (PCL) injuries has been reported to be from 1% to 40% in acute knee injuries.[1-4] This incidence is dependent on the patient population reported, with PCL tears occurring more frequently in trauma patients than in athletic-injury patients.[2,3] These PCL tears result from a variety of injuries. Isolated PCL tears most likely result from a direct blow to the proximal tibia, causing a posteriorly directed force. This occurs with the so-called dashboard knee in motor vehicle accidents, or when the proximal tibia contacts an immovable object. A fall on the flexed knee with the foot in plantar flexion may also induce an isolated PCL tear.[5] Forced flexion plus internal rotation has also been reported to cause isolated PCL tears.[6] Hyperextension, forced varus or valgus, and knee dislocations are associated with PCL tears plus other ligament tears.[2,3,7,8]

Posterior cruciate ligament tears can be isolated or part of a complex of combined knee ligament injuries. PCL injuries can be interstitial disruptions, bony avulsions from the tibia or femur, or nonbony insertion detachments. Patients presenting with PCL tears often give a history of a posteriorly directed force applied to the proximal tibia, or an episode of forced flexion plus internal rotation. More severe injuries involving the PCL plus other ligaments occur with a history of hyperextension, forced varus or valgus, or knee dislocation.

Abrasions of the proximal tibia may be present along with varying degrees of knee effusion (Fig. 7.1). Various physical examination tests have been described that detect PCL tears, and related structural injuries[1,4,9-17]:

- Posterior tibial drop-back
- Lateral thrust gait
- Decreased tibial step off
- Full extension varus-valgus laxity
- False-positive anterior drawer test
- Positive pseudo–Lachman test
- Posterior drawer test
- Reverse pivot shift test
- Posterior sag sign
- Quadriceps active test
- Dynamic posterior shift test
- External rotation thigh-foot angle test

The basic function of each of these tests is to demonstrate proximal posterior tibial displacement relative to the distal femur with the knee flexed 90 degrees. This posterior proximal tibial displacement can occur in a

Fig. 7.1. Anterior tibial abrasions associated with posterior cruciate ligament (PCL) injury in a multiple trauma patient. Left knee has anterior cruciate ligament (ACL), PCL, and medial collateral ligament (MCL) tears. Right knee has ACL and posterolateral corner tears. (From Fanelli et al.,[19] with permission.)

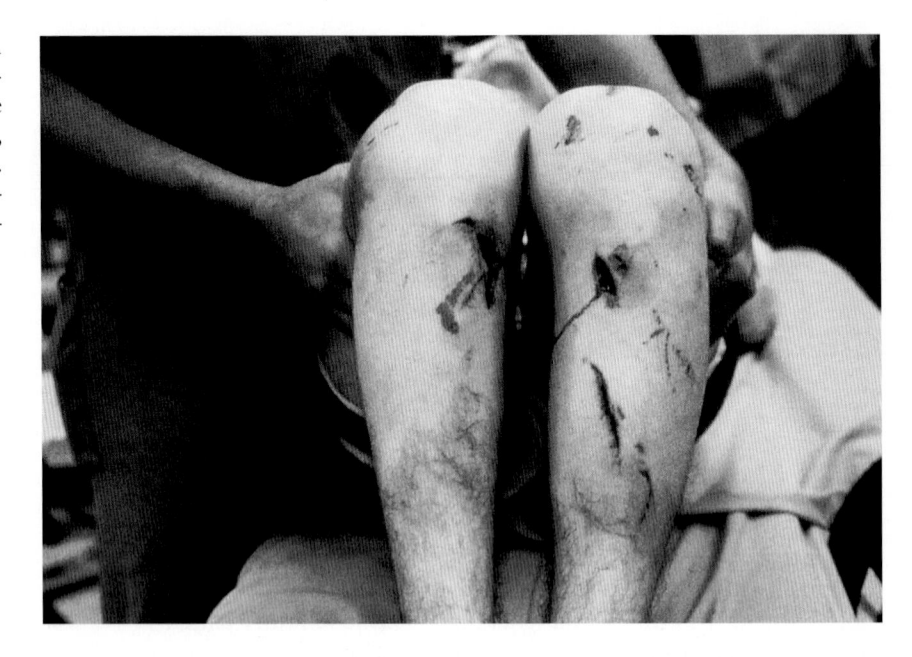

straight anterior-posterior plane, with or without the presence of a rotational component. Exam under anesthesia allows a thorough physical examination of the knee to detect multidirectional instability in the PCL-insufficient knee (especially occult posterolateral instability).

Arthroscopic evaluation of the PCL and related structures in the PCL-injured knee is a useful adjunct to the history, physical examination, arthrometer testing, and imaging studies. Arthroscopic PCL evaluation aids in surgical decision making and planning of reparative or reconstructive procedures.

The Three-Zone Posterior Cruciate Ligament

Arthroscopic evaluation of the PCL has been reported by Lysholm and Guillquist[18] and Fanelli et al.[19] Arthroscopic evaluation of the PCL is a very helpful adjunct to physical examination and imaging studies, especially with respect to surgical planning. At our clinic we have developed and published the three-zone concept of arthroscopic PCL evaluation, and routinely use this method in our treatment of PCL injuries.[19] In this concept, the PCL is divided into three distinct zones. Zone 1 extends from the femoral insertion of the PCL to where that ligament disappears behind the anterior cruciate ligament (ACL). Zone 2 is the portion that lies behind the ACL. Zone 3 is the tibial insertion site (Fig. 7.2).

Surgical Technique

The patient is positioned supine on the operating table, and examination of the normal and injured extremity is performed utilizing the tests outlined above. The extremity is prepped and draped free in the conventional sterile fashion. A lateral arthroscopic post is used instead of a leg holder since this allows more freedom of motion of the lower extremity (Fig. 7.3).

7. Arthroscopic Evaluation of the Posterior Cruciate Ligament and Posterolateral Corner

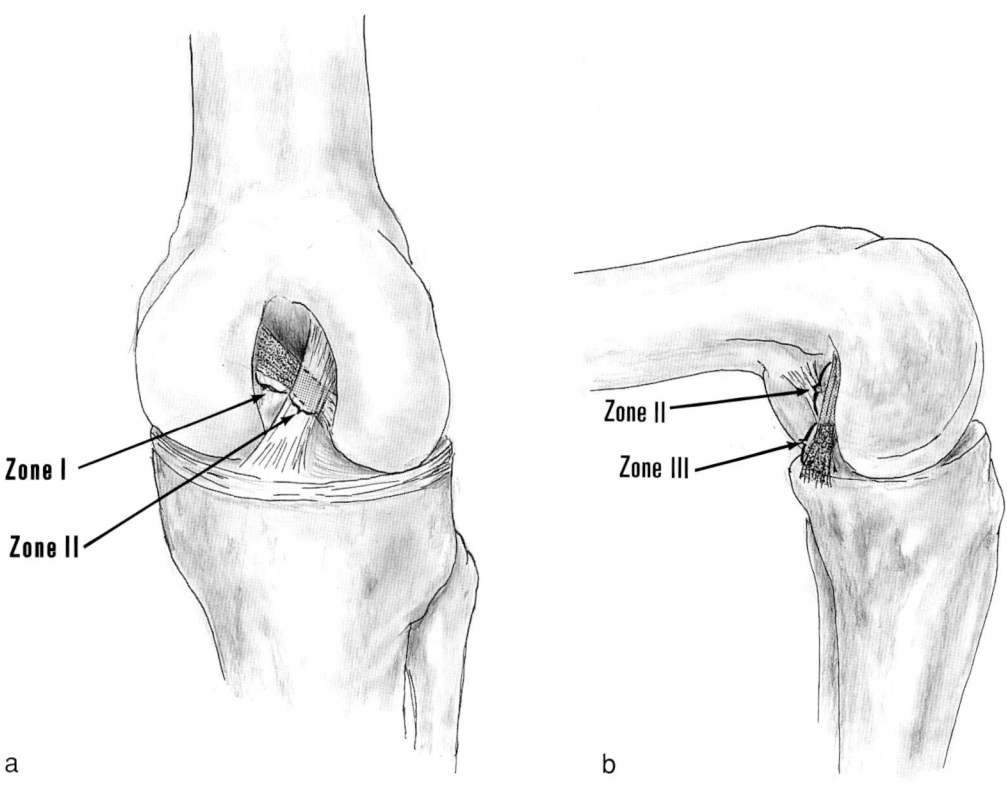

Fig. 7.2. (a) Zones 1 and 2 of the PCL. (b) Zones 2 and 3 of the PCL.

Arthroscopic evaluation of the entire PCL requires a systematic approach, assessing the extent of ligament damage to aid in surgical planning. A 25- or 30-degree arthroscope is utilized through the anterior inferior lateral patellar portal to visualize zone 1 of the PCL, and the 25- or 30-degree arthroscope is placed in the posterior medial arthroscopic portal to visualize zones 2 and 3 (Fig. 7.4). The use of the inferior lateral patellar portal and the posterior medial arthroscopic portals allows complete arthro-

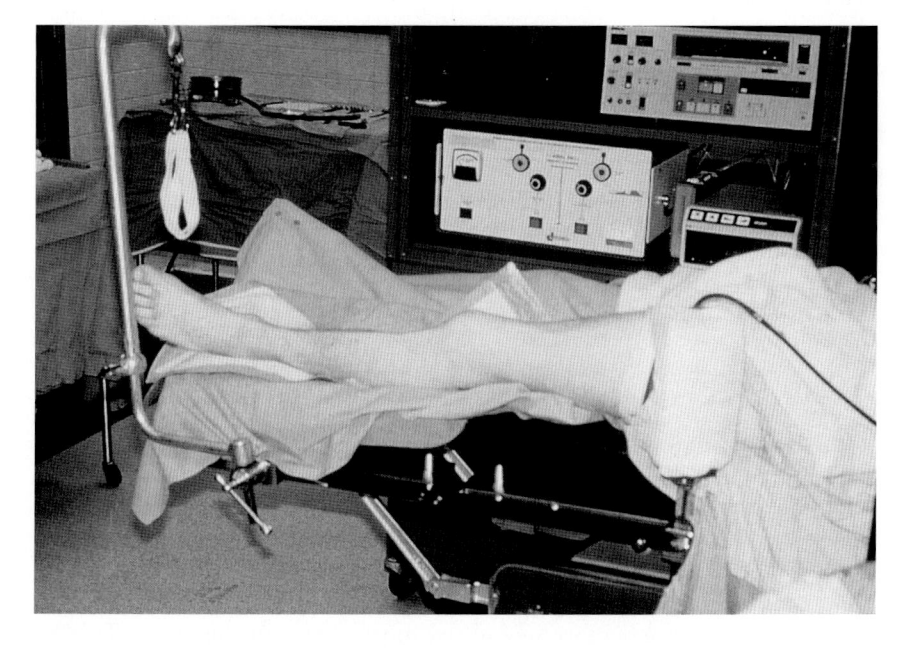

Fig. 7.3. The surgical leg is prepped and draped free, and controlled using a lateral post. The lateral post allows more freedom to maneuver the extremity during surgery.

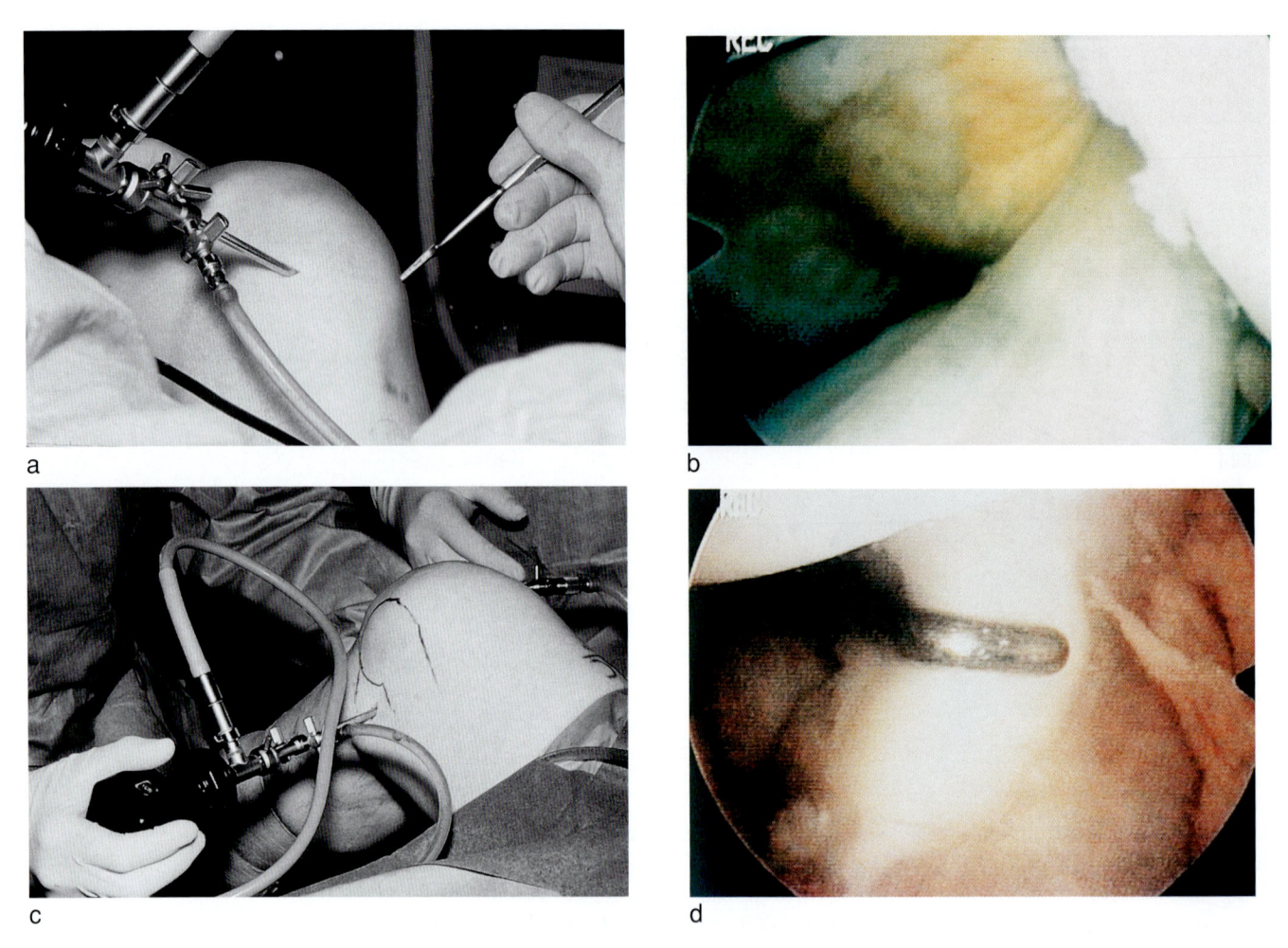

Fig. 7.4. (a) Arthroscope in the anterior-inferior patellar portal viewing zone 1 of the PCL. (b) Arthroscopic view from the anterior-inferior lateral patellar portal of normal PCL zone 1. (c) Arthroscope in the posterior medial portal to view zones 2 and 3 of the PCL. (d) Arthroscopic view from the posterior medial portal of normal PCL zones 2 and 3.

scopic visualization of the entire PCL. Although zones 2 and 3 can be visualized through the intercondylar notch, the posterior medial portal provides a better arthroscopic view of zones 2 and 3. When viewing the PCL from the posterior medial portal, a 25- or 30-degree arthroscope provides excellent visualization. Viewing the PCL through the intercondylar notch requires the use of a 70- or 90-degree arthroscope to fully evaluate the ligament.

Arthroscopic Findings in the PCL-Injured Knee

Arthroscopic findings in PCL injuries are either direct or indirect. Direct findings include damage to the PCL itself, such as midsubstance tears, interstitial tears with ligament stretching, and avulsion of bony insertions. Figure 7.5 shows a zone 1 tear. The 25-degree angled arthroscope is positioned in the anterior-inferior lateral patellar portal. Figure 7.6 depicts a zone 2 interstitial tear, with the arthroscope viewing from the posterior medial portal. Note the hemorrhage in the substance of the ligament. This in-

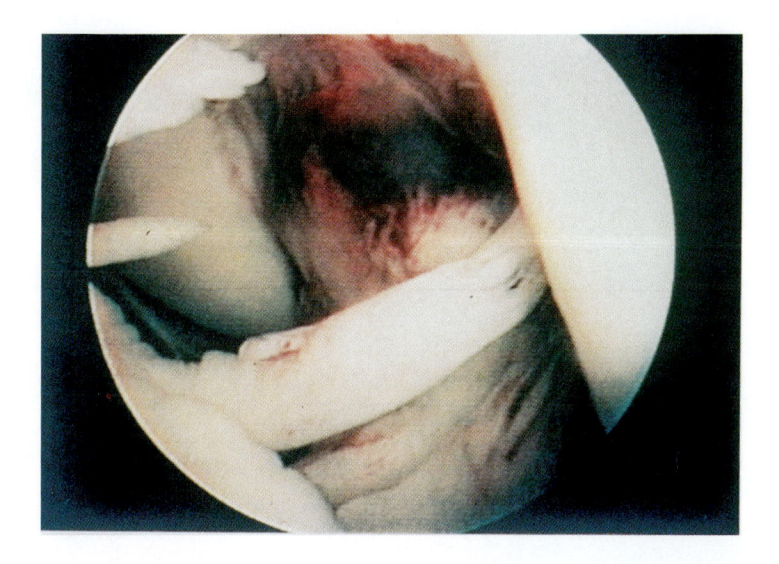

Fig. 7.5. Zone 1 PCL tear. Arthroscope in the anterior-inferior lateral patellar portal.

dicates severe stretching and elongation of the PCL due to disruption of intercollagen cross-linking. Although the PCL is still in continuity, this is a functionally complete PCL tear. Figure 7.7A shows the lateral radiograph of a PCL tibial insertion site with a piece of bone. The expected method of treatment of this injury is direct primary reattachment of the bony piece with a screw and washer. Arthroscopic evaluation of the PCL with the arthroscope in the posterior medial portal demonstrates severe interstitial damage in both zones 2 and 3 (Fig. 7.7B). The amount of interstitial destruction to this PCL contraindicated primary reattachment, and reconstruction was performed.

Indirect arthroscopic findings occur as a result of PCL insufficiency, and include the sloppy ACL sign, altered contact points, and degenerative changes of the patellofemoral joint and medial compartment.[2,3,4,8]

The sloppy ACL sign demonstrates relative laxity of the ACL secondary to posterior tibial drop-back. When the tibia is reduced, the normal ACL tension is restored. Figure 7.8A demonstrates the normal ACL/PCL relationship with a reduced tibiofemoral joint, and Fig. 7.8B demonstrates the "sloppy" ACL sign. The tibial sags posteriorly due to PCL insufficiency, allowing the ACL to become lax or "sloppy." Figure 7.8C and D show an

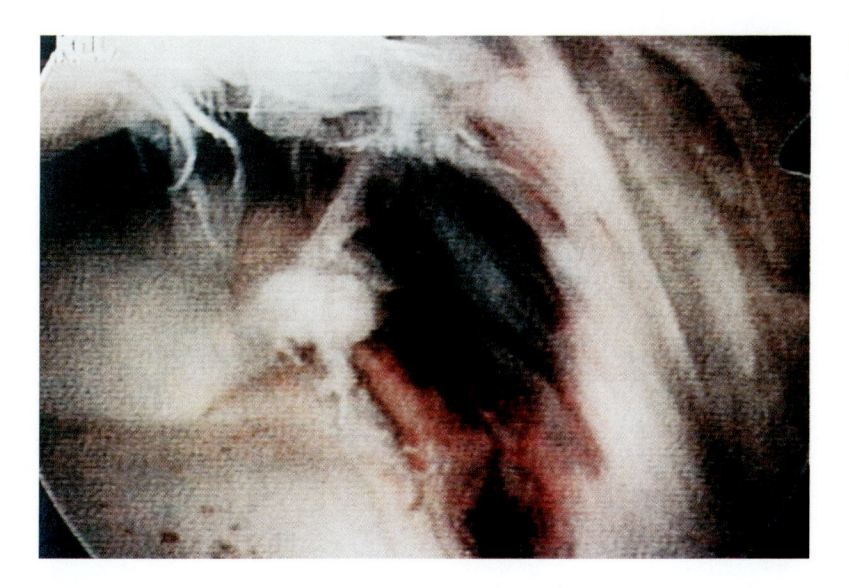

Fig. 7.6. Zone 2 PCL tear. Arthroscope in the posteromedial portal.

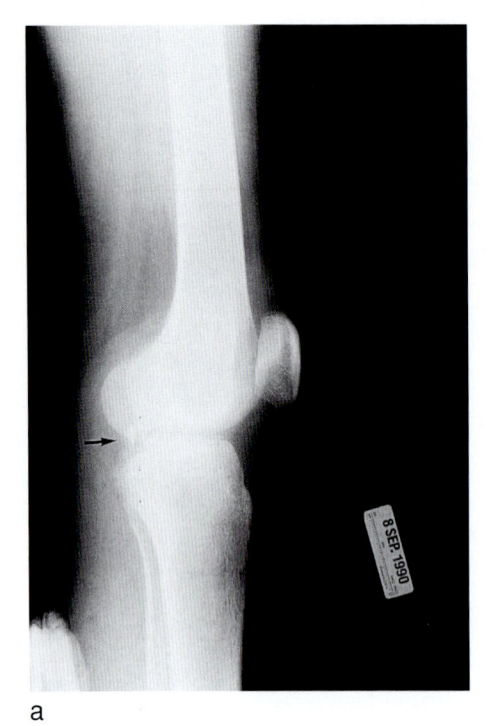

Fig. 7.7. (a) Radiograph of zone 3 PCL tibial avulsion with a piece of bone. (b) Zone 3 PCL tear. Arthroscope in the posteromedial portal. Note interstitial ligament disruption even with bone avulsed.

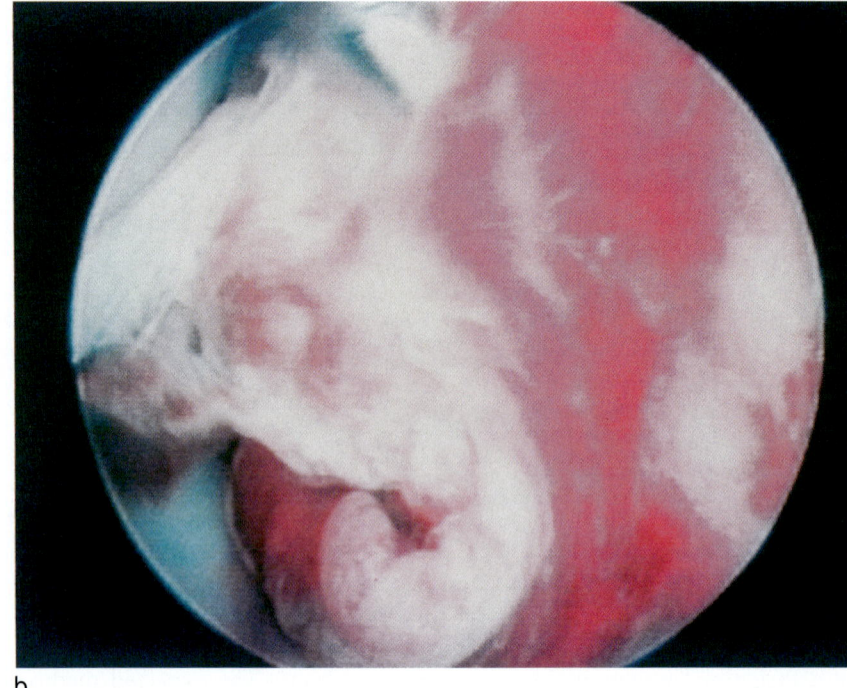

arthroscopic view of the "sloppy" ACL sign. The arthroscope is in the anterior-inferior lateral patellar portal. Notice in Fig. 7.8D the zone 2 damage, which becomes evident when the tibia is subluxed posteriorly.

Altered contact points occur secondary to tibial drop-back with the knee flexed 90 degrees. Clinically this is the posterior sag sign.[14] Placing the arthroscope in the anterior-inferior lateral patellar portal shows closer proximity of the anterior horn of the medial meniscus to the distal femoral condylar articular surfaces than is usually seen (Fig. 7.9).

Degenerative changes of the medial compartment and patellofemoral joint in chronic PCL insufficiency has been described elsewhere.[4,20] These

Fig. 7.8. (a) Drawing of normal ACL/PCL relationship with reduced tibiofemoral joint. (b) Drawing of a knee with a PCL tear. The tibia sags posteriorly allowing the ACL to become lax or "sloppy" (the "sloppy" ACL sign).

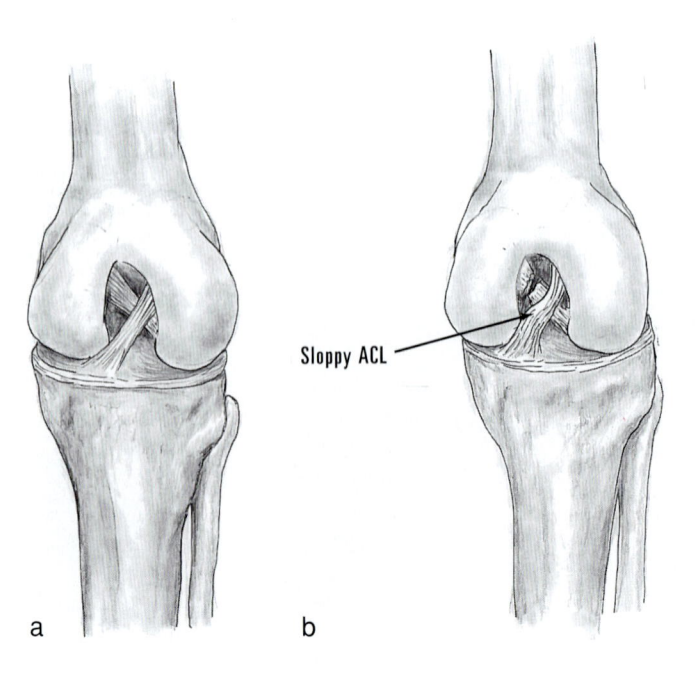

7. Arthroscopic Evaluation of the Posterior Cruciate Ligament and Posterolateral Corner

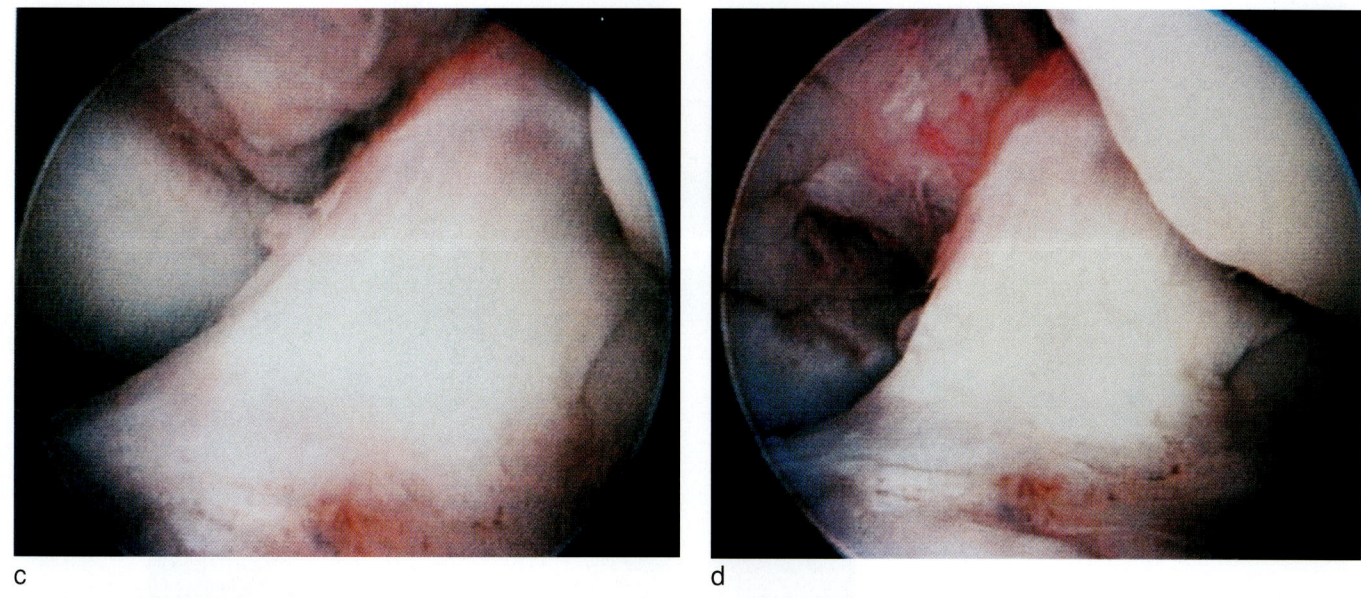

Fig. 7.8 (continued). (c) Reduced tibiofemoral joint with normal tension in the ACL. (d) The tibia is subluxed posteriorly in a knee with a PCL tear. The ACL loses its normal tension and becomes loose or "sloppy" (the "sloppy" ACL sign).

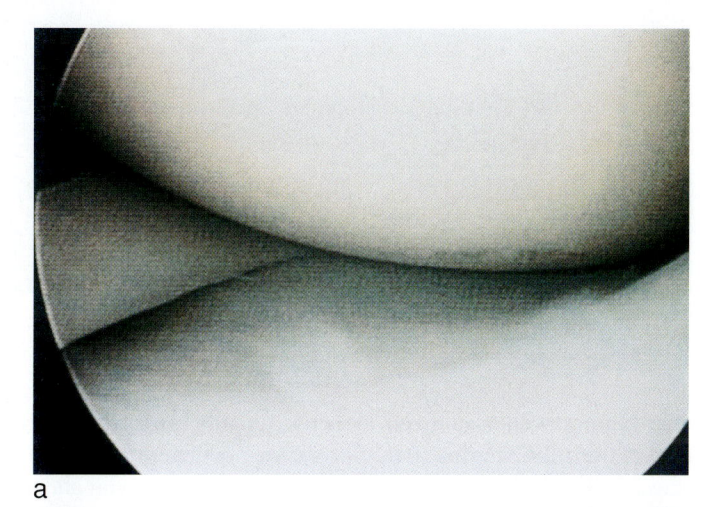

Fig. 7.9. (a) The tibia is reduced in a patient with a PCL tear showing the normal articular surface contact points. The arthroscope is in the anterior-inferior lateral patellar portal viewing the medial compartment. (b) The tibia is subluxed posteriorly in a patient with a PCL tear. Note the impingement of the medial femoral condyle on the anterior medial meniscus due to abnormal posterior tibial translation. These altered articular surface contact points lead to medial compartment degeneration because the medial meniscus does not disperse the joint surface forces. The arthroscope is in the anterior-inferior lateral patellar portal viewing the medial compartment.

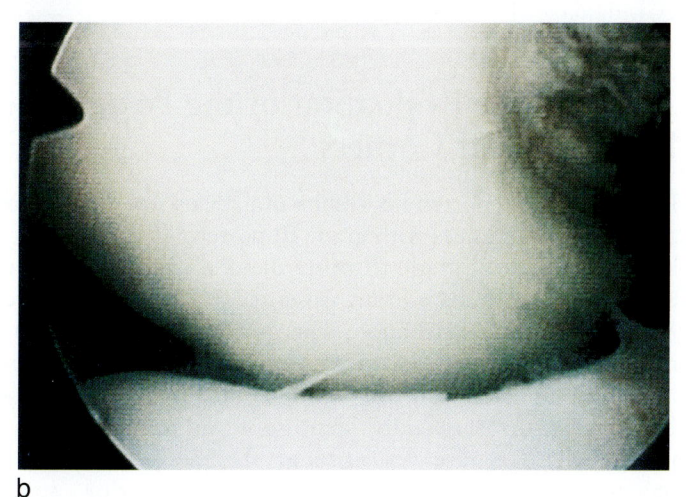

Fig. 7.10. (a) Early medial compartment degenerative changes 2 years after PCL/posterolateral corner tear. (b) Advanced medial compartment degenerative changes 11 years after PCL/posterolateral corner tears.

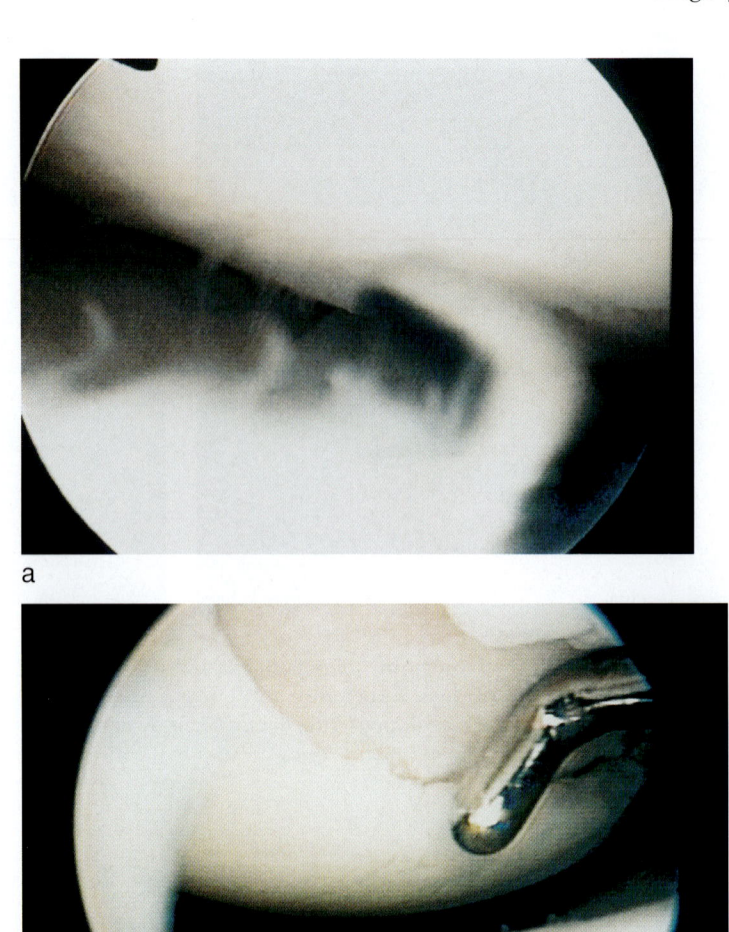

changes are well visualized arthroscopically, and the arthroscopic evaluation is useful in staging articular surface degeneration in chronic PCL tears (Fig. 7.10).[2,3,8] Hemorrhage in the ligament's synovial sheath also indicates PCL damage (Fig. 7.11).

Arthroscopic Evaluation of the Posterolateral and Posteromedial Corners

LaPrade[21] has reported his results of arthroscopic evaluation of the lateral compartment of knees with grade III posterolateral complex injuries. Thirty knees with clinical grade III posterolateral instability underwent examination under anesthesia and arthroscopic evaluation concurrently with open posterolateral reconstruction. The arthroscopic evaluation revealed a significant number of pathologic changes in the lateral compartment that may have gone undetected if only an open reconstruction had been performed. LaPrade's arthroscopically identifiable findings included injuries to the popliteomeniscal fascicles, coronary ligament, mid-third lateral capsular ligament, meniscofemoral portion of the posterior capsule, ligament of

7. Arthroscopic Evaluation of the Posterior Cruciate Ligament and Posterolateral Corner

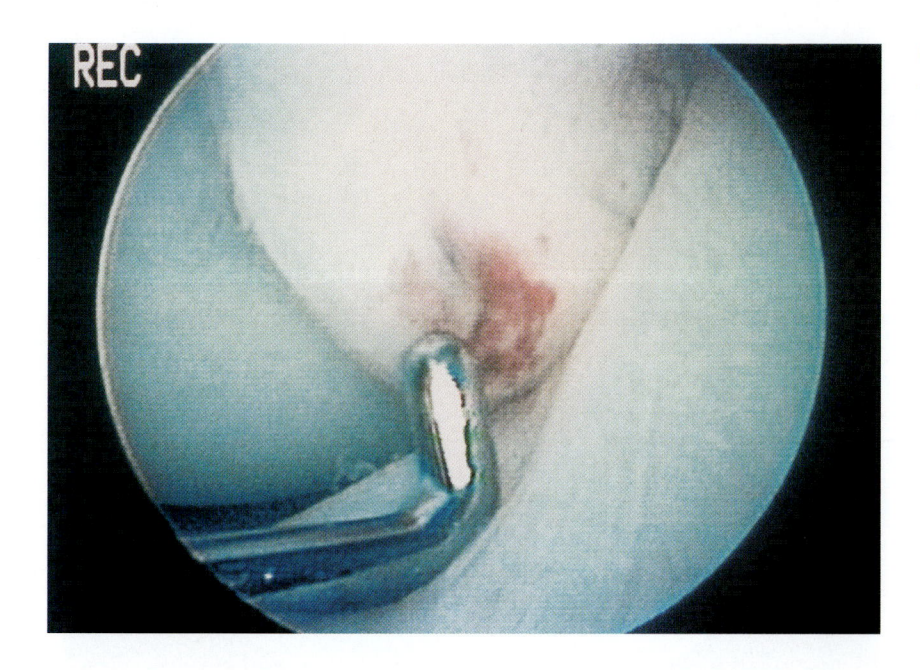

Fig. 7.11. Hemorrhage in the posterior cruciate ligament synovial sheath is a subtle sign indicating damage to the PCL.

Wrisberg, the meniscofemoral portion of the mid-third lateral capsular ligament, and avulsions of the popliteal tendon origin off the femur.

Figure 7.12 demonstrates excessive lateral compartment opening due to severe injury to the posterolateral structures. The arthroscope is positioned in the inferior lateral patellar portal. Note the severe attenuation of the popliteus tendon, and affected capsular structures.

Posteromedial instability has been described by Owens,[22] and can be identified with arthroscopic evaluation. The arthroscope is positioned in the inferior lateral patellar portal, and when valgus stress is applied to the knee there is significant medial compartment opening (Fig. 7.13). There is often obvious attenuation of the posteromedial capsular structures with infrequent disruption of the medial meniscus. We have found that the medial meniscus is most often firmly attached to the tibial side, and the capsular attenuation occurs superior to the meniscotibial attachments.[2,3]

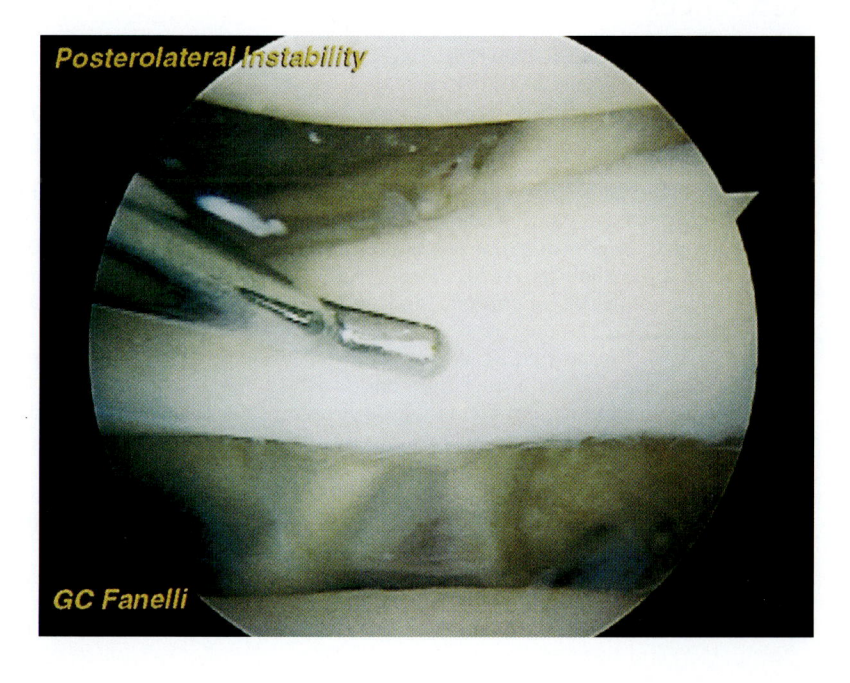

Fig. 7.12. Excessive lateral compartment opening with the knee in the figure-4 position (lateral drive-through sign) in a patient with PCL/posterolateral corner tears. The arthroscope is in the anterior-inferior lateral patellar portal.

Fig. 7.13. Excessive medial compartment opening to valgus stress (medial drive-through sign) in a patient with chronic ACL/PCL/MCL tears. The arthroscope is in the anterior-inferior lateral patellar portal.

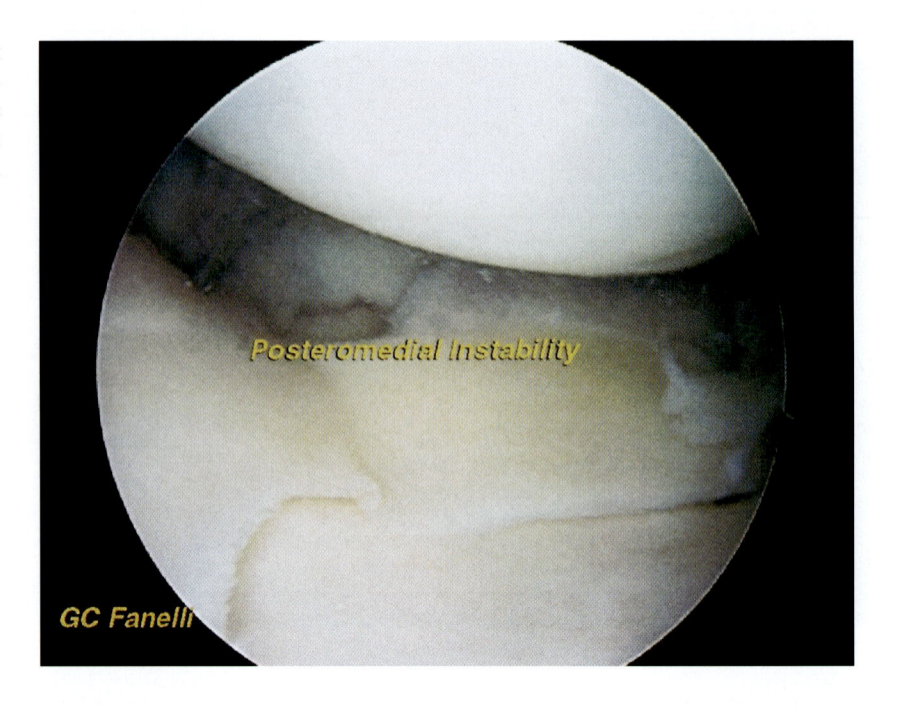

Conclusion

Arthroscopic evaluation of the PCL and related structures in the PCL-injured knee is a useful adjunct to the history, physical examination, arthrometer testing, and imaging studies. Arthroscopic PCL evaluation aids in surgical decision making and planning of reparative or reconstructive procedures. A standard 25- or 30-degree arthroscope and the inferior lateral patellar and posterior medial portals provide excellent visualization of all three zones of the PCL as well as of the posterolateral and posteromedial corners.

References

1. Johnson JC, Bach BR. Current concepts review, posterior cruciate ligament. Am J Knee Surg 1990;3:143–153.
2. Fanelli GC. PCL injuries in trauma patients. Arthroscopy 1993;9(3):291–294.
3. Fanelli GC, Edson CJ. PCL injuries in trauma patients. Part II. Arthroscopy 1995;11:526–529.
4. Clancy WG, Shelbourne KD, Zoellner GB, Keene JS, Reider B, Rosenberg TD. Treatment of knee joint instability secondary to rupture of the posterior cruciate ligament. Report of a new procedure. J Bone Joint Surg 1983;65A:310–322.
5. Girgis FG, Marshall JL, Al Monajem ARS. The cruciate ligaments of the knee joint: anatomical, functional, and experimental analysis. Clin Orthop 1975;106:216–231.
6. Stanish WO, Rubinovich M, Armason T, Lapenskie G. Posterior cruciate ligament tears in wrestlers. Can J Appl Sports Sci 1986;4:173–177.
7. Fanelli GC, Giannotti BF, Edson CJ. Arthroscopically assisted combined anterior and posterior cruciate ligament reconstruction. Arthroscopy 1996;12(1):5–14.
8. Fanelli GC, Giannotti BF, Edson CJ. Arthroscopically assisted combined posterior cruciate ligament/posterior lateral complex reconstruction. Arthroscopy 1996;12(5):521–530.
9. Gollehon DL, Torzilli PA, Warren RF. The role of the posterolateral and cruciate ligaments in the stability of the human knee. J Bone Joint Surg 1987;69A:233–242.

10. Parolie JM, Bergfeld JA. Long term results of nonoperative treatment of isolated posterior cruciate ligament injuries in the athlete. Am J Sports Med 1986;14:35–38.
11. Daniel DM, Stone ML, Barnett P, et al. Use of the quadriceps active test to diagnose PCL disruption and measure posterior laxity of the knee. J Bone Joint Surg 1988;70A:386–391.
12. Grood ES, Hefzy MS, Ledenfeld TN. Factors affecting the region of most isometric femoral attachments. Am J Sports Med 1989;17:197–207.
13. Hughston JC, Baker CL, Norwood LA. Acute combined posterior cruciate and posterolateral instability of the knee. Am J Sports Med 1983;12:204–208.
14. Insall JN, Hood RW. Bone block transfer of the medial head of the gastrocnemius for posterior cruciate insufficiency. J Bone Joint Surg 1982;64:691–699.
15. Jakob RP, Hassler H, Staubli HU. Experimental studies on the functional anatomy and the pathomechanism of the true and reversed pivot shift sign. Acta Orthop Scand 1981;5(suppl):18–32.
16. Loos WC, Fox JM, Blazina ME, Del Pizzo W, Friedman MJ. Acute posterior cruciate ligament injuries. Am J Sports Med 1981;8:86–92.
17. Shelbourne KD, Benedict F, McCarrol JR, et al. Dynamic posterior shift test. Am J Sports Med 1989;17:275–277.
18. Lysholm J, Gillquist J. Arthroscopic examination of the posterior cruciate ligament. J Bone Joint Surg 1981;63A:363–366.
19. Fanelli GC, Giannotti BF, Edson CJ. Current concepts review. The posterior cruciate ligament arthroscopic evaluation and treatment. Arthroscopy 1994;10(6):673–688.
20. Clancy WG. Repair and reconstruction of the posterior cruciate ligament. In: Chapman M, ed. Operative Orthopaedics. Philadelphia: JB Lippincott, 1988:1651–1665.
21. LaPrade RF. Arthroscopic evaluation of the lateral compartment of knees with grade 3 posterolateral knee complex injuries. Am J Sports Med 1997;25(5):596–602.
22. Owens T. Posteromedial pivot shift of the knee: a new test for rupture of the posterior cruciate ligament. A demonstration in 6 patients and a study of anatomical specimens. J Bone Joint Surg 1994;76A(4):532–539.

III

Nonoperative Treatment

Chapter Eight

The Natural History of the Posterior Cruciate Ligament–Deficient Knee

Bradley F. Giannotti

Qualification of the natural history of a disease process is imperative so as to initiate appropriate and specific treatment. Why alter a process if you don't know the unaltered end point? In neoplastic terms, an entity with a "benign" natural history usually can be treated less aggressively, while a "malignant" condition often requires more aggressive modalities. Although posterior cruciate ligament (PCL) injury is not cancer, determining the natural history and treatment is no different.

The natural history of the PCL-deficient knee has not been well defined for a number of reasons. First, the incidence of PCL injury is relatively low compared to other single ligament knee injuries. In several series the incidence has ranged from 1% to 20% of knee ligament injuries.[1,2] Fanelli[3] reported a 42% incidence in PCL injuries in patients presenting to a level I trauma center with a hemarthrosis. Only 7%, however, were isolated PCL tears. In my practice in rural Pennsylvania, PCL injuries account for approximately 3% of single ligament knee injuries. The reported incidence nationally seems to be increasing recently, at least in part due to the increased knowledge, awareness, and improved ability to recognize these sometimes elusive injuries. Parolie and Bergfeld[4] noted that the National Football League predraft examination finds an average of four to six players out of 200 to 250 with chronic PCL-deficient knees. Second, the PCL literature is somewhat of a mosaic reporting on isolated PCL tears, combined anterior cruciate ligament (ACL)/PCL tears, PCL/posterolateral corner insufficiency, and the like. Consequently, results and conclusions regarding treatment and the natural history, respectively, are somewhat varied. Third, the PCL-deficient knee is a time-dependent phenomenon. Long-term studies of the PCL-deficient knee are needed to establish its natural history, as has been done by hip and knee arthroplasty reports.

Despite these shortcomings, there are now sufficient data to draw some conclusions about both long- and short-term effects of the PCL-deficient knee. This chapter discusses PCL/combined ligament injuries and isolated PCL insufficiency. The combined ligament injuries represent PCL insufficiency concomitantly with ACL insufficiency, medial collateral ligament (MCL) insufficiency, or posterolateral complex insufficiency. With a few exceptions, the literature does not distinguish between different ligament combination injuries. Certainly, there is assumed to be a difference between PCL insufficiency combined with a grade I MCL sprain and the more devastating combined ACL/PCL tear.

PCL/Combined Ligament Injury

Posterior cruciate ligament tears combined with other ligament tears (ACL, MCL, posterolateral complex) represent a functional problem both acutely and chronically.[1,5-8] When division of the PCL is combined with division of lateral structures, complex abnormal movement results.[8] Trickey[8] points out that not only does the tibia subluxate posteriorly, but the lateral tibial condyle *rotates* posteriorly as well, proportionate to the degree of posterolateral insufficiency. This represents a tremendously unstable situation in the acute setting, and it puts secondary and tertiary structures at risk in the subacute time frame. If left untreated, this initial instability has a high likelihood of leading to later functional disability. Abnormal shear forces are undoubtedly present, all but ensuring future degenerative changes (Fig. 8.1).

In a similar fashion, when the PCL and medial structures are divided, the tibia may subluxate posteriorly, and the medial femoral condyle can rotate posteriorly (posteromedial rotatory subluxation) and/or anteriorly (anteromedial subluxation). As in the PCL/posterolateral combined setting, this is an acutely unstable situation, and it predisposes the patient to chronic degenerative disability. Furthermore, combined PCL/multiligament deficiency and its inherent instability risks the secondary and tertiary supporting structures. Whereas meniscal tears are unusual in isolated PCL injuries, Johnson and Bach[1] note that in combined ligament injuries such as these, meniscal pathology is common. This finding should not be dismissed as minor. If the shock absorber function of the meniscus is compromised, degenerative changes may be imminent. If the stability function of the meniscus is lost, a vicious cycle of recurrent injury and instability is possible (Fig. 8.2).

The PCL is thought to be a primary knee stabilizer providing a central axis about which normal and abnormal internal and external rotation occur.[9] Hughston et al.[9,10] described this analogy: When the PCL is intact, all instabilities are rotational about this axis, and when torn, all instabilities are straight, because the tibia does not rotate on the femur about a central axis and the knee joint opens like a door on a hinge when varus or valgus stress is applied. One can appreciate the dramatic resultant instability as well as the susceptibility of further ligamentous compromise.

In 1971 Meyers and Harvey[11] reported on 18 knee dislocations. All but one knee that did not undergo surgical repair of all torn ligaments were unstable and symptomatic at follow-up. In 1975, Meyers et al.[12] reported on 33 traumatic knee dislocations available for minimum 1-year reevalua-

Fig. 8.1. Untreated posterior cruciate ligament (PCL)/posterolateral instability presented with chronic medial compartment and patellar pain with radiographic evidence of severe degenerative joint disease at age 40. (From Johnson and Bach,[1] with permission.)

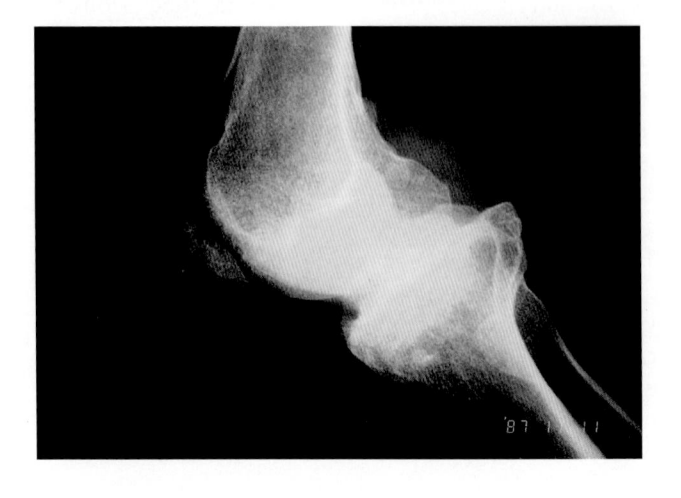

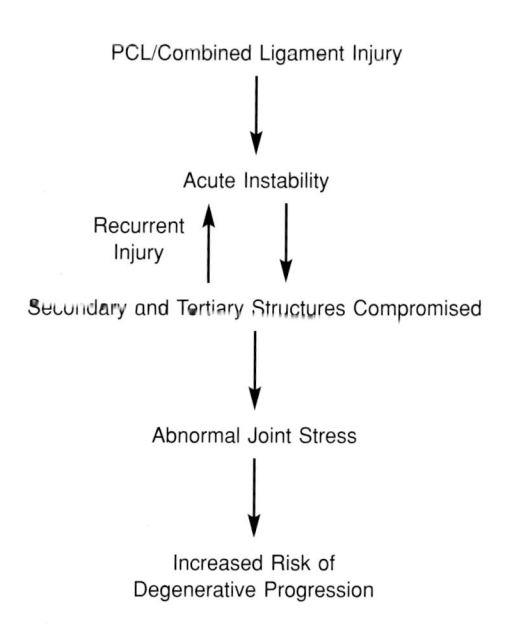

Fig. 8.2. Cycle of projected increased risk of degenerative joint disease following PCL/combined ligament injury.

tion. Of this group, 13 were treated by manipulation and immobilization. The results were graded good in one, fair in one, and poor in 10.

It is not the purpose of this chapter to recommend specific treatment. Certainly, however, many patients with combined PCL/multiligament knee injuries do poorly in both the acute and chronic setting when treated non-operatively. To alter this otherwise dismal outcome of the combined PCL/multiligament–deficient knee, many authors recommend reconstruction in the acute setting with seemingly good results.[5,7,8,13–16]

The Isolated Posterior Cruciate Ligament–Deficient Knee

Much has been written regarding conservative care in the ACL-deficient knee. Certainly, in the young, active patient with ACL deficiency, muscle rehabilitation is of limited value. Cross and Powell[17] studied 116 patients with PCL-deficient knees. The authors noted that 14 patients required ligamentous reconstruction. Though no information was given on degree of instability, the authors noted that 80% did well with nonoperative care. They suggested that quadriceps muscle rehabilitation can protect the PCL-deficient knee far better than those with ACL deficiency. Furthermore, they indicated that those knees with multiple episodes of instability are predisposed to develop osteoarthritis.[17]

For a large number of patients with combined PCL/multiligament knee insufficiency, acute, subacute, and chronic disability results when treated nonoperatively. The earlier literature on the isolated PCL-deficient knee fails to reach this consensus and has led to some confusion.

Dandy and Pusey[18] reported on 20 patients with apparent isolated midsubstance PCL tears treated nonoperatively and followed for a mean interval of 7.2 years. The authors advocate a conservative approach, noting that the majority of patients were able to pursue sports activities.[18] Subjectively, they noted 10 fair results, while objectively eight were fair and 12 were poor. Despite this noncorrelation between objective and subjective results, 14 continued to have pain while walking and nine had episodic giving way.[18]

Four years after Dandy and Pusey's report, Parolie and Bergfeld[4] published their results on 25 patients with isolated PCL injuries treated nonoperatively. The authors noted an 80% subjective satisfaction rate with a mean follow-up of 6.2 years. Additionally, they found a positive correlation between quadriceps torque capacity and ability to return to athletic activity. Despite this, they also noted that 50% of patients studied had persistent knee pain and stiffness. Nine patients had radiograhic evidence of arthritis, usually involving the medial compartment, though the authors indicated that this is not inevitable in the patient with less than 5 to 10 mm of posterior instability.

Clancy,[19] in a review article, suggested that patients with chronic, isolated, PCL laxity often develop a "giving way" of the joint, or what he terms "pseudoinstability," in that they suffer sharp and sudden pain from making contact with an eroded area on the medial femoral condyle. He further notes subjective complaints of pain or achiness with work and occasional effusions, consistent with the previously mentioned findings of Parolie and Bergfeld.[4]

As suggested earlier, PCL injury is more commonly a trauma event as seen in motor vehicle deceleration accidents.[3] Athletes, especially football players, also have a degree of risk. Unlike ACL injuries, which are often noncontact injuries, PCL injuries are usually the result of contact with a posteriorly directed force at the level of the tibial tubercle.

Fowler and Messieh[20] prospectively followed for an average of 2.6 years 13 patients with isolated PCL tears sustained while involved in various athletic activities. All were treated with physical therapy. Hyperflexion was the most common mechanism of injury. The authors noted that all patients were able to return to their previous activities without limitations. They, like Dandy and Pusey,[18] found no correlation between subjective and objective results. Objectively, only three patients had good results and 10 had fair results. Moreover, KT-1000 testing at 25 degrees revealed that 12 of the 13 patients had a side-to-side difference of 5 mm or less. Clinically, over half demonstrated a trace to 1+ posterior drawer test. The results of this and other studies suggest that return to sports is possible, even likely, in the active patient with isolated PCL deficiency. Interestingly, however, in the Fowler and Messieh study, 5 of 13 patients were found to have partial PCL tears, and most had less than a 5-mm side-to-side difference. Furthermore, no inference can be made as to further degenerative change, given the relatively short follow-up.

It is important to remember that technically, the natural history of PCL injury implies that no form of treatment is administered. To the author's knowledge, no results have been published with this strict absence of treatment. Rather, published results inform us of a modified natural history. Nonoperative treatment implies a rehabilitation program of varying intensity and duration. Certainly, the literature so far reviewed supports quadriceps strengthening as a mode of treatment for the isolated PCL-deficient knee. It is apparent thus far that quadriceps strength correlates somewhat positively with stability and function in the PCL-deficient knee in the short-term or acute setting. But this seems to be somewhat dependent on the degree of initial instability.[21]

Hughston and Degenhardt[13] in 1982 reported on their results utilizing the medial gastrocnemius tendon as a PCL substitute in patients with chronic PCL insufficiency. All patients in the study were initially placed on an unspecified rehabilitation program for an average of 5 years before presenting for surgery; 42% failed to achieve functional recovery. It is unclear in the study how many patients initially had combined ligamentous injury, and the degree of posterior instability was not quantified. Nevertheless, many patients noted subjective difficulty with simple stair climbing as well as with recreational and occupational activities. Some required the use of a cane. Interestingly, these patients were quite young, with an average age of 23 at the time of injury.

Degenhardt and Hughston[21] documented the results of 23 patients with chronic PCL insufficiency treated nonoperatively with an average follow-up of 7.5 years. Once again, a muscle rehabilitation program was instituted emphasizing quadriceps strengthening. They noted that only two patients with isolated posterior instability required reconstruction. It was apparent that the severity of the posterior drawer test was important prognostically. The average posterior drawer sign in the nonoperative group was only 1.3, whereas in the operative group it averaged 2.6. In addition to quadriceps weakness, poor prognostic signs were genu varum, a history of previous menisectomy, and the presence of multiple instability.

Thus far, two findings can be gleaned from the literature regarding isolated PCL deficiency. First, those with low-grade instability can achieve good functional results and are probably not predisposed to future arthritic change. However, the greater the instability or posterior drawer, the greater the propensity to develop arthritic change, though qualification is difficult to specify. Second, quadriceps rehabilitation is protective in the isolated PCL-deficient knee with low-grade instability. It is of questionable benefit in the uncommon, highly unstable, isolated PCL-deficient knee.

Keller et al.[22] evaluated 40 patients with isolated PCL injuries treated nonoperatively for an average of 6 years. On a questionnaire, 65% reported that their activity level after injury was significantly limited. Almost half of the patients in the study stated that they had not fully recovered despite participation in a quadriceps rehabilitation program. As a group, all demonstrated excellent muscular strength, but this did not correlate with subjective results. Ninety percent continued to experience knee pain with walking. Notably, the longer the interval from injury, the lower the knee questionnaire score. Additionally, patients with a greater degree of posterior ligamentous laxity had more subjective complaints about their involved knee and lower scores on a modified Noyes knee questionnaire.

Keller et al.[22] further found that radiographic evidence of degenerative changes worsened with increasing interval from injury (Fig. 8.3). The degenerative changes occurred first in the medial compartment and, with increasing time, the lateral and patellofemoral compartments became involved. Again, contrary to other studies, the subjective and radiographic changes occurred despite excellent quadriceps strength.

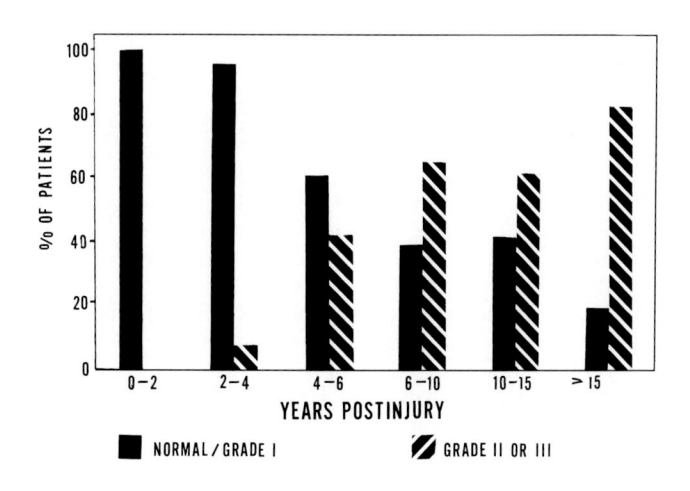

Fig. 8.3. Radiographic changes following PCL injury by years since injury. (From Keller et al.,[22] with permission.)

Dejour et al.[23] reviewed 45 cases of isolated, chronic PCL insufficiency (5 to 44 years postinjury) both functionally and radiographically. Twenty-two of the 45 patients continued to have varying degrees of instability but did not significantly inhibit return to sports. All but five continued to complain of pain. Patellofemoral arthritis was present in 62% of cases. The authors further characterize the natural history of isolated PCL rupture in three stages:

- Stage I (3–18 months): Functional adaptation.
- Stage II (18 months–20 years): Functional tolerance.
- Stage III (20 years and beyond): Arthritic deterioration.

Conclusions

1. Combined PCL/posterolateral complex and PCL/ACL injuries do poorly both acutely and chronically. This appears to be so in terms of stability, functionality, and degenerative change.
2. Isolated PCL tears with 1+ side-to-side difference do well when treated nonoperatively, with rehabilitation, but may be predisposed to future arthritic change.
3. Isolated PCL tears with 2+ side-to-side injury may do well treated nonoperatively with aggressive rehabilitation, especially in the inactive or lightly active population. Future arthrosis is a factor.
4. Isolated PCL insufficiency patients with 3+ side-to-side difference in general do poorly acutely and chronically. Acutely, the knee is unstable and degenerative joint disease is inevitable. These isolated injuries are uncommon, and occult concomitant injury should be relentlessly sought.

References

1. Johnson JC, Bach BR. Current concepts review posterior cruciate ligament. Am J Knee Surg 1990;3(3):143–153.
2. Clendenin MB, DeLee JC, Heckman JD. Interstitial tears of the posterior cruciate ligament of the knee. Orthopedics 1980;3:764–772.
3. Fanelli GC. Posterior cruciate ligament injuries in trauma patients. Arthroscopy 1993;9:291–294.
4. Parolie JM, Bergfeld JA. Long-term results of nonoperative treatment of isolated posterior cruciate ligament injuries in the athlete. Am J Sports Med 1986;14:35–38.
5. Fanelli GC, Giannotti BF, Edson CJ. Current concepts review: the posterior cruciate ligament arthroscopic evaluation and treatment. Arthroscopy 1994;10:673–688.
6. Fanelli GC, Edson CJ, Foster J. PCL injuries in acute traumatic hemarthrosis of the knee. Paper presented at the American Academy of Orthopaedic Surgeons annual meeting, New Orleans, 1994.
7. Fanelli GC, Edson CJ, Foster J, et al. Arthroscopically assisted combined anterior and posterior cruciate ligament reconstruction. Paper presented at the American Academy of Orthopaedic Surgeons annual meeting, New Orleans, 1994.
8. Trickey EL. Injuries to the posterior cruciate ligament: diagnosis and treatment of early injuries and reconstruction of late instability. Clin Orthop 1980;147:76–81.
9. Hughston JC, Bowden JA, Andrews JR. Acute tears of the posterior cruciate ligament. J Bone Joint Surg 1980;62A:438–450.

10. Hughston JC. The posterior cruciate ligament in knee joint stability. Proceedings of the American Academy of Orthopaedic Surgeons. J Bone Joint Surg 1969;51A:1045–1046.
11. Meyers MH, Harvey JP. Traumatic dislocation of the knee: a study of eighteen cases. J Bone Joint Surg 1971;53A:16–29.
12. Meyers MH, Moore TM, Harvey JP. Follow-up notes on articles previously published in the journal: traumatic dislocation of the knee joint. J Bone Joint Surg 1975;57A:430–433.
13. Hughston JC, Degenhardt TC. Reconstruction of the posterior cruciate ligament. Clin Orthop 1982;164:59–77.
14. Sisto DJ, Warren RF. Complete knee dislocation: a follow-up study of operative treatment. Clin Orthop 1985;198:94–101.
15. Fanelli GC, Giannotti BF, Edson CJ. Arthroscopically assisted combined anterior and posterior cruciate ligament reconstruction. Arthroscopy 1996;12(1):5–14.
16. Fanelli GC, Giannotti BF, Edson CJ. Arthroscopically assisted combined posterior cruciate ligament/posterior lateral complex reconstruction. Arthroscopy 1996;12(5):521–530.
17. Cross MJ, Powell J. Natural history of posterior cruciate ligament disruption. Orthop Trans 1981;5(1):486.
18. Dandy DJ, Pusey RJ. The long-term results of unrepaired tears of the posterior cruciate ligament. J Bone Joint Surg 1982;64B:92–94.
19. Clancy WG. Knee ligamentous injury in sports: the past, present, and future. Med Sci Sports Exerc 1983;15(1):9–14.
20. Fowler PJ, Messieh, SS. Isolated posterior cruciate ligament injuries in athletes. Am J Sports Med 1987;15(6):553–557.
21. Degenhardt TC, Hughston JC. Chronic posterior cruciate instability: nonoperative management. Orthop Trans 1981;5(1):486–487.
22. Keller PM, Shelbourne KD, McCarroll JR, et al. Nonoperatively treated isolated posterior cruciate ligament injuries. Am J Sports Med 1993;21(1):132–136.
23. Dejour H, Walch G, Peyrot J. The natural history of rupture of the posterior cruciate ligament. Orthop Trans 1987;11(1):146.

Chapter Nine

Nonoperative Treatment of Posterior Cruciate Ligament Injuries

John A. Bergfeld, Manuel Leyes, and Gary J. Calabrese

The posterior cruciate ligament (PCL) injury is one of the most controversial areas in sports medicine.[1] There is no consensus in the literature with respect to its incidence, natural history, and indications for conservative or surgical treatment. Unlike anterior cruciate ligament (ACL) reconstruction, the results of surgical reconstruction of the PCL are inconsistent, and there is little agreement among surgeons as to surgical techniques. Therefore, it is critical to identify those patients who can be treated nonoperatively and who will do as well, if not better, than the patients undergoing surgical reconstruction.

This chapter addresses the diagnosis of PCL injuries, the indications for nonoperative treatment, the biomechanical principles, specific guidelines of the rehabilitation protocol at our clinic, and the results of nonoperative treatment.

Diagnosis of Posterior Cruciate Ligament Injuries

Successful treatment of PCL injuries depends on an early and accurate diagnosis. A thorough precise history and physical examination can be considered diagnostic in the majority of the cases.[2] The mechanism of injury often suggests the diagnosis. A frequent mechanism is a violent anteroposterior force on the upper front of the flexed knee. This commonly occurs in motor vehicle accidents in which the knee hits the dashboard. In sports injuries the athlete receives a blow to the anterior aspect of the tibia when falling on the flexed knee. Not uncommonly, the athlete's knee is further forced into hyperflexion. A PCL injury should be suspected in every patient with a knee injury associated with an abrasion on the anterior aspect of the proximal tibia and a mild bloody effusion.[3-6] In motor vehicle trauma, injury to the PCL is often overlooked. In sports injuries, the isolated PCL injury may also not be recognized because of the benign appearance of the knee, 1+ to 2+ effusion, and mild pain. Other mechanisms of injury include hyperextension of the knee and forceful valgus or varus stress combined with leg rotation.[4,7-12]

In the presence of a tense hemarthrosis, skin abrasions, muscle spasm, or concomitant injuries, the evaluation of an acute PCL injury may be difficult.[13] If the diagnosis is not clear in the first visit, the patient must be scheduled for a clinical reexamination within the next 7 to 10 days. The posterior sag sign, the apparent disappearance of the tibial tubercle in lateral inspection when the knee is flexed 90 degrees, is a characteristic finding of the PCL-deficient knee.[3,14-16] In some cases, however, quadriceps muscle spasm, effusion, or anterior tibial induration may obscure this finding.[17] A frequent, but nonspecific finding in chronic PCL injuries is an abnormal hyperextension of the knee.[17,18] The most sensitive and specific clinical test for PCL injuries is the posterior drawer test at 90 degrees of flexion.[2,19,20] The different clinical tests for PCL, posteromedial, and posterolateral corner injuries have been described in Chapter 4.

Posterior cruciate ligament injuries are usually divided into two broad categories: isolated injuries to the PCL and injuries to the PCL combined with injuries to other capsuloligamentous structures. The distinction between these two separate entities is essential, since they require a different treatment and bear a different prognosis. In general, isolated PCL injuries are more common in the athletic population.[3,21–24] The isolated PCL injury may be treated nonoperatively and has an excellent prognosis. Conversely, in trauma patients, 95% of PCL injuries are combined with other ligament injuries.[25] These injuries usually require surgical treatment and have a more guarded prognosis.

Successful treatment of PCL injuries depends on an accurate diagnosis. Patients with isolated injuries to the PCL may be identified by an accurate physical examination. The knee will be stable to medial and lateral stress as well as rotational stress. With the knee at 90 degrees of flexion and the tibia at neutral rotation, abnormal posterior laxity is less than 10 mm. The posterior translation will decrease with the posterior drawer test when the tibia is taken from neutral to internal tibial rotation.[21,26] Biomechanical studies demonstrated a 4-mm difference in posterior translation between neutral and internal rotation of the tibia.[27] This is easily detectable on physical examination and it is very useful to help differentiate knees that can be treated nonoperatively from those that require operative treatment. The presence of decreased posterior laxity with internal rotation indicates a PCL injury that will respond favorably to nonoperative treatment. A PCL-injured knee with a positive drawer test (>10 mm) that does not decrease with internal tibial rotation usually requires PCL reconstruction.

Bergfeld, in association with Ritchie, Kambic and Manning,[27] performed an experimental study in 20 cadaveric knee specimens to determine the anatomic structure responsible for this clinical observation. The TestStar device was used to perform single-plane posterior drawer tests with the knee in neutral tibial rotation and in 20 degrees of internal tibial rotation. The intact knee was tested, and then the injured knee was tested after sequential sectioning of the meniscofemoral ligaments, the PCL, the posteromedial capsule, and the superficial medial collateral ligament (MCL). With the knee in internal rotation, posterior displacement was significantly less than in neutral rotation for each state until the superficial MCL was sectioned; posterior translation was increased after its sectioning. The authors' data suggest that the superficial MCL, not the meniscofemoral ligaments, is the structure responsible for this decrease in posterior tibial translation with tibial internal rotation in the PCL-deficient knee. It should be noted that the investigators did not study the lateral and posterolateral structures of the knee.

The symptoms and appearance of the injured knee may also help to differentiate between the two entities. Acute isolated PCL injuries are usually associated with moderate anterior and posterior pain that increases with flexion, mild to moderate effusion, and lack of 10 to 20 degrees of flexion.[28] Often the isolated PCL injury is passed off as minor unless an accurate physical examination is performed. Combined injuries are more disabling. Typically they cause severe pain, moderate to tense effusion, and peripheral ecchymosis.

Isolated PCL injuries may be classified into partial (grade I or II) and complete (grade III) tears. Grade I injuries have a palpable but diminished step-off (0 to 5 mm) between the tibial plateaus and the femoral condyles. Grade II injuries have lost their step-off, but the tibial plateaus cannot be pushed beyond the femoral condyles (5 to 10 mm). In grade III or complete PCL injuries, the tibia can be displaced beyond the femoral condyles (>10 mm).[28]

Posterolateral corner involvement should be ruled out when the posterior tibial translation is greater than 10 mm. In this setting an increase of >10 to 15 degrees of tibial external rotation at 30 degrees of flexion indicates a posterolateral corner injury. If the increased rotation occurs at 30 and 90 degrees of flexion, both the PCL and the posterolateral corner are involved.[24] Examination under anesthesia can be helpful to confirm the presence or absence of a posterolateral corner injury.

Ancillary Studies

Anteroposterior, lateral, and oblique knee radiographs are recommended to evaluate bone avulsions of the PCL. The PCL is extremely strong and injuries often avulse the bony tibial attachment rather than rupture the ligament.[29] Plain radiographs may also detect small tibial plateau fractures that in the setting of a PCL-injured knee suggest a severe combined ligament injury.[28] Forty-five-degree flexion weight-bearing views and patellar views are helpful in the evaluation of degenerative changes in the chronic PCL-deficient knee.

The KT-1000 arthrometer is considered a moderately reliable tool for the measurement of tibial translation in patients with PCL tears and reconstructions.[30] In a recent study, Eakin and Cannon[31] reported that the overall accuracy of arthrometry for detection of PCL injury was 96% for 40 pounds of posterior force and 94% for total anterior-posterior translation at 40 pounds. The accuracy of the arthrometer was not affected by concomitant ACL injury.

Stress radiography is a simple, accurate, and reliable diagnostic method that is particularly useful in the grossly swollen and multiple ligament-injured knee.[16,32] Hewett et al.[33] have shown that stress radiographs are superior to arthrometric (KT-1000) and posterior drawer testing in the evaluation of the injured PCL. These authors determined that 8 mm or more of increased posterior translation on stress radiographs indicated a complete rupture of the PCL.

Magnetic resonance imaging (MRI) is not necessary to make the diagnosis of PCL injury but it is useful to determine associated injuries, in particular, meniscal injuries and bone bruising. The PCL injury is easily identified with MRI by using simple signal intensity and structural characteristics.[34] MRI has a sensitivity and a specificity of 100% in PCL and MCL injuries.[35] The "hockey stick" configuration of the normal PCL can be seen entirely in a single MRI section or in the composite of two consecutive sections. The normal PCL has a very low signal intensity. On the other hand, an acute torn PCL may appear in continuity but with an increased signal. Tears are more obvious on T2-weighted images due to the high signal intensity of fluid within the tear.[36] MRI is also useful in determining the location of the PCL tear—femoral, midsubstance, or tibial.[28,34] Differentiation between complete and partial-thickness PCL tears by MRI criteria alone is problematic.[37] Follow-up MRIs have also been used to confirm the healing of the injured PCL.[22,38] As MRI becomes more sophisticated, its value increases. In the future we may be able to predict which knees will go on to arthritic changes based on the initial MRI findings.

Radionuclide imaging is a useful technique to identify early articular cartilage degeneration that can be associated with chronic PCL injury.

Indications for Nonoperative Treatment

Early reports documented that although patients with isolated PCL tears had clinical laxity on physical examination, they were satisfied with the results of an aggressive rehabilitation program and were able to return to sports activities.[3,21,24,26,29,39–42] Some studies, however, have reported longer follow-up of these patients and have documented the progressive deterioration of not only the objective physical findings but also the patients' subjective symptoms and radiologic findings.[43–45]

Despite this controversy, we believe, as do many other authors,[3,16,21,29,39–41,45] that isolated PCL injuries should be treated nonoperatively, since abnormal posterior residual laxity in most of these knees is consistent with functional stability and minimal symptoms (Fig. 9.1). The return to work and sports activities is much faster without surgery, and the risks and costs of conservative treatment are much lower than those of surgery. We should bear in mind that reconstruction of the PCL is a major surgical procedure that, along with the subsequent rehabilitation, requires a significant commitment from the patient of time and expense. Unfortunately, the results of surgical reconstruction of the PCL are not predictable. Despite PCL reconstruction, many patients still have residual abnormal posterior laxity and posterior sag at the final follow-up.[12,15,46–50] We recommend nonoperative treatment in acute isolated PCL injuries, newly diagnosed chronic isolated PCL injuries with no history of prior rehabilitation, and acute or chronic isolated or combined PCL injury in a patient incapable of complying with the postoperative rehabilitation. Patient selection remains crucial. We want to emphasize the value of the posterior drawer test at 90 degrees in neutral and internal tibial rotation to differentiate the knee that can be treated nonoperatively from the knee that requires operative treatment. Our indications for operative treatment are described in Chapter 12.

Biomechanical Principles of Rehabilitation

The biomechanics of the PCL have been described in detail in Chapter 1. Therefore, we are going to focus here only on those aspects relevant to the rehabilitation process following a PCL tear. This process presents a challenge to the injured athlete and to the rehabilitation team.

An understanding of the posterior tibiofemoral shear forces during activities of daily living and during rehabilitation exercises is important to minimize the amount of stress applied to the injured PCL. During level walking, forces of up to one-half body weight occur in the PCL.[51,52] In other activities that require a greater knee flexion angle, such as descending stairs[53] or squatting,[54] the posterior tibiofemoral forces are much higher, 1.7 and 3 times body weight, respectively. On the other hand, activities such as cycling place little stress on the PCL. Ericson and Nisell[55] estimated forces as low as 0.05 body weight during ergometer cycling.

Rehabilitation of the lower extremities is classified into two basic cate-

9. Nonoperative Treatment of Posterior Cruciate Ligament Injuries

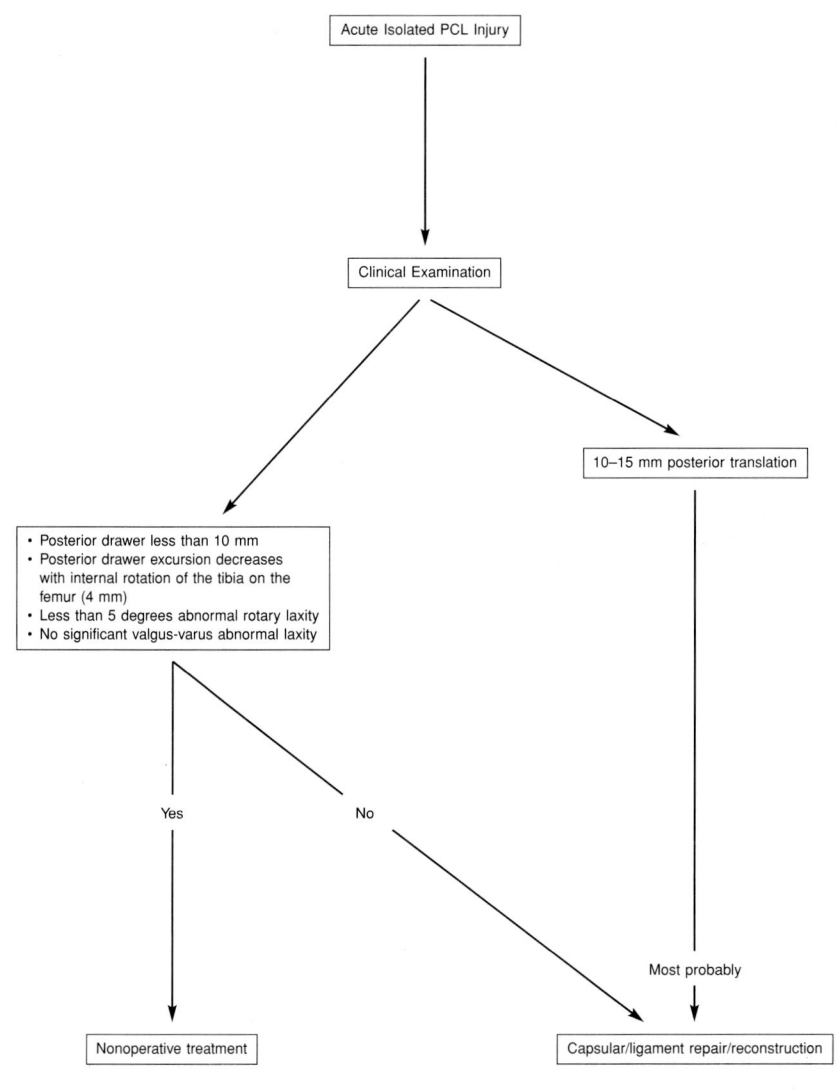

Fig. 9.1. Algorithm for the treatment of isolated posterior cruciate ligament (PCL) injuries.

gories known as open (OKCEs) and closed (CKCEs) kinetic chain exercises.[56] In OKCEs the distal segment is free to move and an isolated movement of a joint occurs. Conversely, in CKCEs the distal segment is fixed, and movement at one joint results in simultaneous movement of the rest of the joints in the kinetic chain. OKCEs and CKCEs produce different effects on tibial translation and PCL load. The CKCEs more closely duplicate the normal forces across the knee joint with daily living and sports activities.

Biomechanics of Open Kinetic Chain Exercises

Open kinetic chain (OKC) resisted knee flexion places a significant load on the PCL.[57–59] Electromyographic studies have shown that during OKC resisted knee flexion, the hamstring to quadriceps muscle activity ratio is 10 to 1 at 60 degrees and 90 degrees of knee flexion.[57] This excessively high hamstring muscle activity may contribute to PCL stretching.[57] Pandy and Shelburne[60] have noted that for isolated contractions of the hamstrings, the PCL force increased as hamstrings force increased for all flexion angles greater than 10 degrees. The PCL was unloaded at flexion angles less than 10 degrees. When hamstrings force was kept constant, PCL force increased with increasing knee flexion. OKC resisted knee flexion places deleterious stress on the PCL and should be avoided in the early rehabilitation phase.

OKC resisted knee extension at an angle greater than the quadriceps neutral (QN) angle (QN Q-angle = 60 to 75 degrees) results in a posterior tibiofemoral shear force, and at an angle lower than the QN Q-angle results in an anterior shear force. Wilk et al.[61] noted an anterior shear force (ACL stress) during OKC knee extension from 40 degrees to full extension, and Escamilla et al.[62] have found that the quadriceps muscle activity is greatest in OKC knee extension when the knee is near full extension. On the other hand, Lutz et al.[58] have quantified a maximum posterior shear force of 1,780 newtons at 90 degrees of knee flexion. Carlin et al.[63] reported that at 30 degrees of knee flexion, approximately 45% of the resistance to posterior tibial loading was caused by contact between the tibia and the femoral condyles, whereas at 90 degrees of knee flexion, no resistance was caused by such contact.

We should also bear in mind that the point where the resistance is located significantly affects the direction and magnitude of tibial displacement during resisted knee extension. Generally, a more proximally placed center of resistance will result in a posterior tibial displacement; therefore, the resistance to the tibia should be placed distally in the presence of a PCL injury.[64]

Based on the previously mentioned studies, OKC extension is a safe exercise after a PCL injury. To minimize PCL stress, the extensions should be initially performed from 60 to 0 degrees. Care should be taken to avoid irritation of the patellofemoral joint.

Biomechanics of Closed Kinetic Chain Exercises

During closed kinetic chain (CKC) exercises there is also a posterior shear force placed on the PCL that increases linearly as the knee flexion angle increases.[58,65,66] O'Connor[67] used a computer model of the knee to study the forces in cruciate ligaments induced by co-contraction of the extensor and flexor muscles, in the absence of external loads. His calculations showed that co-contraction of the quadriceps and hamstrings muscles loads the ACL from full extension to 22 degrees of flexion, and loads the PCL at higher flexion angles. Zheng et al.[65] compared the forces during OKCEs (seated knee extension) and CKCEs (leg press and squat). Peak PCL tensions were approximately twice as great in CKCEs, and increased with knee flexion.

Wilk et al.[61] investigated the tibiofemoral joint kinetics and electromyographic activity of the quadriceps, hamstring, and gastrocnemius muscles during OKC knee extension and CKC leg press and squat. During the CKCEs, a posterior shear force occurred throughout the range of motion, with the peak occurring from 85 to 105 degrees of knee flexion. Electromyographic data indicated greater hamstring and quadriceps muscle co-contraction during the squat compared with the other two exercises. During the leg press, the hamstring muscle activity was minimal, 12% of maximal velocity isometric contraction. Among the different CKCEs, leg presses seem to be safer than squats for patients with PCL injury[62] because the squats generate approximately twice as much hamstring activity.

Stuart et al.[66] analyzed the intersegmental forces at the tibiofemoral joint and the muscle activity during the power squat, the front squat, and the lunge. The mean tibiofemoral shear force was posterior throughout the cycle of all three exercises, and its magnitude increased with knee flexion during the descent phase of each exercise. Increased quadriceps muscle activity and decreased hamstring muscle activity was required to perform the lunge as compared with the power squat and the front squat.

Based on these studies, to minimize PCL stress, CKC front squats or leg press should be initially performed from 0 to 60 degrees of knee flexion.[57] The parity in the quadriceps/hamstrings contraction during the first 30 degrees of the squat may contribute to a stabilizing effect to the tibiofemoral joint.[57]

Biomechanical Changes and Compensatory Mechanisms in PCL-Deficient Knees

The kinematics of the knee are significantly altered when the PCL is injured.[68] The PCL is the primary restraint to posterior translation of the tibia on the femur. In PCL-deficient knees the tibial tuberosity displaces posteriorly and the angle between the quadriceps and patellar tendons decreases, causing an increase in patellofemoral joint reaction forces.[57,59]

Skyhar et al.[69] reported a 16% increase in patellofemoral pressure after sectioning the PCL. These increased contact pressures have potential harmful effects on the patellofemoral articular cartilage, leading some patients to a patellofemoral pain syndrome and eventually to patellofemoral osteoarthritis.[14,18,24,39,45,70]

MacDonald et al.[71] used a static cadaveric knee model with simulated physiologic loads to study the biomechanical changes after cutting the PCL. They demonstrated a statistically significant increase in contact pressure and pressure concentration on the medial compartment of the knee at 60 degrees of flexion. This finding correlates with the long-term degenerative changes observed in the medial femoral condyle in PCL-deficient knees.[21,29,41,43,72]

Functional stability in many of the PCL-deficient knees is achieved because the dysfunctional PCL can be compensated by quadriceps activity.[17] Isolated quadriceps contraction has been shown to produce an anterior drawer force to the tibia from full extension to about 70 degrees of flexion.[73,74] In PCL-deficient knees the voluntary quadriceps activity corrects the posterior sag to the balanced neutral position.[17] With strong quadriceps, the function of the knee is very good, even if the knee has remained posteriorly unstable.[3,21,29,39–41]

Cain and Schwab[18] evaluated an athlete with a ruptured PCL who was successfully competing in professional football. They used isokinetic strength evaluation and high-speed cinematography gait analysis combined with electromyography of the quadriceps muscles to document the compensatory mechanisms involved. The athlete could fully compensate for the PCL insufficiency of his right knee by contracting his quadriceps prior to heel strike, much earlier in the gait cycle than the left quadriceps. This early contraction increased the dynamic stability by setting the knee in full extension to prevent the posterior displacement of the tibia in relation to the femur. This mechanism requires an extremely strong quadriceps and a considerable degree of muscular coordination.

Treatment/Rehabilitation Protocol

Our protocol is based on the previous biomechanical observations (Table 9.1). Acute PCL disruption is initially managed by splinting or bracing in full extension for pain relief. Cold and compression are also beneficial to decrease pain and swelling. Range of motion exercises from 0 to 60 degrees, with the knee in a removable knee immobilizer or drop-lock brace, are initiated immediately to counteract the harmful effects of immobilization and joint effusion. The patient is advised to ambulate with two crutches, and weight bearing is progressed as tolerated. If no operative treatment is contemplated, nonsteroidal antiinflammatory drugs (NSAIDs) are begun to counteract the inflammatory effusion.

The early rehabilitation emphasizes quadriceps and hip flexor strengthening. Isometric quadriceps sets and straight leg raises (SLRs) are initiated. These isometric exercises have a benign effect on the injured tissue and are second only to eccentric exercises in generating intramuscular tension.[75] Concentric resisted knee extension exercises from 0 to 60 degrees are also started. Active resisted knee extension in that arc of motion produces an anterior force on the tibiofemoral joint,[55,64,76,77] and it is therefore safe for

Table 9.1. Nonoperative management of posterior cruciate ligament (PCL) injuries

Acute phase: protection phase (day 1–week 2)	
Goals	Reduce pain and inflammation
	Promote tissue healing
	Restore range of motion
	Early weight bearing
	Early strengthening
Range of motion	0–60 degrees, progress as inflammation and pain subside
	Drop-lock brace/removable knee immobilizer
Weight bearing	As tolerated; use crutches if needed for normal gait pattern
Exercises	Isometric quadriceps, straight leg raise, hip adduction and abduction
	Knee extension (60–0 degrees)
	Multiangle isometrics (quadriceps)
	Minisquats, leg press (0–45 degrees)
	Electric stimulation to quadriceps if necessary
Modalities	Ice, compression
Medication	Nonsteroidal antiinflammatory drugs
Avoid	Gravity induced posterior tibial sag
	OKC knee flexion exercises (hamstrings)
Subacute phase: moderate protection phase (weeks 2–6)	
Goals	Progressive muscle strengthening, emphasizing endurance and stress of the quadriceps
Range of Motion	As tolerated; discontinue brace
Weight bearing	Discontinue crutches when there is normal gait pattern and full extension of the knee without a quadriceps lag
Exercises	Light resistance and high repetitions OKC and CKC quadriceps strengthening
	CKC hip flexion
	Bicycling, leg press (0–60 degrees)
	Aquatic exercises, Stairmaster, rowing, NordicTrack
	Step-ups and calf raises, sport-cord progression
	Slideboard
	Proprioception and balance training
Avoid	OKC knee flexion exercises and heavy resistance extension exercises that cause patellofemoral pain
Functional recovery phase: minimal protection phase (weeks 8–12)	
Goal	Gradual return to sports activities
Exercises	Continue strengthening exercises
	Begin gentle hamstring exercises
	Acceleration/deceleration
	Sprint, jumping, cutting, pivoting
Functional bracing	Enhances proprioceptive awareness
	Attempt to prevent posterior tibial sag and repeated stress to posterior capsular structures

CKC, closed kinetic chain; OKC, open kinetic chain.

the PCL-injured knee. On the other hand, resisted knee extension from greater flexion angles should be initially avoided since it produces a posteriorly directed shear force that may overstress the healing PCL.[55,64,76] Hamstring exercises are avoided in the early phase of rehabilitation because of the significant posterior tibiofemoral shear force that they generate.

When the patient demonstrates good leg control, ambulates without gait deviation, and the knee joint effusion is decreasing, the crutches are discontinued and progression to more aggressive rehabilitation is instituted. The rehabilitation program is designed to strengthen the quadriceps to dynamically stabilize the knee joint, while protecting the patellofemoral articulation from compressive forces. The rehabilitation team should adjust the patient's program if patellofemoral symptoms develop. Combined OKCEs and CKCEs in pain-free and crepitus-free range of motion protects the patellofemoral joint.

The rehabilitation program is advanced to include progressive resistive exercises (PREs) utilizing light resistance and high repetitions to avoid inducing patellofemoral pain symptoms. Biking, unilateral balance and proprioceptive training, aquatic therapy, and ambulation without assistive devices are also initiated. Proprioceptive treatment should emphasize recruitment of the quadriceps muscle to dynamically stabilize the knee.[78]

Closed kinetic chain exercises continue, incorporating calf raises, step-up/down exercises starting at 2 inches and progressing to 8- to 10-inch boxes, and Stairmaster and Versa climber progression from seated to standing positions. These CKCEs produce a stabilizing effect on the tibiofemoral joint by coactivation of the quadriceps and the hamstrings.[58,67,79] At 6 to 8 weeks the patient is progressed to a walk/jog program, running/straight sprint program, jumping vertical to horizontal program, cutting and pivoting, and acceleration-deceleration functional maneuvers. The hamstring muscles are strengthened by performing CKCEs and open chain hip extension with the knee near terminal extension.[78]

Swelling of the knee should be monitored during the entire course of rehabilitation.[71] Increased swelling may cause a significant reflex inhibition of the quadriceps,[80] which will slow the progression of the rehabilitation. Effusion may be controlled with short periods of cessation of exercises, ice, elevation, compression, and antiinflammatory medication.[81]

Return to competitive sports is allowed when the patient has full range of motion, no change in laxity, no pain, no swelling, and strength equal to the opposite side, and demonstrates confidence in the knee. In our series the range of time of return to competition was 3 weeks to 6 months, averaging approximately 6 weeks.

A functional brace may assist in preventing the posterior tibial sag and enhance proprioception, but its use in the PCL-deficient knee remains controversial. If a functional brace it utilized, a custom-contoured combined instability design is applied (Don Joy, Smith & Nephew Corp., Carlsbad, CA). Patients do not generally complain of functional instability during activity but are encouraged to continue indefinitely with a preventative exercise program.

Activities that require frequent high-intensity loading of the patellofemoral joint are discouraged due to the potential contribution to the articular degeneration of that joint surface.

Results of Conservative Treatment

Sports Activity

The percentage of athletes that return to sports participation after conservative treatment of isolated PCL injuries is consistently high in the different series. Parolie and Bergfeld[21] reported that 84% of the athletes returned to their preinjury sports, 68% at the same level. All the patients who were seen acutely and started early motion and a vigorous rehabilitation program returned to full sports participation. The importance of an early diagnosis is also evident in Fowler and Messleh's[3] series. They studied 13 athletes with isolated midsubstance tears of the PCL who were seen acutely and placed on a vigorous physical therapy program. All patients returned to their previous activity with no limitations. A long-term review by Cross and Powell[29] found that 86% of the patients who injured their PCLs while participating in athletics had good to excellent results, while only 8% had good to excellent results following motor vehicle accidents or falls. The authors attributed this difference, in part, to the greater readiness of the athletes to participate in vigorous quadriceps rehabilitation.

In Degenhardt and Hughston's[82] series, 78% of the patients were able to participate in sports that required running and cutting maneuvers, and 13% were more limited by age than by the knee. In the San Diego Kaiser PCL study, the mean total hours of participation in the patients' two favorite sports fell from 361 hours per year before the PCL injury to 141 hours per year at 27 months of average follow-up.[83] Shelbourne et al.[22] reported that of 133 patients with isolated PCL tears who were followed for an average of 5.4 years, half returned to the same sports, one-third returned to the same sport at a lower level, and one-sixth took up a different sport.

In Parolie and Bergfeld's[21] series, the outcome of conservatively treated isolated PCL tears depended more on the quadriceps status of the knee than on the amount of residual posterior laxity. Those patients who fully returned to sports and were fully satisfied with their knees had a mean torque for three velocities of testing (Cybex II), which was greater than 100% of the uninvolved quadriceps. Those findings corroborated Cross and Powell's[29] and Degenhart and Hughston's[82] observation that good results correlated well with good quadriceps strength.

The degree of abnormal laxity determined by clinical examination as well as by the KT-1000 arthrometer does not appear related to the patients' ability to return to sports activities.[21,22,29,41] Absolute static stability is not required to achieve good functional results.[3] In our athletes the KT-1000 showed an average 6.7-mm excursion in the knees and an 8.5-mm excursion in the unsatisfactory knees. This 1.8-mm difference was not statistically significant.

Long-Term Results

Patients, and particularly athletes, with nonoperatively treated PCL injuries should have a close follow-up evaluation. The most common complaint of patients with chronic PCL laxity is aching in the knee when walking long distances, going up or down stairs, or squatting.[41] A revision of the literature would indicate a high incidence of symptoms, with knee pain in 52% to 89% of the patients[29,41,43,45] and swelling in 20% to 50%.[41,44,45] These figures may be misleading because many patients in these studies do not meet the criteria of isolated PCL injury as defined earlier in this chapter. In Parolie and Bergfeld's[21] series, 48% of the patients were free of pain, 24% complained of occasional knee pain that was unrelated to exercise, 20% had pain after exercise, and 8% complained intermittently of pain prior to and after exercise. None of the patients had consistent pain in the knee, and the majority could not localize the pain in one compartment. Twenty percent complained of pain with patellofemoral compression, and 12% had medial joint line tenderness.

Boynton and Tietjens[44] evaluated 38 patients with isolated PCL injuries at a mean follow-up of 13.4 years, and found a positive correlation between the degree of posterior laxity, as graded on manual examination, and the knee scores for both symptoms and function. On the other hand, Shelbourne et al.[22] have not found a correlation between pain score and the amount of PCL laxity.

Patients with chronic isolated PCL injuries may develop articular changes in the injured knee. These changes usually affect the medial compartment and the patellofemoral joint.[14,18,21,22,40,44,45] The incidence is higher with combined ligament injuries, active lifestyle, repeated episodes of instability, and associated meniscal or osteochondral injuries.

Some studies report an incidence of arthritis and disability that is higher than the one we have observed in our patients.[39,44,45] We attribute this discrepancy to the different criteria used to diagnose isolated PCL injuries. Parolie and Bergfeld[21] reported a 36% incidence of degenerative changes after a mean follow-up of 6.2 years. Similar figures have been reported by Degenhardt and Hughston and Kennedy and Grainger.[4]

In a recent study by Shelbourne et al.,[22] after a mean follow-up of 5.4 years, 28 of the 68 patients (41.2%) showed degenerative changes on weight-bearing posteroanterior views. There seems to be no correlation between the degree of posterior drawer laxity and the radiographic evidence of osteoarthritis.[22,44]

At this time, we cannot say that surgical reconstruction of isolated PCL injuries will prevent the development of osteoarthritis. In a series by Lipscomb et al.,[84] 15 of 25 patients with isolated PCL injuries who underwent PCL reconstruction showed radiographic evidence of degenerative changes after an average follow-up of 7 years. The osteoarthritis predominantly involved the medial and patellofemoral compartments.

Bone scan is most sensitive in the diagnosis of early arthritic changes.[85] In patients with chronic PCL injuries, an increased uptake in the bone scan may be an early sign of articular cartilage damage and an indication for the need of surgical treatment.[28]

References

1. Miller MD. Management of the PCL-injured knee. Instructional Course Lecture 103, AOSSM annual meeting, Vancouver, 1998.
2. Rubinstein RA Jr, Shelbourne KD, McCarroll JR, VanMeter CD, Rettig AC. The accuracy of the clinical examination in the setting of posterior cruciate ligament injuries. Am J Sports Med 1994;22(4):550–557.
3. Fowler PJ, Messieh SS. Isolated posterior cruciate ligament injuries in athletes. Am J Sports Med 1987;15(6):553–557.
4. Kennedy JC, Grainger RW. The posterior cruciate ligament. J Trauma 1967;7(3):367–377.
5. McMaster WC. Isolated posterior cruciate ligament injury: literature review and case reports. J Trauma 1975;15(11):1025–1029.
6. Trickey EL. Injuries to the posterior cruciate ligament: diagnosis and treatment

of early injuries and reconstruction of late instability. Clin Orthop 1980;147:76–81.
7. Kannus P, Jarvinen M. Knee instabilities and their clinical examination. Finnish Sports Exerc Med 1985;4:79–89.
8. Mayer PJ, Micheli LJ. Avulsion of the femoral attachment of the posterior cruciate ligament in an eleven-year-old boy. Case report. J Bone Joint Surg 1979;61A(3):431–432.
9. Sanders WE, Wilkins KE, Neidre A. Acute insufficiency of the posterior cruciate ligament in children. Two case reports. J Bone Joint Surg 1980;62A(1):129–131.
10. Stanish WD, Rubinovich M, Armason T, Lapenskie G. Posterior cruciate ligament tears in wrestlers. Can J Appl Sport Sci 1986;11(4):173–177.
11. Trickey EL. Rupture of the posterior cruciate ligament of the knee. J Bone Joint Surg 1968;50B(2):334–341.
12. Hughston JC, Bowden JA, Andrews JR, et al. Acute tears of the posterior cruciate ligament. Results of operative treatment. J Bone Joint Surg 1980;62A(3):438–450.
13. Andrews JR, Edwards JC, Satterwhite YE. Isolated posterior cruciate ligament injuries. History, mechanism of injury, physical findings and ancillary tests. Clin Sports Med 1994;13:519–530.
14. Clancy WG. Repair and reconstruction of the posterior cruciate ligament. Instructional Course, AOSSM annual meeting, Traverse City, MI, 1989.
15. Clendenin MB, DeLee JC, Heckman JD. Interstitial tears of the posterior cruciate ligament of the knee. Orthopedics 1980;3:764–772.
16. Satku K, Chew CN, Seow H. Posterior cruciate ligament injuries. Acta Orthop Scand 1984;55(1):26–29.
17. Kannus P, Bergfeld J, Jarvinen M, et al. Injuries to the posterior cruciate ligament of the knee. Sports Med 1991;12(2):110–131.
18. Cain TE, Schwab GH. Performance of an athlete with straight posterior knee instability. Am J Sports Med 1981;9(4):203–208.
19. Clancy WG Jr, Shelbourne KD, Zoellner GB, et al. Treatment of knee joint instability secondary to rupture of the posterior cruciate ligament. Report of a new procedure. J Bone Joint Surg 1983;65A(3):310–322.
20. Covey CD, Sapega AA. Injuries of the posterior cruciate ligament. J Bone Joint Surg 1993;75A(9):1376–1386.
21. Parolie JM, Bergfeld JA. Long-term results of nonoperative treatment of isolated posterior cruciate ligament injuries in the athlete. Am J Sports Med 1986;14(1):35–38.
22. Shelbourne KD, Patel DV, Davis TJ. The natural history of acute, isolated, nonoperatively treated posterior cruciate ligament injuries: a prospective study. Paper presented at the AOSSM annual meeting. Vancouver, 1998.
23. Shino K, Horibe S, Nakata K, Maeda A, Hamada M, Nakamura N. Conservative treatment of isolated injuries to the posterior cruciate ligament in athletes. J Bone Joint Surg 1995;77B(6):895–900.
24. Veltri DM, Warren RF. Anatomy, biomechanics, and physical findings in posterolateral knee instability. Clin Sports Med 1994;13(3):599–614.
25. Fanelli GC, Edson CJ. Posterior cruciate ligament injuries in trauma patients: Part II. Arthroscopy 1995;11(5):526–529.
26. Shelbourne KD, Patel DV. Natural history of acute isolated nonoperatively treated posterior cruciate ligament injuries of the knee: a prospective study. Paper presented at the American Academy of the Orthopaedic Surgeons annual meeting, San Francisco, 1997.
27. Ritchie JR, Bergfeld JA, Kambic H, Manning T. Isolated sectioning of the medial and posteromedial capsular ligaments in the posterior cruciate ligament-deficient knee. Influence on posterior tibial translation. Am J Sports Med 1998;26(3):389–394.
28. Harner CD, Höher J. Evaluation and treatment of posterior cruciate ligament injuries. Am J Sports Med 1998;26(3):471–482.
29. Cross MJ, Powell JF. Long-term followup of posterior cruciate ligament rupture: a study of 116 cases. Am J Sports Med 1984;12(4):292–297.
30. Huber FE, Irrgang JJ, Harner C, Lephart S. Intratester and intertester reliability of the KT-1000 arthrometer in the assessment of posterior laxity of the knee. Am J Sports Med 1997;25(4):479–485.

31. Eakin CL, Cannon WD Jr. Arthrometric evaluation of posterior cruciate ligament injuries. Am J Sports Med 1998;26(1):96–102.
32. Staubli HU, Jakob RP. Posterior instability of the knee near extension. A clinical and stress radiographic analysis of acute injuries of the posterior cruciate ligament. J Bone Joint Surg 1990;72B(2):225–230.
33. Hewett TE, Noyes FR, Lee MD. Diagnosis of complete and partial posterior cruciate ligament ruptures. Stress radiography compared with KT-1000 arthrometer and posterior drawer testing. Am J Sports Med 1997;25(5):648–655.
34. Sonin AH, Fitzgerald SW, Hoff FL, Friedman H, Bresler ME. MR imaging of the posterior cruciate ligament: normal, abnormal, and associated injury patterns. Radiographics 1995;15(3):551–561.
35. Gross ML, Grover JS, Bassett LW, Seeger LL, Finerman GA. Magnetic resonance imaging of the posterior cruciate ligament. Clinical use to improve diagnostic accuracy. Am J Sports Med 1992;20(6):732–737.
36. Grover JS, Bassett LW, Gross ML, et al. Posterior cruciate ligament: MR imaging. Radiology 1990;174(2):527–530.
37. Patten RM, Richardson ML, Zink-Brody G, Rolfe BA. Complete vs partial-thickness tears of the posterior cruciate ligament: MR findings. J Comput Assist Tomogr 1994;18(5):793–799.
38. Bellelli A. Cicatrization of complete traumatic lesions of the posterior cruciate ligament. Magnetic resonance follow-up of 10 cases and a proposal for modification of Gross classification. Radiol Med (Torino) 1998;95(4):286–292.
39. Torg JS, Barton TM, Pavlov H, Stine R. Natural history of the posterior cruciate ligament-deficient knee. Clin Orthop 1989;246:208–216.
40. Barton TM, Torg JS, Das M. Posterior cruciate ligament insufficiency. A review of the literature. Sports Med 1984;1:419–430.
41. Dandy DJ, Pusey RJ. The long-term results of unrepaired tears of the posterior cruciate ligament. J Bone Joint Surg 1982;64B(1):92–94.
42. Tietjens BB. Posterior cruciate ligament injuries. J Bone Joint Surg 1985;67B:674.
43. Keller PM, Shelbourne KD, McCarroll JR, Rettig AC. Nonoperatively treated isolated posterior cruciate ligament injuries. Am J Sports Med 1993;21(1):132–136.
44. Boynton MD, Tietjens BR. Long-term followup of the untreated isolated posterior cruciate ligament-deficient knee. Am J Sports Med 1996;24(3):306–310.
45. Dejour H, Walch G, Peyrot J, Eberhard P. The natural history of rupture of the posterior cruciate ligament. Rev Chir Orthop 1988;74(1):35–43.
46. Bianchi M. Acute tears of the posterior cruciate ligament: clinical study and results of operative treatment in 27 cases. Am J Sports Med 1983;11(5):308–314.
47. Eriksson E, Haggmark T, Johnson RJ. Reconstruction of the posterior cruciate ligament. Orthopedics 1986;9(2):217–220.
48. Loos WC, Fox JM, Blazina ME, Del Pizzo W, Friedman MJ. Acute posterior cruciate ligament injuries. Am J Sports Med 1981;9(2):86–92.
49. Moore HA, Larson RL. Posterior cruciate ligament injuries. Results of early surgical repair. Am J Sports Med 1980;8(2):68–78.
50. Strand T, Molster AO, Engesaeter LB, et al. Primary repair in posterior cruciate ligament injuries. Acta Orthop Scand 1984;55(5):545–547.
51. Morrison JB. The mechanics of the knee joint in relation to normal walking. J Biomech 1970;3(1):51–61.
52. Denham RA, Bishop RE. Mechanics of the knee and problems in reconstructive surgery. J Bone Joint Surg 1978;60B(3):345–352.
53. Morrison JB. Function of the knee joint in various activities. Biomed Eng 1969;4(12):573–580.
54. Dahlkuits NJ, Mago P, Seedholm BB. Forces during squatting and rising from a deep squat. Eng Med 1982;11:69–79.
55. Ericson MO, Nisell R. Tibiofemoral joint forces during ergometer cycling. Am J Sports Med 1986;14(4):285–290.
56. Chu DA. Rehabilitation of the lower extremity. Clin Sports Med 1995;14(1):205–222.
57. Wilk KE. Rehabilitation of isolated and combined posterior cruciate ligament injuries. Clin Sports Med 1994;13(3):649–677.

58. Lutz GE, Palmitier RA, An KN, Chao EY. Comparison of tibiofemoral joint forces during open-kinetic-chain and closed-kinetic-chain exercises. J Bone Joint Surg 1993;75A(5):732–739.
59. Kaufman KR, An KN, Litchy WJ, Morrey BF, Chao EY. Dynamic joint forces during knee isokinetic exercise. Am J Sports Med 1991;19(3):305–316.
60. Pandy MG, Shelburne KB. Dependence of cruciate-ligament loading on muscle forces and external load. J Biomech 1997;30(10):1015–1024.
61. Wilk KE, Escamilla RF, Fleisig GS, Barrentine SW, Andrews JR, Boyd ML. A comparison of tibiofemoral joint forces and electromyographic activity during open and closed kinetic chain exercises. Am J Sports Med 1996;24(4):518–527.
62. Escamilla RF, Fleisig GS, Zheng N, Barrentine SW, Wilk KE, Andrews JR. Biomechanics of the knee during closed kinetic chain and open kinetic chain exercises. Med Sci Sports Exerc 1998;30(4):556–569.
63. Carlin GJ, Livesay GA, Harner CD, Ishibashi Y, Kim HS, Woo SL. In-situ forces in the human posterior cruciate ligament in response to posterior tibial loading. Ann Biomed Eng 1996;24(2):193–197.
64. Jurist KA, Otis JC. Anteroposterior tibiofemoral displacements during isometric extension efforts. The roles of external load and knee flexion angle. Am J Sports Med 1985;13(4):254–258.
65. Zheng N, Fleisig GS, Escamilla RF, Barrentine SW. An analytical model of the knee for estimation of internal forces during exercise. J Biomech 1998;31(10):963–967.
66. Stuart MJ, Meglan DA, Lutz GE, Growney ES, An KN. Comparison of intersegmental tibiofemoral joint forces and muscle activity during various closed kinetic chain exercises. Am J Sports Med 1996;24(6):792–799.
67. O'Connor JJ. Can muscle co-contraction protect knee ligaments after injury or repair? J Bone Joint Surg 1993;75B(1):41–48.
68. Muller W. The Knee, Form, Function and Ligament Reconstruction. New York: Springer, 1983.
69. Skyhar MJ, Warren RF, Ortiz GJ, Schwartz E, Otis JC. The effects of sectioning of the posterior cruciate ligament and the posterolateral complex on the articular contact pressures within the knee. J Bone Joint Surg 1993;75A(5):694–699.
70. Insall JN, Hood RW. Bone-block transfer of the medial head of the gastrocnemius for posterior cruciate insufficiency. J Bone Joint Surg 1982;64A(5):691–699.
71. MacDonald P, Miniaci A, Fowler P, Marks P, Finlay B. A biomechanical analysis of joint contact forces in the posterior cruciate deficient knee. Knee Surg Sports Traumatol Arthrosc 1996;3(4):252–255.
72. Clancy WG. Repair and reconstruction of the posterior cruciate ligament. In: Chapman MW, ed. Operative Orthopaedics. Philadelphia: JB Lippincott, 1988:1651–1666.
73. Beynnon B, Yu J, Huston D, Fleming, et al. A sagittal plane model of the knee and cruciate ligaments with application of a sensitivity analysis. J Biomech Eng 1996;118(2):227–239.
74. Yasuda K, Sasaki T. Exercise after anterior cruciate ligament reconstruction. The force exerted on the tibia by the separate isometric contractions of the quadriceps or the hamstrings. Clin Orthop 1987;220:275–283.
75. Davies GJ. A Comparison of Isokinetics in Clinical Usage, 3rd ed. La Crosse, WI: S&S Publishers, 1987.
76. Daniel DM, Stone ML, Barnett P, et al. Use of the quadriceps active test to diagnose posterior cruciate ligament disruption and measure posterior laxity of the knee. J Bone Joint Surg 1988;70A(3):386–391.
77. Mangine RE, Eifert-Mangine MA. Postoperative posterior cruciate ligament reconstruciton rehabilitation. In: Engle RP, ed. Knee Ligament Rehabilitation. New York: Churchill Livingstone, 1991:165–176.
78. Irrgang JJ. Rehabilitation for nonoperative and operative management of knee injuries. In: Fu FH, Harner CD, Vince KG, eds. Knee Surgery. Baltimore: Williams & Wilkins, 1994:485–502.
79. Wilk KE, Andrews JR. Current concepts in the treatment of anterior cruciate ligament disruptions. J Orthop Sports Phys Ther 1992;15:279–293.
80. Spencer JD, Hayes KC, Alexander IJ. Knee joint effusion and quadriceps reflex inhibition in man. Arch Phys Med Rehabil 1984;65(4):171–177.

81. Huegel M, Indelicato PA. Trends in rehabilitation following anterior cruciate ligament reconstruction. Clin Sports Med 1988;7(4):801–811.
82. Degenhardt TC, Hughston JC. Chronic posterior cruciate instability: nonoperative management. Orthop Trans 1981;5:486–487.
83. Hirshman HP, Daniel MD, Miyasaka K. The fate of unoperated knee ligament injuries. In: Daniel DM, Akeson WH, O'Connor, eds. Knee Ligaments. Structure, Function, Injury and Repair. New York. Raven Press.
84. Lipscomb AB Jr, Anderson AF, Norwig ED, Hovis WD, Brown DL. Isolated posterior cruciate ligament reconstruction. Long-term results. Am J Sports Med 1993;21(4):490–496.
85. Dye SF, Chew MH. The use of scintigraphy to detect increased osseous metabolic activity about the knee. Instr Course Lect 1994;43:453–469.

IV

Surgical Treatment

Chapter Ten

Graft Selection in Posterior Cruciate Ligament Surgery

Walter R. Shelton

Reconstruction of the posterior cruciate ligament (PCL) is one of the most challenging problems of all knee surgery. Early attempts at repair alone proved unsatisfactory due to a lack of understanding of the complex anatomy of the PCL, and the tremendous stress placed on the ligament during normal knee motion. These failures led to a nonoperative treatment philosophy for most PCL tears, since results were equal to or better than surgical repair.

The use of stronger grafts and fixation techniques along with precise tunnel placement, arthroscopic techniques, and a better understanding of rehabilitation principles have greatly improved the results of PCL surgery in recent years. While nonoperative treatment is still recommended for isolated tears with mild instability, repair using an augmenting graft or reconstruction using a graft to replace the entire PCL has evolved as a more reliable procedure for the treatment of severe and combined instabilities.

Properties of an Ideal Graft

Despite significant advances in graft choices for PCL reconstruction, the perfect graft does not exist. The anatomy of the PCL is complex.[1,2] Its midsubstance cross-sectional width averages 13 mm, spreading out to a femoral attachment of 32 mm. It can be divided into at least two separate bundles. The large anterolateral bundle tightens as the knee goes into flexion, while the smaller posterior-medial bundle is taunt in extension.[3,4] More accurately, instead of one bundle tightening in flexion and the other in extension, there is probably a cascade of tightening and relaxing of individual fiber bundles from anterior to posterior as the knee goes from extension into flexion. The intraarticular length of the PCL averages 38 mm. Passage and fixation of grafts are usually more difficult than in anterior cruciate ligament (ACL) reconstruction, and a longer graft of 125 to 130 mm is needed. This length allows graft fixation either inside or outside the tunnels. PCL graft choice must take into account the anatomic issues of length and cross-sectional area to closely reduplicate PCL function.

The normal PCL has a strength of $1{,}627 \pm 491$ N[5] and experiences shear forces of three to six times that seen by the ACL during active knee flexion.[6] A graft must be strong enough to withstand these forces. Grafts also tend to lose strength during revascularization and remodeling phases of ligament healing[7]; thus, a graft that exceeds normal PCL strength would be preferable. During the healing phase, a graft of insufficient strength will tend to stretch out when subjected to the forces seen by the PCL.

Adequate length and thickness are necessary when using two femoral insertions to reproduce the broader attachment of the two bands of the PCL. This technique can produce a more anatomically correct ligament, but requires a more versatile graft.

Graft fixation at tunnel entrance into the joint rather than distally in the tunnels has been shown to be advantageous.[8] Especially in the tibia, fixation of the graft posteriorly can reduce the stress on the graft by the so-called killer angle. When the graft employed has soft tissue bridging the entire posterior aspect of the tibia, the tibial plateau acts as a wedge or fulcrum on the graft causing a mechanical disadvantage. The ability to withstand this stress is necessary to prevent graft stretch during healing, and the use of an onlay graft or fixation of a bone plug posteriorly in the tibial tunnel can reduce these forces. A graft with at least one bone plug is necessary when either technique is employed.

Graft Choices for PCL Reconstruction

Semitendinosus/Gracilis

The hamstring tendons may be used to augment repair of the PCL or to fully reconstruct it.[9,10] It can be used as an allograft or autograft, the latter being more common (Fig. 10.1).

Hughston et al.[11] described repair of the PCL, but repair alone without augmentation has been reported to be unreliable.[12] Repair of a torn PCL with augmentation has produced good results.[13] Repair and augmentation is most efficacious when the tear is either proximal or distal in the ligament and not midsubstance. The object is to reinforce and protect the repaired PCL with the least amount of damage possible to attachment sites.

The semitendinosus alone or in combination with the gracilis[14–16] has been the most common augmentation graft used in PCL repair. It has sufficient length and may be placed through a small 6- to 7-mm tunnel, which minimizes ligament damage at the attachment sites.

The semitendinosus and gracilis may be doubled to produce a four-strand graft for reconstruction of the PCL. This technique increases the bulk and strength of the graft sufficiently for use in PCL replacements. Its major weakness lies in not having bone plugs for rigid attachment. The ligamentous graft is exposed to the posterior killer angle, and stretching of

Fig. 10.1. Quadrupled strand hamstring graft.

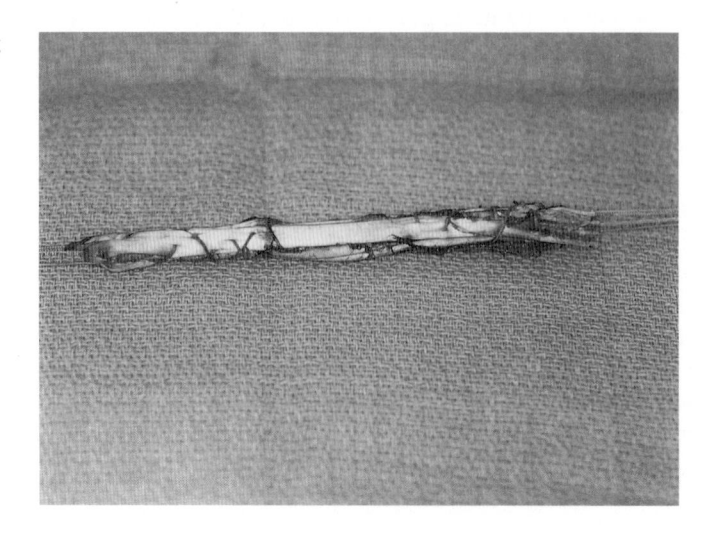

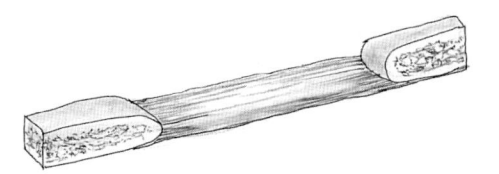

10. Graft Selection in Posterior Cruciate Ligament Surgery

Fig. 10.2. Patella tendon graft.

the graft is a concern. The semitendinosus/gracilis graft is most often a secondary choice for PCL reconstruction, and is often needed to augment other areas such as the posterior lateral corner during PCL surgery.

Patella Tendon

The patella tendon may be used as an allograft or autograft for PCL reconstruction (Fig. 10.2). Clancy et al.[17] were the first to show good results with its use, and subsequent reports have confirmed its use either with fixation of the bone plugs in tunnels[18] or as a posterior onlay graft.[19] The patella tendon has sufficient strength[20] to withstand PCL forces. Its size (10 × 4 mm) does not closely reproduce that of a normal PCL and especially on the femur does not reduplicate a broad attachment site. The small cross-sectional area does not lend itself to dividing into two bundles, if this form of femoral fixation is desired. The bone-to-bone distance is usually 10 mm longer than needed; therefore, one bone plug must be fixed within a tunnel, which is less secure than tunnel entrance fixation.

The presence of two bone plugs means that at least one must be passed from one tunnel to the other. This passage can be a difficult technical feat, an undesirable aspect to the use of this graft.

Quadriceps Tendon

The quadriceps tendon may be used as an allograft or autograft in PCL reconstruction. Its popularity has increased recently due to a large cross-sectional area (12 × 8 mm) and adequate length (up to 130 mm) (Fig. 10.3). This size closely approaches the normal PCL size and may be easily split to allow for two femoral attachment sites. The bone plug on one end may be used in either the femur or the tibia. The superior pole of the patella slopes at a 45-degree angle, fixed at the posterior cortex of the tibia. This slope allows the ligament fibers to parallel the posterior tibial cortex,

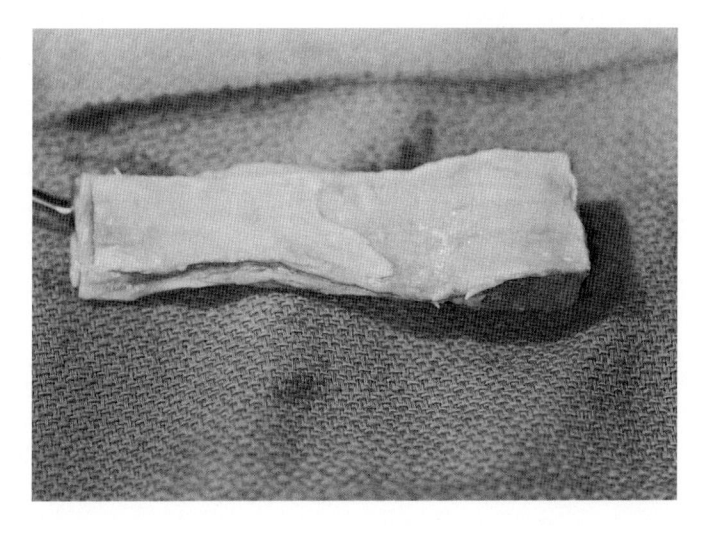

Fig. 10.3. Allograft quadriceps tendon with attached patella bone plug. Note 45-degree slope of proximal patella.

closely duplicating a posterior onlay graft. The reduction of the killer-angle force on the graft is accomplished without a posterior incision.

The major disadvantage of the quadriceps tendon is the absence of a bone plug on the proximal end. While soft tissue fixation techniques have improved and gained acceptance, it does not produce as strong a fixation as bone-to-bone fixation on both ends of the graft.

Achilles Tendon

The Achilles tendon used as an allograft has been a very popular graft choice for PCL reconstruction (Fig. 10.4). Its size, 12×8 mm, can reduplicate the size of the PCL and it can also be split for two femoral attachment sites. Fanellia et al.[3,21,22] have published good results with this graft when used for PCL reconstruction. The bone plug may be used for either femoral or tibial attachment with soft tissue fixation on the proximal graft. It may be obtained in lengths in excess of 130 mm. The main disadvantage of the use of the Achilles tendon is that it only has one bone plug, so fixation strength is less than for grafts with two bone plugs. It can only be used as an allograft with the inherent disadvantage of slower graft maturation and the risk of disease transmission.

Synthetic Grafts

The use of a purely synthetic replacement of the PCL is an appealing idea since problems with length, strength, disease transmission, stretching out, maturation time, and donor-site morbidity could be eliminated. Materials such as Goretex,[23,24] Dacron,[25,26] and others have been used in the knee. Unfortunately, attachment site problems, cyclical wear, and fatigue failure have proven too difficult to overcome.[27]

Reconstruction of the PCL with a biologic graft augmented with a synthetic one has been reported by Noyes and Barber-Westin.[28] They used a ligament augmentation device to augment a patella tendon graft, but they found it did not improve their results when compared to using the patella tendon alone. At present there is no acceptable synthetic graft for use in PCL reconstruction.

Other Grafts

Many structures have been tried as PCL grafts, but, due to poor results or excessive morbidity resulting from their harvest, are not recommended. The medial head of the gastrocnemius seemed to be an ideal graft with size, strength, and an attached bone plug. The graft was located posteriorly and did not have to be separated from its blood supply. Poor results have universally been reported with its use because of an inability to firmly fix the tibial attachment. Likewise, the use of the popliteal tendon or the meniscus[29] sacrifices too important a structure to consider using as a PCL graft. Results with these grafts have been uniformly poor, and their use is not recommended.

Conclusion

The choice of a graft for PCL reconstruction or repair is an important aspect of the surgical reconstruction. High stress forces on the graft, complex anatomic makeup of the PCL, and difficult graft passage and fixation must

Fig. 10.4. Achilles tendon allograft.

all be considered when choosing a graft. The patient should participate in the decision, especially when choosing between an allograft or an autograft.

Although most tendons about the knee, at one time or another, have been used to reconstruct the PCL, only a quadrupled semitendinosus/gracilis as an allograft or autograft, the patellar tendon as an allograft or autograft, the quadriceps tendon as an allograft or autograft, or the Achilles tendon as an allograft have proved successful. Synthetic ligaments have met with failure and are presently not indicated in PCL reconstruction either for complete replacement or augmentation.

References

1. Girgis FG, Marshall JL, Monajem ARS. The cruciate ligaments of the knee joint. Clin Orthop 1975;106:216–231.
2. Harner CD. Transactions. Orthop Res Soc 1992;17:123.
3. Fanelli GC, Giannotti BF, Edson CJ. The posterior cruciate ligament: arthroscopic evaluation and treatment. Arthroscopy 1994;10(6):673–688.
4. Covey DC, Sapega AA. Anatomy and function of the posterior cruciate ligament. Clin Sports Med 1994;13(3):509–518.
5. Prietto MP, Bain JR, Stonebrook SN, Settlage RA. Tensile strength of the human posterior cruciate ligament (PCL). Trans Orthop Res Soc 1988;13:195.
6. Lutz GE, Palmitier RA, Chao EYS. Comparison of tibiofemoral joint forces during open-kinetic-chain and closed-kinetic-chain exercises. J Bone Joint Surg 1993;75A:732–739.
7. Cordrey LJ, McCorkle H, Hilton E. A comparative study of fresh autogenous and preserved homogenous tendon grafts in rabbits. J Bone Joint Surg 1963;45A:182–195.
8. Burns WC, Draganich LF, Pyevich M, Reider B. The effect of femoral tunnel position and graft tensioning technique on posterior laxity of the posterior cruciate ligament-reconstructed knee. Am J Sports Med 1995;23(4):424–430.
9. Bianchi M. Acute tears of the posterior cruciate ligament: clinical study and results of operative treatment in 27 cases. Am J Sports Med 1983;11(5):308–314.
10. Lipscomb AB, Johnston RK, Snyder RB: The technique of cruciate ligament reconstruction. Am J Sports Med 1981;9:77–81.
11. Hughston JC, Bowden JA, Andrews JR, Norwood LA. Acute tears of the posterior cruciate ligament. J Bone Joint Surg 1980;62A(3):438–450.
12. Stand T, Molster A, Engesaeter LB, Raugstad TS, Alho A. Primary repair in posterior cruciate ligament injuries. Acta Orthop Scand 1984;55:545–547.
13. Clancy WG, Shelbourne KD, Zoellner GB, Keene JS, Reider B, Rosenberg TD. Treatment of knee joint instability secondary to rupture of the posterior cruciate ligament. J Bone Joint Surg 1983;65A:310–322.
14. Roth JH, Bray RC, Best TM, Cunning LA, Jacobson RP. Posterior cruciate ligament reconstruction by transfer of the medial gastrocnemius tendon. Am J Sports Med 1988;16(1):21–28.
15. Kennedy JC, Galpin RD. The use of the medial head of the gastrocnemius muscle in the posterior cruciate-deficient knee. Am J Sports Med 1982;10(2):63–74.
16. Insall JN, Hood RW. Bone-block transfer of the medial head of the gastrocnemius for posterior cruciate insufficiency. J Bone Joint Surg 1982;64A:691–699.
17. Clancy WG, Shelbourne KD, Zoellner GB, Keene JS, Reider B, Rosenberg TD. Treatment of knee joint instability secondary to rupture of the posterior cruciate ligament. J Bone Joint Surg 1983;65A:310–322.
18. Kim SJ, Min BY. Arthroscopic intraarticular interference screw technique of posterior cruciate ligament reconstruction: one-incision technique. Arthroscopy 1994;10(3):319–323.
19. Berg EE. Posterior cruciate ligament tibial inlay reconstruction. Arthroscopy 1995;11(1):69–76.
20. Noyes FR, Butler DL, Grood ES, Zernicke RF, Hefzy MS. Biomechanical analysis of human ligament grafts used in knee-ligament repairs and reconstructions. J Bone Joint Surg 1984;66A:344–352.

21. Fanelli GC, Giannotti BF, Edson CJ. Arthroscopically assisted combined ACL/PCL reconstruction. Arthroscopy 1996;12(1):5–14.
22. Fanelli GC, Giannotti BF, Edson CJ. Arthroscopically assisted PCL/posterolateral complex reconstruction. Arthroscopy 1996;12(5):521–530.
23. Ferkel RD, Fox JM, Del Pizzo W, Freidman MJ, Snyder SJ. Arthroscopic "second-look" at the GORETEX ligament. Am J Sports Med 1989;17(2):147–153.
24. Jenkins DHR. The repair of cruciate ligaments with flexible carbon fibre: a long term study of the induction of new ligaments and of the fate of the implanted carbon. J Bone Joint Surg 1978;60B:520–522.
25. Chiu F-Y, Wu J-J, Hsu H-C, Lin L, Lo W-H. Management of insufficiency of posterior cruciate ligaments. Chin Med J 1994;53:282–287.
26. Barrett GR, Savoie FH. Operative management of acute PCL injuries with associated pathology: long term results. Orthopedics 1991;14:687–692.
27. Good L, Tarlow SD, Odensten M, Gillquist J. Load tolerance, security, and failure modes of fixation devices for synthetic knee ligaments. Clin Orthop 1990;253:190–196.
28. Noyes FR, Barber-Westin SD. Posterior cruciate allograft reconstruction with and without a ligament augmentation device. Arthroscopy 1994;10(4):371–382.
29. Tillberg B. The late repair of torn cruciate ligaments using menisci. J Bone Joint Surg 1977;59B(1):15–19.

Chapter Eleven

Arthroscopically Assisted Posterior Cruciate Ligament Reconstruction: Transtibial Tunnel Technique

Gregory C. Fanelli

The incidence of posterior cruciate ligament (PCL) injuries, reported to be from 1% to 40% of acute knee injuries, seems to be patient population dependent—3% in the general population and 38% in regional trauma centers.[1-10] The rate of PCL injuries in athletics is not specifically known. We see a 38% incidence of PCL injuries in acute knee injuries at our regional trauma center; 96.5% are PCL/multiple ligament injuries, and only 3.5% are isolated PCL injuries.[9,10] The most frequently occurring are combined anterior cruciate ligament (ACL)/PCL injuries (45.9%) and combined PCL/posterolateral corner tears (41.2%). We have performed all of our PCL reconstructions using the transtibial tunnel technique. This chapter presents our indications, graft selection, surgical technique, postoperative rehabilitation program, and results of PCL reconstruction using the transtibial tunnel technique.

Mechanism of Injury

The PCL may be injured by abrupt posterior translation of the proximal tibia to the 90-degree flexed knee such as with a dashboard injury, or a fall on a flexed knee with the foot in plantar flexion.[7,11] The PCL can also be injured when excessive tension is produced in the ligament at extremes of flexion or extension—forced hyperflexion, forced hyperextension (knee dislocation), or forced varus or valgus.[7,9,10,12]

Evaluation

Knowing the mechanism of the PCL injury will help to determine what the associated structural injuries may be. The most accurate physical examination test in our hands is the decreased tibial step-off test. When the tibial step-off is decreased by 0 to 5 mm, the injury is grade I; 6 to 10 mm, the injury is grade II; and 11 mm or greater (negative step-off), the injury is grade III. When a high-energy mechanism of injury is present, it is critical to document the status of the neurovascular system and the presence of structural injuries occurring in addition to the PCL tear. Combined PCL/posterolateral instability is frequently misinterpreted as an isolated PCL tear, and failure to address the posterolateral instability at the time of PCL reconstruction will lead to failure of the PCL reconstruction.

Surgical Indications

Posterior cruciate ligament treatment considerations include whether the injury is acute or chronic, of isolated or multiple ligaments, bony avulsion or interstitial disruption, and what the additional structural injuries are. Nonsurgical treatment is utilized for acute isolated PCL injuries (grade I or II) with 10 mm or less pathologic posterior tibial translation tested at 90 degrees of knee flexion, and for asymptomatic chronic isolated PCL injuries. Close follow-up is recommended to detect unrecognized posterolateral instability with progressive functional instability.

Surgical treatment is recommended for acute PCL injuries with insertion-site avulsions, tibial step-off decreased greater than 10 mm, the PCL/multiple ligament–injured knee, and when other structural injuries (ligaments, articular surface, meniscus) are present. Surgical treatment in chronic PCL-injured knees is recommended when an isolated PCL tear becomes symptomatic. Progressive functional instability develops because of unrecognized posterolateral, posteromedial, or anterior instabilities, and early posttraumatic arthrosis develops. The treatment decision in the chronic PCL-injured knee is between ligament reconstruction or high tibial osteotomy.

Surgical Timing

The timing of surgery in acute PCL injuries depends on how and where the PCL is torn, what the associated collateral ligament or ACL injuries are, and certain "treatment modifiers." Insertion-site avulsions, whether bony or soft tissue, should be repaired early if the PCL substance/midbody is intact, as determined by the three-zone arthroscopic evaluation of the PCL (see Chapter 7).[12]

For PCL/posterolateral corner tears, it is recommended to proceed with surgical reconstruction within 2 to 3 weeks, or as soon as is safely possible. This time frame will enable capsular sealing to occur so the PCL reconstruction can be performed as an arthroscopically assisted procedure, and this time frame also facilitates primary repair of the posterolateral structures. At our clinic we recommend augmented posterolateral primary repair in acute posterolateral injuries, since this provides a much stronger posterolateral corner than primary repair alone. We have reported excellent results with PCL/posterolateral surgical reconstruction in the acute injuries 6 to 8 weeks postinjury.[12,13]

Most acute PCL/medial collateral ligament (MCL) tears can be successfully treated with full extension MCL brace treatment for 4 to 6 weeks followed by PCL reconstruction. The degree of medial side damage determines whether the medial ligament complex needs to be addressed surgically. These PCL/MCL injured knees are at particularly high risk for postoperative stiffness after open medial ligament surgery.

The acute ACL/PCL–injured knee is assumed to be a dislocated knee even if the patient did not present with a documented tibiofemoral dislocation. The neurovascular status of the extremity must be documented, and arteriography obtained as indicated. Surgery must proceed immediately in irreducible dislocations. When the tibiofemoral joint reduction can be maintained, then the timing for ACL/PCL reconstruction and collateral ligament surgery is determined by the collateral ligament complex injured and the patient's overall health. We have reported excellent results with delayed reconstruction in acute ACL/PCL–injured knees (see Chapter 15).[12,14–16]

"Surgical timing modifiers" are special considerations and conditions

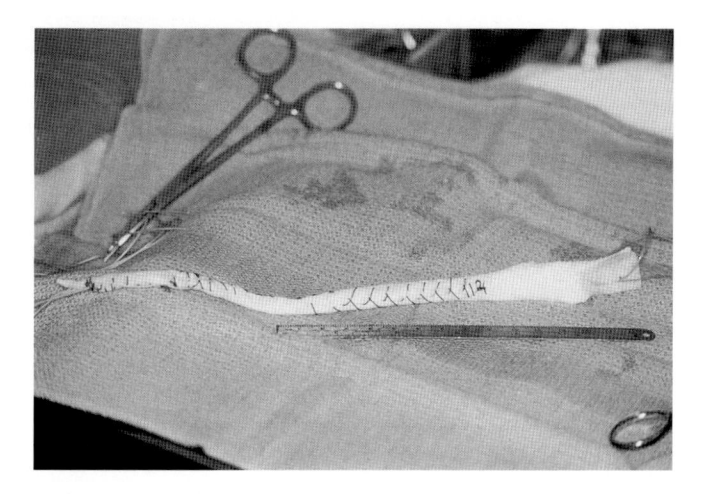

Fig. 11.1. Achilles tendon allograft prepared for posterior cruciate ligament (PCL) reconstruction. (From Fanelli et al.,[15] with permission.)

that will alter the timing of surgery in the acute PCL-injured knee, and they must be considered on an individual basis. These modifiers include the vascular status of the extremity, reduction stability of the tibiofemoral and/or patellofemoral joints, the skin condition around the knee, concomitant multiple system injuries, open vs. closed injuries, fractures, and meniscal and articular surface injuries. The presence of these modifiers may necessitate surgery being performed earlier or later than the operating surgeon would prefer.

Graft Selection

The ideal graft material should be strong, provide secure fixation, be easy to pass, be readily available, and have low donor-site morbidity. The available options in the United States are autograft and allograft sources. Our preferred graft for the PCL is the Achilles tendon allograft because of its large cross-sectional area and strength, absence of donor-site morbidity, and easy passage with secure fixation (Fig. 11.1).

Surgical Technique

PCL Reconstruction Principles

The principles of PCL reconstruction are:

- Identify and treat all pathology.
- Ensure the safety of the procedure.
- Ensure accurate tunnel placement.
- Use anatomic graft insertion sites.
- Utilize strong graft material.
- Minimize graft bending.
- Utilize graft tensioning to restore normal tibial step off.
- Secure the graft fixation.
- Prepare the patient for slow postoperative rehabilitation.

Our preferred approach to PCL reconstruction is a transtibial tunnel arthroscopic technique utilizing gravity fluid inflow. Collateral ligament surgery is performed through a medial or lateral hockey-stick incision as indicated. The posteromedial extracapsular/extraarticular safety incision is used in all PCL reconstructions. Here is an outline of the order of PCL reconstruction, as discussed in the following sections:

- Patient positioning and preparation.
- Posteromedial safety incision.
- Capsule elevation.
- Drill guide positioning.
- Tibial tunnel drilling.
- Femoral tunnel drilling.
- Tunnel preparation and graft passage.
- Graft tensioning and fixation.
- Associated ligament surgery.

Patient Positioning and Preparation

The patient is placed on the operating table in the supine position, and after satisfactory induction of anesthesia, the operative and nonoperative lower extremities are carefully examined. A tourniquet is applied to the upper thigh of the operative extremity, and that extremity is prepped and draped in a sterile fashion. When allograft tissue is used, it is prepared prior to bringing the patient into the operating room. Autograft tissue is harvested prior to beginning the arthroscopic portion of the surgical procedure.

The arthroscopic instruments are inserted with the inflow through the superolateral patellar portal. Instrumentation and visualization are achieved through the inferomedial and inferolateral patellar portals, and are interchanged as necessary. Exploration of the joint consists of evaluation of the patellofemoral joint, the medial and lateral compartments, medial and lateral menisci, the intercondylar notch, and the posteromedial and posterolateral corners.

Posteromedial Safety Incision

The PCL tear is identified, and the intact ACL is confirmed. The residual stump of the PCL is debrided with hand tools and a synovial shaver. An extracapsular/extraarticular posteromedial safety incision is made by cre-

Fig. 11.2. (a) Extraarticular/extracapsular posteromedial safety incision used during PCL reconstruction to confirm instrument position, and to aid in protecting the neurovascular structures. (b) Intraoperative photograph of posteromedial safety incision.

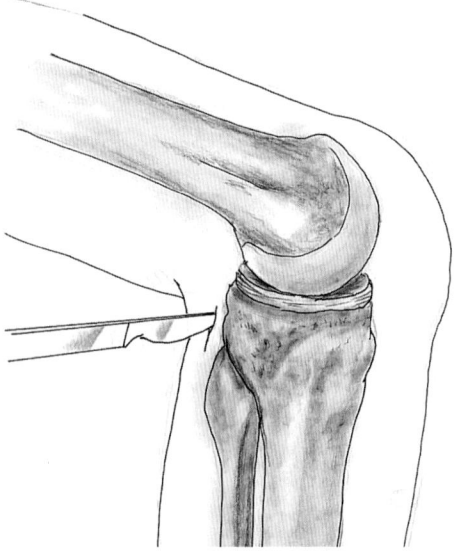

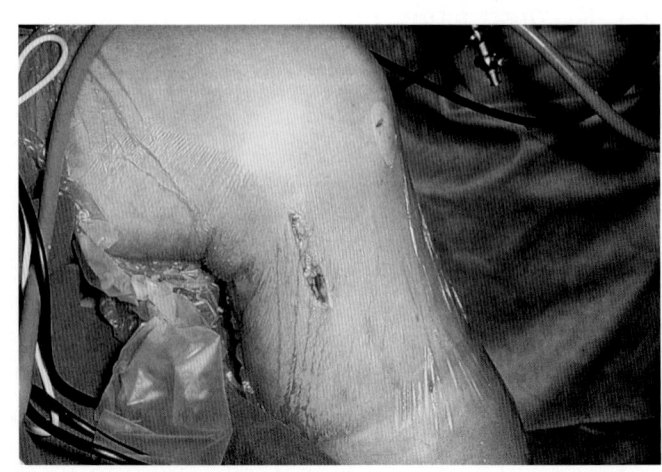

a

b

ating an incision approximately 1.5 to 2.0 cm in length starting at the posteromedial border of the tibia at the level of the joint line and extending distally (Fig. 11.2). Dissection is carried down to the crural fascia, which is incised longitudinally. Care is taken to protect the neurovascular structures. An interval is developed between the medial head of the gastrocnemius muscle posteriorly and the capsule of the knee joint anteriorly. The surgeon's gloved finger is positioned to have the neurovascular structures posterior to the finger, and the knee joint posterior capsule anterior to the finger. This is so that the surgeon can monitor hand tools, shavers, and the PCL/ACL drill guide positioned in the posterior aspect of the knee for safety. This also allows for accurate placement of the tibial tunnel guide wire both in a mediolateral and proximal distal direction.

Capsule Elevation

The curved over-the-top PCL instruments are used to sequentially lyse adhesions in the posterior aspect of the knee, and to elevate the capsule from the tibial ridge posteriorly. This will allow accurate placement of the PCL/ACL drill guide with correct placement of the tibial tunnel (Fig. 11.3).

Fig. 11.3. (a) Curved PCL over-the-top instruments (Arthrotek, Inc., Warsaw, IN). (b) Posterior capsular elevation using curved over-the-top PCL instruments. (c) Model demonstrating posterior capsular elevation using curved over-the-top PCL instrumentation. Note surgeon's finger in extracapsular/extraarticular position to monitor instrument position by feeling the capsule, protecting the neurovascular structures. (d) Intraoperative photograph demonstrating capsular elevation technique using posteromedial safety incision. (From Fanelli et al.,[12] with permission.)

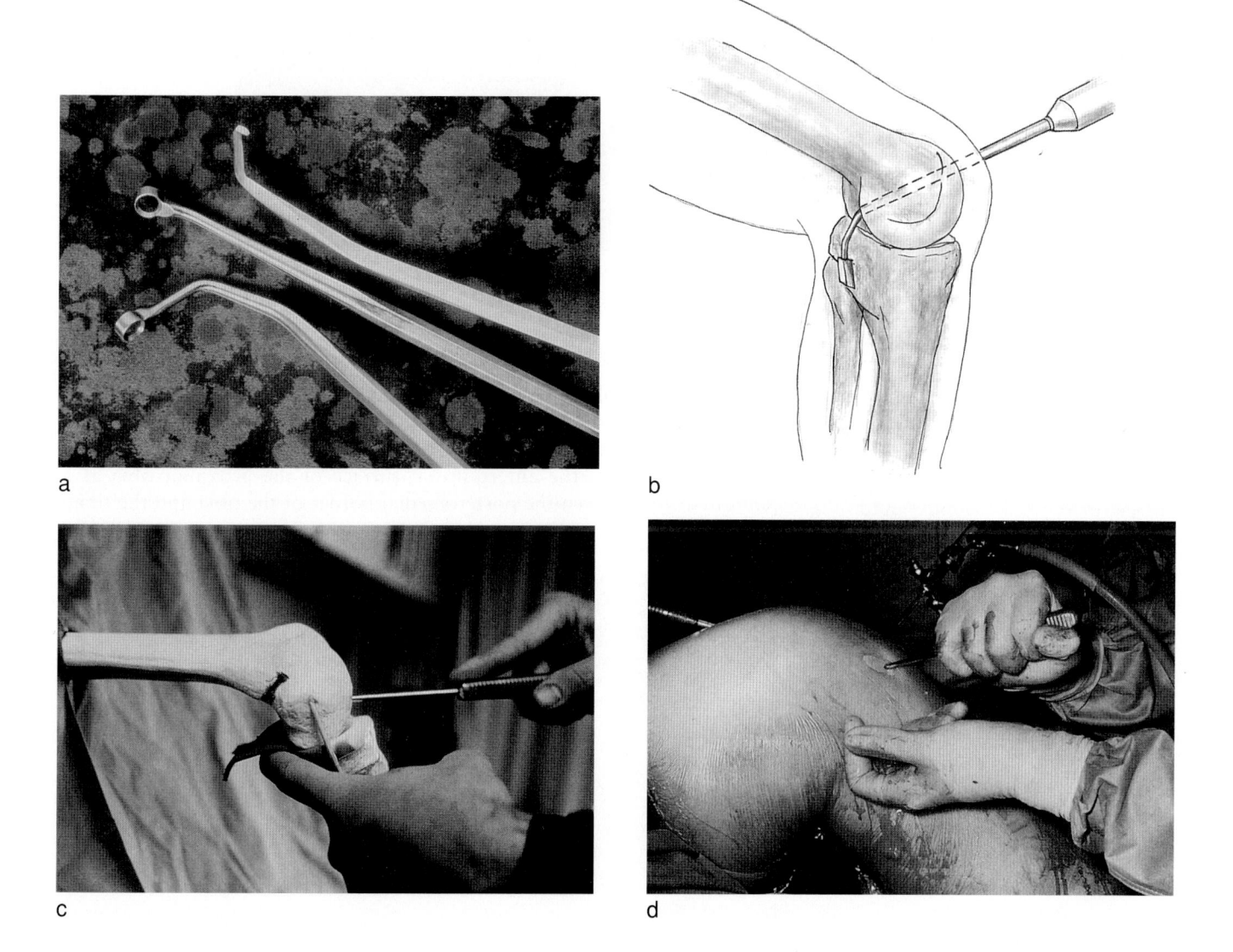

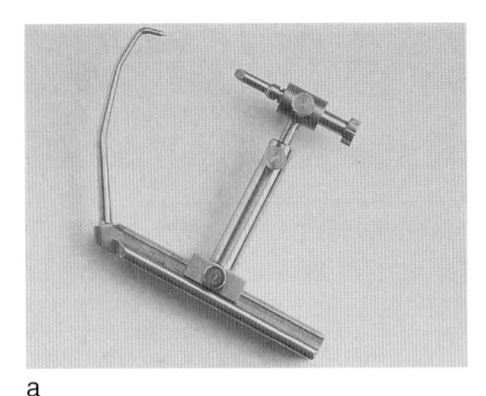

Fig. 11.4. (a) Fanelli PCL/ACL drill guide (Arthrotek, Inc., Warsaw, IN). (b) PCL/ACL drill guide positioned for creation of the tibial tunnel. (c) Intraoperative photograph of PCL/ACL drill guide positioned for tibial tunnel creation. (From Fanelli et al.,[12] with permission.)

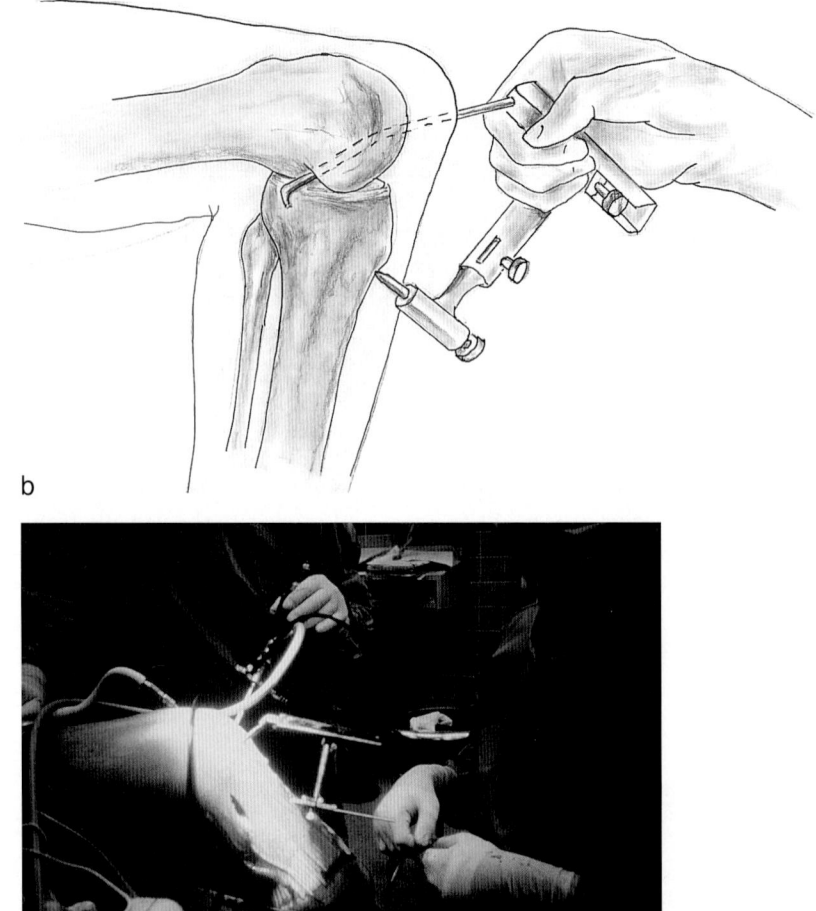

Fig. 11.5. (a) Drawing of tibial tunnel orientation. A properly positioned tibial tunnel allows a smooth pathway of graft material around the posterior aspect of the tibia. The two 45-degree angle turns eliminate the posterior tibial "killer turn." (b) Intraoperative radiograph demonstrating the smooth transitions the PCL graft material will make. (From Fanelli et al.,[13] with permission.)

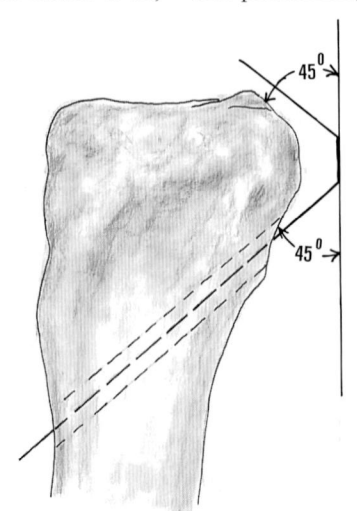

Drill Guide Positioning

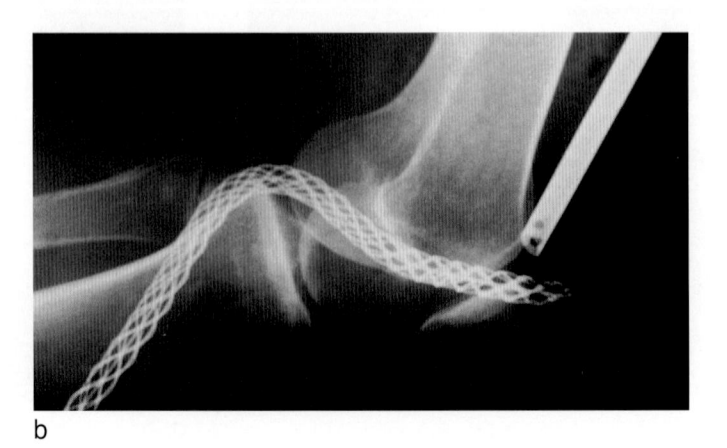

The PCL/ACL drill guide is inserted into the knee joint through the inferomedial patella portal. The tip of the guide is positioned at the inferolateral aspect of the PCL anatomic insertion site. The bullet portion of the drill guide contacts the anteromedial surface of the proximal tibia at a point midway between the posteromedial border of the tibia and the tibial crest anteriorly, approximately 1 cm below the tibial tubercle (Fig. 11.4). This will provide an angle of graft orientation such that the graft will turn two very smooth 45-degree angles on the posterior aspect of the tibia, and will not have an acute 90-degree angle turn, which may cause pressure necrosis of the graft (Fig. 11.5).

The tip of the guide in the posterior aspect of the tibia is confirmed with the surgeon's finger through the extraarticular/extracapsular posteromedial safety incision. When the PCL/ACL drill guide is positioned in the desired area, a blunt spade-tipped guide wire is drilled from anterior to posterior. The surgeon's finger confirms the position of the guide wire through the posteromedial safety incision. This is the author's preferred method. Intraoperative anteroposterior and lateral radiographs and arthroscopic visualization may also be utilized; however, it is the author's opinion that the safest and most accurate method for drill guide placement, guide wire drilling, and tunnel reaming is when the surgeon directly confirms the position of these instruments through the posteromedial safety incision (Fig. 11.6). This is because the shape of the posterior aspect of the proximal tibia may allow guide wire and reamer posterior cortical penetration to occur before these instruments are visualized arthroscopically, and because radiography may endanger the neurovascular structures[17] (Fig. 11.7).

Fig. 11.6. (a) Surgeon's finger confirming the position of the blunt spade-tipped guide wire through the capsule. The surgeon's finger is in the extracapsular/extraarticular posteromedial safety incision. (b) Intraoperative photograph demonstrating the use of the posteromedial safety incision to confirm guide wire and reamer placement. (c,d) Intraoperative radiographs demonstrating proper placement of tibial tunnel guide wire. Guide wire exits at the inferior and lateral aspect of the PCL tibial anatomic insertion site.

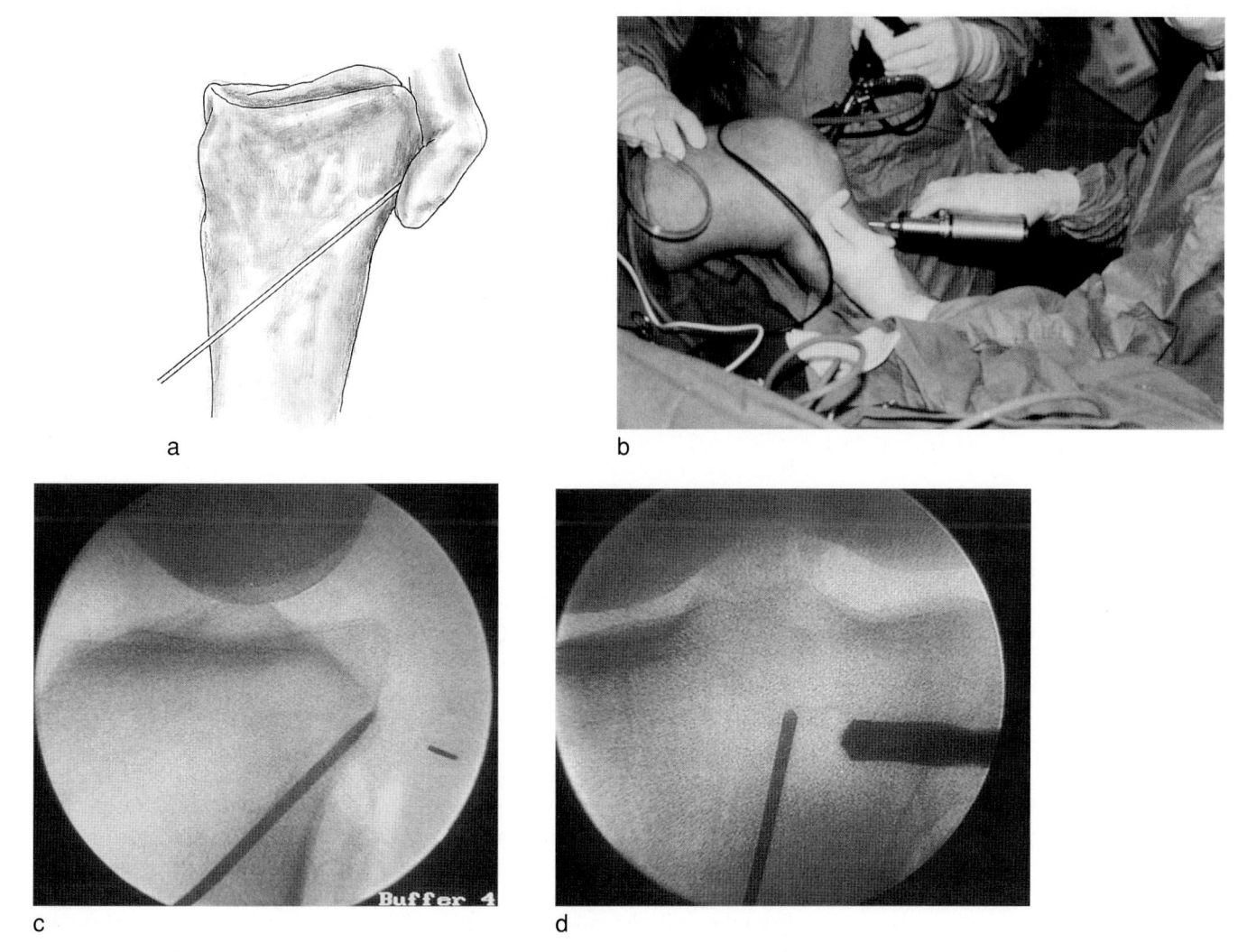

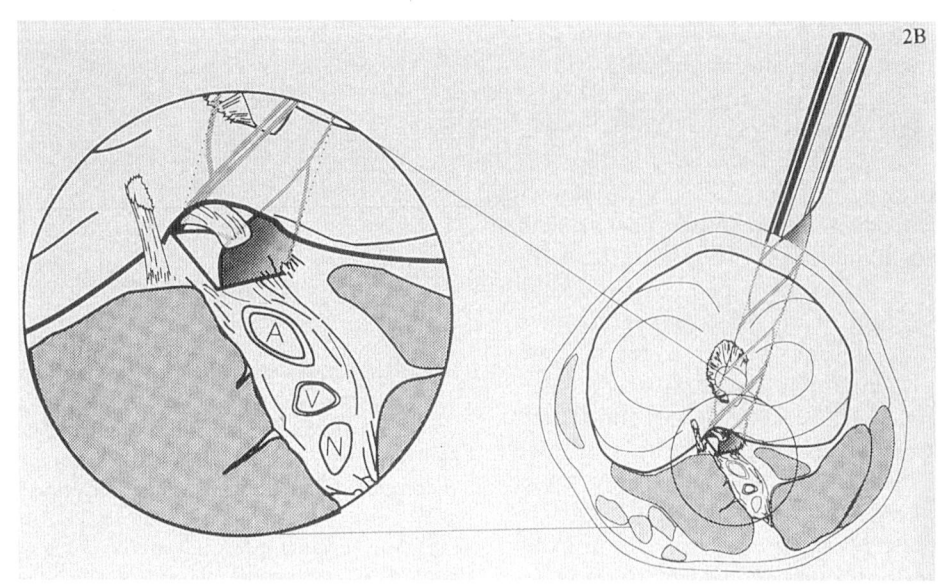

Fig. 11.7. (a) Axial section at the level of the PCL insertion demonstrating the relationship of the neurovascular structures and their proximity to the exit point of the drill bit. (b) Sagittal section demonstrating the relationship of the neurovascular structures to the drill bit that has prematurely exited the posterior cortex. (From Jackson et al.,[17] with permission.)

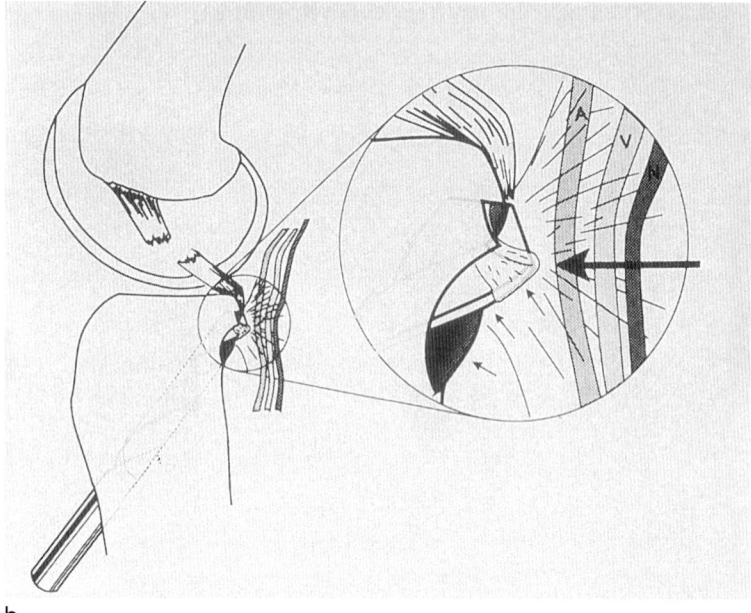

Tibial Tunnel Drilling

The appropriately sized standard cannulated reamer is used to create the tibial tunnel. The author's preferred method is to monitor the position of the guide wire and the reamer with the surgeon's finger in the extracapsular/extraarticular posteromedial safety incision for the reasons outlined above. Other methods include using the PCL closed curette to cup the tip of the guide wire while visualizing with the arthroscope in the posteromedial portal. The author recommends always using the posteromedial safety incision, and feeling the instruments, guide wire, and reamers with the surgeon's finger. The guide wire is advanced to the posterior cortex of the tibia just under the periosteum, and the position confirmed with the surgeon's finger. The standard cannulated reamer is advanced over the guide wire until it comes to the posterior cortex of the tibia. The chuck is disengaged from the drill, the guide wire removed, and completion of the tibial tunnel

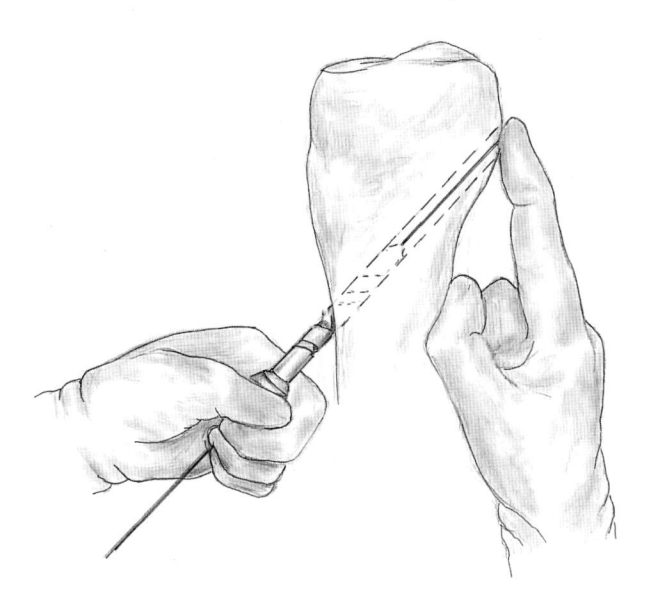

11. Arthroscopically Assisted PCL Reconstruction: Transtibial Tunnel Technique

is performed by hand (Fig. 11.8). The tunnel edges are chamfered and rasped with the PCL/ACL system rasp (Fig. 11.9).

Femoral Tunnel Drilling

The PCL/ACL drill guide is positioned to create the femoral tunnel. The arm of the guide is introduced through the inferomedial patella portal, and is positioned so that the guide wire will exit through the center of the stump of the anterolateral bundle of the PCL (Fig. 11.10). The bullet of the PCL/ACL drill guide contacts the medial surface of the distal femur halfway between the medial femoral condyle articular surface and the medial epicondyle of the femur. We prefer to have a minimum 1- to 2-cm bone bridge between the PCL femoral tunnel and the distal femoral articular surface. This will decrease the chances of subchondral collapse and/or avascular necrosis.[18] The blunt spade-tipped guide wire is drilled through the

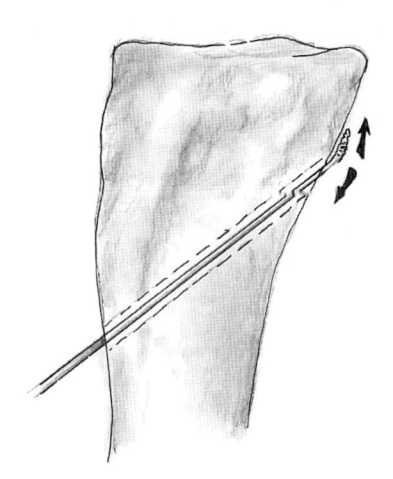

Fig. 11.8. PCL tibial tunnel drilling. The final stages of the PCL tibial tunnel are completed by hand for an additional margin of safety.

Fig. 11.9. PCL tibial tunnel edges are chamfered and rasped to eliminate sharp edges.

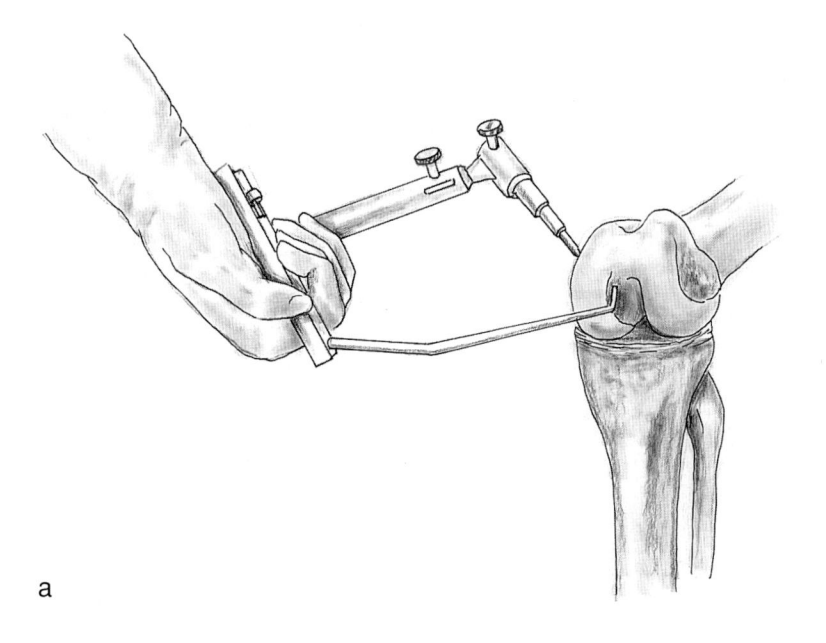

Fig. 11.10. (a) PCL/ACL drill guide positioned for creation of the femoral tunnel.

a

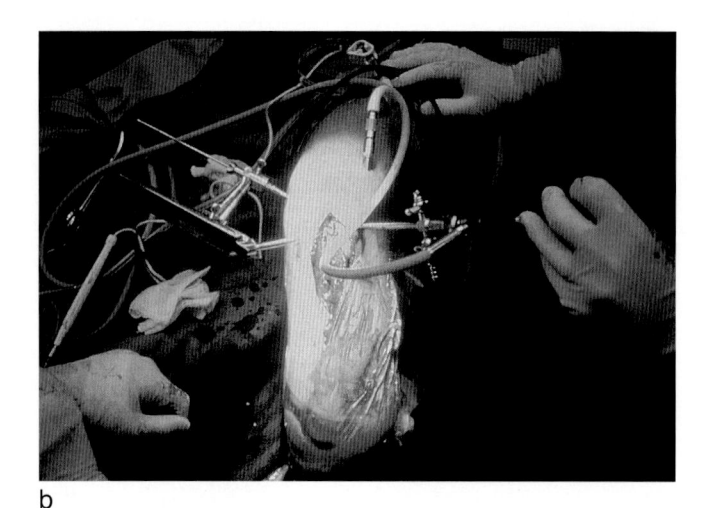

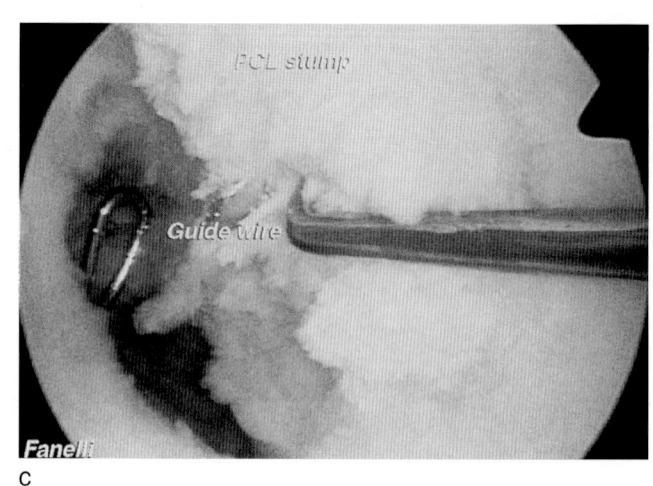

Fig. 11.10 (continued). (b) Intraoperative photograph demonstrating position of PCL/ACL drill guide for creation of femoral tunnel. (From Fanelli et al.,[12] with permission.) (c) Arthroscopic view of guide wire placement for PCL femoral tunnel creation. Guide wire exits through the center of the anterolateral fiber bundle region of the posterior cruciate ligament.

guide, and just as it begins to emerge through the center of the stump of the PCL anterolateral bundle, the drill guide is disengaged. The accuracy of the guide wire placement is confirmed arthroscopically with probing and visualization. Care must be taken to ensure the patellofemoral joint has not been violated by arthroscopically examining the patellofemoral joint prior to drilling.

The appropriately sized standard cannulated reamer is used to create the femoral tunnel. A curette is used to cap the tip of the guide wire to prevent inadvertent guide wire advancement into the joint with subsequent ACL or articular surface damage. As the reamer is about to penetrate into the joint, the reamer is disengaged from the drill, and the final reaming is completed by hand (Fig. 11.11). This adds an additional margin of safety, minimizing the potential for ACL and articular surface damage. The reaming debris is evacuated with a synovial shaver to minimize fat pad inflammatory response with subsequent risk of arthrofibrosis. The tunnel edges are chamfered and rasped.

Tunnel Preparation and Graft Passage

A 16-gauge curved passing wire is introduced through the tibial tunnel, into the joint, and retrieved through the femoral tunnel to facilitate graft passage (Figs. 11.12 and 11.13). A 7.9-mm flexible tunnel rasp/smoother is attached to the passing wire and pulled into position. The flexible rasp is gently moved back and forth, chamfering and rasping the tunnel edges at 90, 60, 30, and 0 degrees of knee flexion (Fig. 11.14). Care is taken to avoid excessive rasp pressure, which would alter tunnel configuration.

Traction sutures of the graft material are attached to the loop of the flexible rasp, and the graft is pulled into position (Fig. 11.15). The graft is secured on the femoral or tibial side using interference screw fixation, press fit fixation, and post and washer or spiked ligament washer backup fixation as indicated (Fig. 11.16). With the PCL graft secured on one side, additional ligament injuries are addressed at this time.

Graft Tensioning and Fixation

Fig. 11.11. Final reaming of the femoral tunnel by hand. This minimizes the chance of ACL and/or articular surface injuries due to inadvertent drill advancement.

The PCL graft is in place and has been secured on either the femoral or the tibial end with bioabsorbable interference screw primary fixation, and suture and post or spiked ligament washer backup fixation. Traction is placed on the PCL graft traction sutures, and the knee is cycled through 25 full flexion-extension cycles to allow graft settling and graft pretensioning. The knee is placed in approximately 70 degrees of flexion, a firm anterior drawer force is applied to restore the normal tibial step-off, and traction is placed on the graft traction sutures setting the tension in the graft. Final fixation is achieved with interference screw fixation, post and suture, or spiked ligament washer as indicated (Fig. 11.17).

When final fixation of the PCL reconstruction is complete, and posterolateral and/or posteromedial reconstruction has been performed, the tension is set and final fixation of the posterolateral and/or posteromedial corner reconstruction is performed with the knee in 30 degrees of flexion. In cases of combined ACL/PCL reconstruction, the ACL reconstruction is secured after final fixation of the PCL and posterolateral complex reconstructions (discussed in detail in Chapter 15 and ref. 14).

Associated Ligament Surgery

Our preferred technique for posterolateral reconstruction is the split biceps tendon transfer to the lateral femoral epicondyle.[15,16] The requirements for this procedure include an intact proximal tibiofibular joint, intact posterolateral capsular attachments to the common biceps tendon, and an intact biceps femoris tendon insertion into the fibular head. This technique restores the function of the popliteofibular ligament and lateral collateral ligament, tightens the posterolateral capsule, and provides a post of strong autogenous tissue to reinforce the posterolateral corner.

A lateral hockey-stick incision is made. The peroneal nerve is dissected free and protected throughout the procedure. The long head and common biceps femoris tendon is isolated, and the anterior two-thirds is separated from the short head muscle. The tendon is detached proximal and left attached distally to its anatomic insertion site on the fibular head. The strip of biceps tendon should be 12 to 14 cm long. The iliotibial band is incised in line with its fibers, and the fibular collateral ligament and popliteus ten-

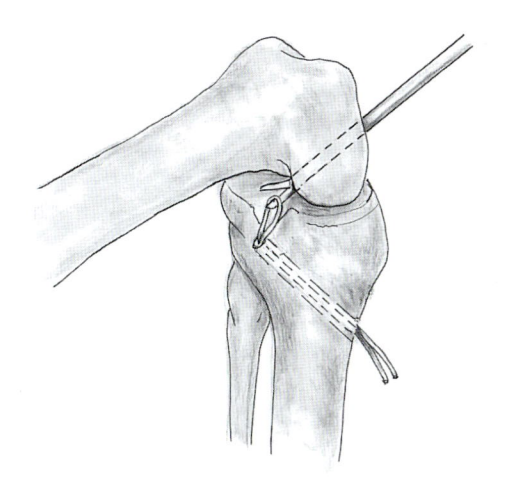

Fig. 11.12. A 16-gauge looped wire facilitates graft passage.

Fig. 11.13. (a) A 16-gauge passing wire loop is passed through the tibial tunnel, into the joint, and retrieved through the femoral tunnel. The flexible rasp is attached with suture to the passing wire loop. (b) Intraoperative arthroscopic view demonstrating position of 16-gauge wire loop to facilitate graft passage.

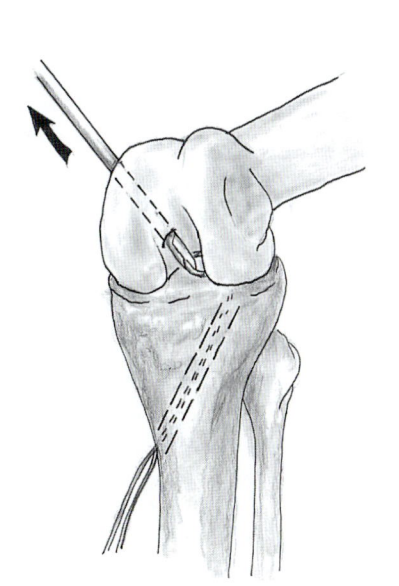

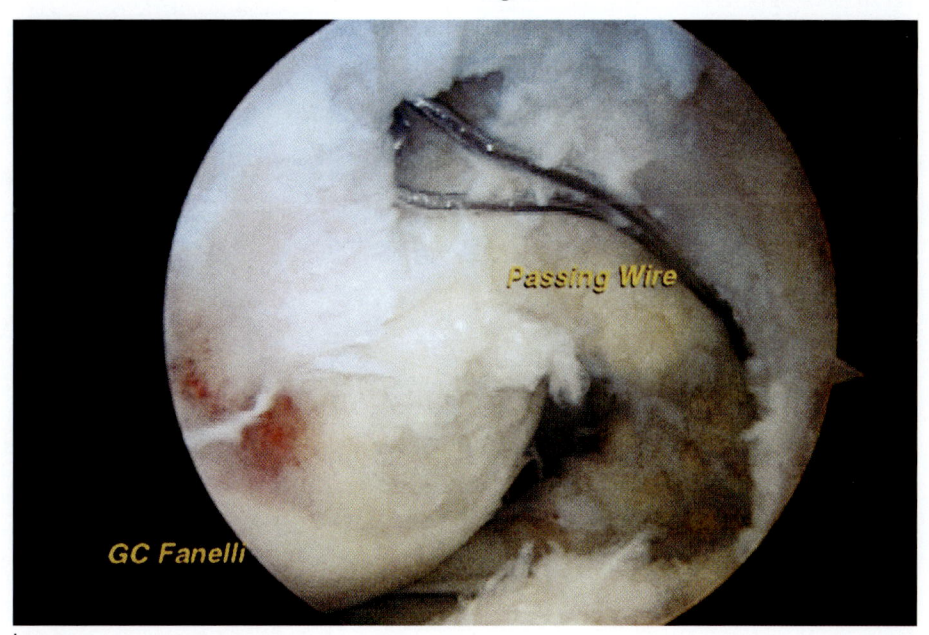

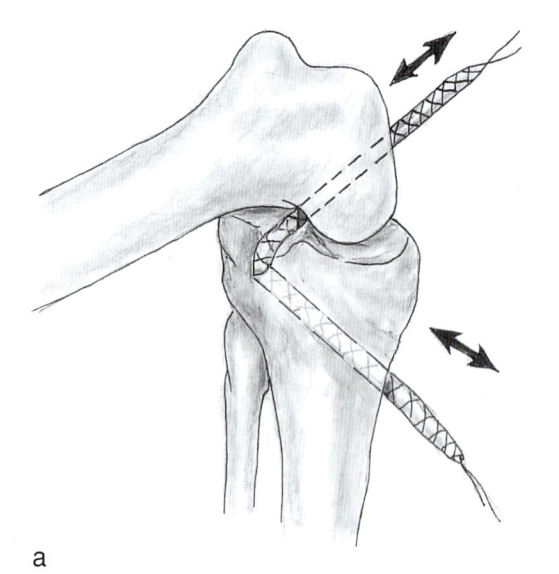

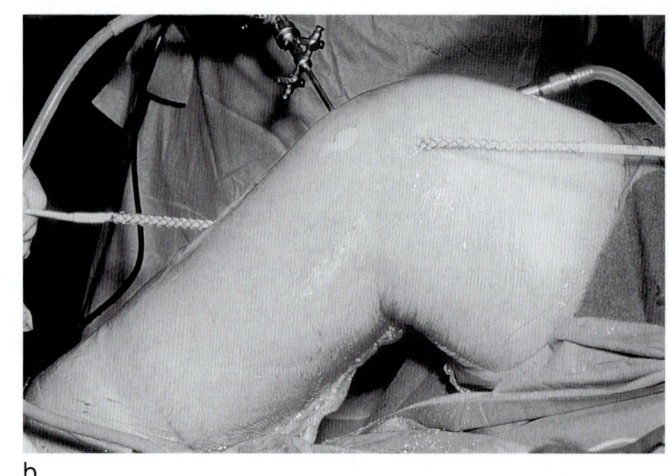

Fig. 11.14. (a) Flexible rasp in position to chamfer tunnel edges at 90, 60, 30, and 0 degrees of knee flexion. (b) Intraoperative photograph. Flexible rasp smooths tunnel edges and removes excess soft tissue from tunnel openings. Flexible rasp also helps to pull tendon graft material into position. (From Fanelli et al.,[12] with permission.)

don are exposed. A drill hole is made 1 cm anterior to the fibular collateral ligament femoral insertion.[19] A longitudinal incision is made in the lateral capsule just posterior to the fibular collateral ligament. The split biceps tendon is passed medial to the iliotibial band, and secured to the lateral femoral epicondylar region with a screw and spiked ligament washer at the above-mentioned point. The residual tail of the transferred split biceps tendon is passed medial to the iliotibial band, and secured to the fibular head. The posterolateral capsule that had been previously incised is then shifted and sewn into the strut of transferred biceps tendon to eliminate posterolateral capsular redundancy (Fig. 11.18).

Posteromedial and medial reconstructions are performed through a medial hockey-stick incision (Fig. 11.19). Care is taken to maintain adequate skin bridges between incisions. The superficial MCL is exposed, and a longitudinal incision is made just posterior to its posterior border. Care is taken not to damage the medial meniscus during the capsular incision. The

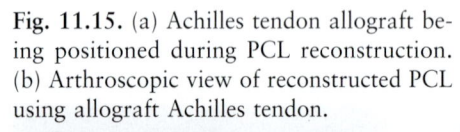

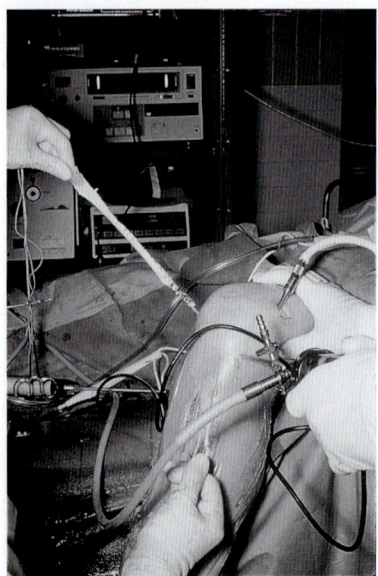

Fig. 11.15. (a) Achilles tendon allograft being positioned during PCL reconstruction. (b) Arthroscopic view of reconstructed PCL using allograft Achilles tendon.

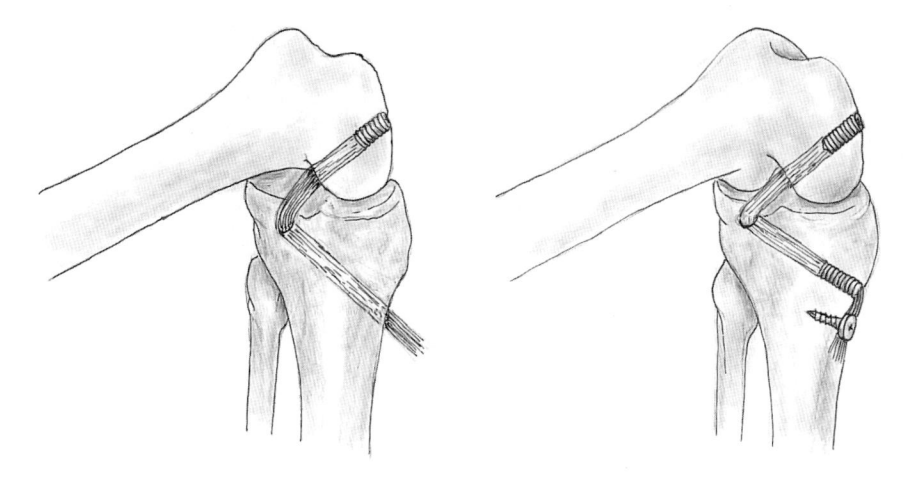

Fig. 11.16. Femoral-side PCL graft fixation using interference screw.

Fig. 11.17. Final fixation of PCL reconstruction demonstrating interference screw fixation on the femoral side, and interference screw and spike washer fixation on the tibial side.

interval between the posteromedial capsule and the medial meniscus is developed. The posteromedial capsule is shifted anterosuperiorly. The medial meniscus is repaired to the new capsular position, and the shifted capsule is sewn into the MCL. When superficial MCL reconstruction is indicated, it is performed with allograft tissue or semitendinosus free autograft. The graft material is attached to the anatomic insertion sites of the superficial MCL on the femur and the tibia. The posteromedial capsular advancement is performed, and sewn into the newly reconstructed MCL.

Postoperative Rehabilitation

Postoperative rehabilitation proceeds slowly. The knee is locked in full extension in a postoperative long leg brace for 6 weeks. The patient is non–weight bearing with crutches for this 6-week period. In postoperative weeks 7 through 10, the brace is unlocked, and the patient progresses to full range of motion and full weight bearing. When weight bearing is full and there is good control of the surgical leg, the crutches are discontinued. Progressive range of motion occurs slowly, and strength training entails closed kinetic chain exercises. Return to sports and heavy work occurs at the end of postoperative month 9, provided adequate range of motion and strength are present. It should be noted that a loss of approximately 10 to 15 degrees of terminal flexion can be expected in these complex knee lig-

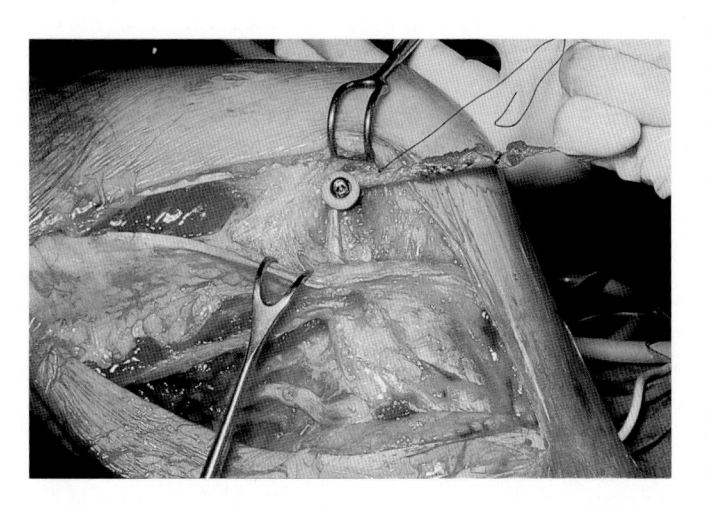

Fig. 11.18. The posterolateral reconstruction is performed using a split biceps tendon transfer combined with a posterolateral capsular shift. This technique re-creates the function of the popliteofibular ligament, and lateral collateral ligament, and eliminates the posterolateral capsular redundancy. The anterior two-thirds of the long head and common biceps femoris tendon are isolated from the short head muscle. The tendon is detached proximally, and left attached distally to its anatomic insertion site. The peroneal nerve is protected. The split biceps tendon is passed medial to the iliotibial band, and secured to the lateral femoral epicondyle approximately 1 cm anterior to the fibular collateral ligament femoral insertion using a cancellous screw and spiked ligament washer.

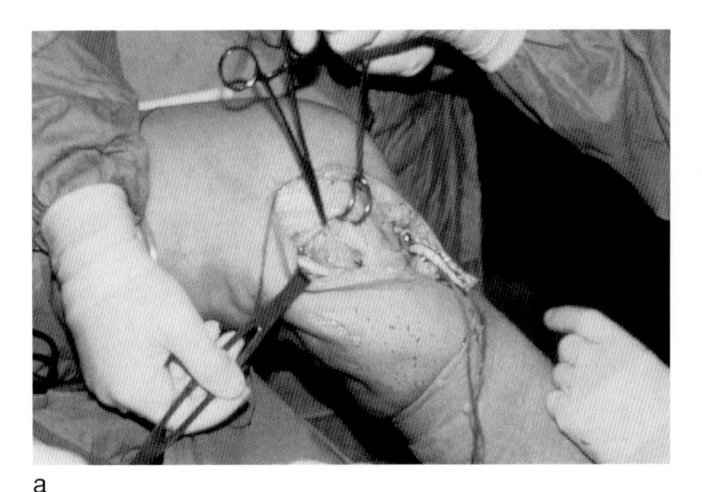

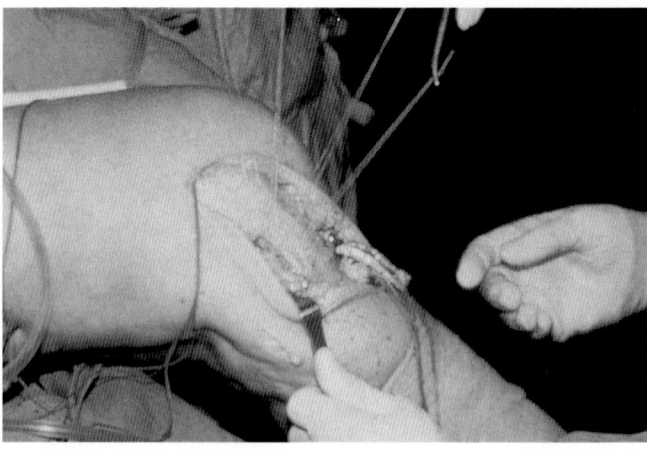

Fig. 11.19. Posteromedial reconstruction using posteromedial capsular shift procedure. (From Fanelli and Feldman,[16] with permission.)

ament reconstructions. This does not cause a functional problem for these patients, and is not a cause for alarm. Here is a summary of the postoperative rehabilitation plan:

0–6 weeks:	Long leg brace locked in full extension.
7–10 weeks:	Begin range of motion; progress to full weight bearing with crutches.
11–24 weeks:	Progressive range of motion; initial strength training.
25–36 weeks:	Advanced strength training.
37 weeks:	Return to sports and heavy labor if strength and range of motion are appropriate.

Fanelli Sports Injury Clinic Results

We have previously published the results of our arthroscopically assisted combined ACL/PCL and PCL/posterolateral complex reconstructions using the reconstructive technique described in this chapter.[12–14] Our results indicated that we were able to restore functional stability in all knees, and all patients in these series were able to return to their preinjury level of activity. In the combined ACL/PCL series,[14] we were able to predictably restore normal tibial step-off and posterior drawer in 9 of 20 of the reconstructed knees, and to achieve grade I posterior drawer with 5-mm decreased tibial step-off in 11 of 20 reconstructed knees. All knees had grade III posterior drawer tests preoperatively, and negative step-off. Fifteen of 20 knees had normal postoperative Lachman tests improved from grade III preoperative Lachman tests, and 20 of 20 knees had elimination of the preoperative pivot shift. Posterolateral and medial instability was corrected in all cases.

In the combined PCL/posterolateral complex reconstruction series,[13] we were able to predictably restore normal tibial step-off and posterior drawer tests in 10 of 21 patients (48%), and grade I posterior drawer with 5-mm decreased tibial step-off in 10 of 21 knees (48%). Arthrometer measurements showed a statistically significant improvement from preoperative to postoperative measurements using the PCL screen and corrected posterior measurements.[8] All cases of posterolateral instability were corrected.

We have recently presented our results of 3- to 8-year follow-up of PCL reconstructions analyzed with stress radiography demonstrating the longevity of these reconstructions[20]; 31 arthroscopically assisted PCL reconstructions were performed in PCL injured knees with grade III posterior drawer tests and negative tibial step-off using the transtibial tunnel PCL reconstruction technique. There were 16 PCL/posterolateral and 15 ACL/PCL reconstructions. Mean follow-up was 56.1 months, with a range of 36 to 96 months. In the PCL/PLC group there were 11 men and five women, and six right, 10 left, seven acute, and nine chronic knees. In the ACL/PCL group there were 11 men and four women, and six right, nine left, nine acute, and six chronic knees; 25 Achilles tendon allografts, five bone–patellar tendon–bone autografts, and one quadriceps tendon autograft were used for the PCL reconstructions. All patients were evaluated with lateral stress radiography with the knee in 90 degrees of flexion using the Telos device with 32 lb of posterior directed force applied at the level of the tibial tubercle. Posterior tibial displacement was measured on the lateral radiograph in millimeters. The reconstructed knee was compared to the normal knee, and the difference in posterior tibial displacement was compared between the surgical and normal knees. Results were reported as the difference measured in millimeters between the surgical and normal knees.

The mean postreconstruction side-to-side differences between the surgical and normal knees in the PCL/PLC group were 3.9 mm (range 0 to 7 mm), and in the ACL/PCL group 3.5 mm (range 0 to 9 mm). The mean side-to-side differences in the PCL/PLC group for allograft and autograft were 4.1 mm (0 to 7 mm) for allografts, and 3.3 mm (2 to 5 mm) for autografts. Postsurgical mean side-to-side differences in the PCL/PLC group for acute and chronic knees were 4.4 mm (1 to 7 mm) for acute, and 3.5 mm (0 to 6 mm) for chronic injuries. The mean side-to-side differences in the ACL/PCL group for allograft and autograft were 3.0 mm (0 to 6 mm) for allografts, and 5.3 mm (3 to 9 mm) for autografts. Postsurgical mean side-to-side differences in the ACL/PCL group for acute and chronic knees were 4.0 mm (0 to 9 mm) for acute, and 2.7 mm (0 to 6 mm) for chronic injuries.

We concluded that arthroscopic PCL reconstruction using the transtibial tunnel technique is a predictable and reproducible procedure. Analysis with Telos stress radiography demonstrated side-to-side differences of 0 to 3 mm in 45.2% ($n = 14$) of knees, 4 to 5 mm in 32.3% (10) of knees, and 6 to 7 mm in 19.4% (6) of knees in this series. We were able to restore a normal posterior drawer test and normal tibial step-off on clinical exam in 45.2% of the PCL reconstructed knees, and to restore a grade I posterior drawer and 5 mm decreased tibial step off in 51.6% of the knees in this series. Preoperatively, all knees had a grade III posterior drawer, and negative tibial step-off. There appeared to be no significant difference between acute and chronic reconstructions, and between autograft and allograft reconstructions.

Conclusion

Posterior cruciate ligament injuries occur more frequently in trauma patients than in athletic injury patients, and the isolated PCL tear is relatively rare compared to the PCL/multiple ligament–injured knee in our clinical setting. Careful clinical examination is essential to determine the associated structural injuries occurring with PCL tears, especially posterolateral instability. The arthroscopically assisted transtibial tunnel technique using strong graft material, reproducing the anterolateral component of the PCL, using the posteromedial safety incision, and requiring a slow deliberate postoperative rehabilitation program, in our experience, provides reproducible documented results, with a low complication rate.

References

1. Parolie JM, Bergfeld JA. Long term results of nonoperative treatment of isolated posterior cruciate ligament injuries in the athlete. Am J Sports Med 1986;14:35–38.
2. Fanelli GC. PCL tears—who needs surgery? Paper presented at the AANA Annual Meeting, Palm Desert, CA, 1993.
3. Torg JS, Barton JM. Natural history of the posterior cruciate deficient knee. Clin Orthop 1989;246:208–216.
4. Dandy DJ, Pusey RJ. The long term results of unrepaired tears of the posterior cruciate ligament. J Bone Joint Surg 1982;64B:92–94.
5. Keller PM, Shelbourne KD, McCarroll JR, Rettig AC. Nonoperatively treated isolated posterior cruciate ligament injuries. Am J Sports Med 1993;12:132–136.
6. Trickey EL. Injuries to the posterior cruciate ligament. Clin Orthop 1980;147:76–81.
7. Johnson JC, Back BR. Current concepts review, posterior cruciate ligament. Am J Knee Surg 1990;3:143–153.
8. Daniel DM, Akeson W, O'Conner J, eds. Knee Ligaments—Structure, Function, Injury, and Repair. New York: Raven Press, 1990.
9. Fanelli GC. PCL injuries in trauma patients. Arthroscopy 1993;9:291–294.
10. Fanelli GC, Edson CJ. PCL injuries in trauma patients. Part II. Arthroscopy 1995;11:526–529.
11. Clancy WG. Repair and reconstruction of the posterior cruciate ligament. In: Chapman M, ed. Operative Orthopaedics. Philadelphia: JB Lippincott, 1988:1651–1665.
12. Fanelli GC, Giannotti BF, Edson CJ. Current concepts review. The posterior cruciate ligament arthroscopic evaluation and treatment. Arthroscopy 1994;10(6):673–688.
13. Fanelli GC, Giannotti BF, Edson CJ. Arthroscopically assisted combined posterior cruciate ligament/posterior lateral complex reconstruction. Arthroscopy 1996;12(5):521–530.
14. Fanelli GC, Gianotti BF, Edson CJ. Arthroscopically assisted combined anterior and posterior cruciate ligament reconstruction. Arthroscopy 1996;12(1):5–14.
15. Fanelli GC, Feldmann DD. Management of combined anterior cruciate ligament/posterior cruciate ligament/posterolateral complex injuries of the knee. Oper Tech Sports Med 1999;7(3):143–149.
16. Fanelli GC, Feldmann DD. The dislocated/multiple ligament injured knee. Oper Tech Orthop 1999;9(4):1–12.
17. Jackson DW, Proctor CS, Simon TM. Arthroscopic assisted PCL reconstruction: a technical note on potential neurovascular injury related to drill bit configuration. Arthroscopy 1993;9(2):224–227.
18. Athanasian EA, Wickiewicz TL, Warren RF. Osteonecrosis of the femoral condyle after arthroscopic reconstruction of a cruciate ligament. J Bone Joint Surg 1995;77A(9):1418–1422.
19. Wascher DC, Grauer JD, Markoff KL. Biceps tendon tenodesis for posterolateral instability of the knee. An in vitro study. Am J Sports Med 1993;21(3):400–406.
20. Fanelli GC, Maish D, Edson CJ. Stress radiographic analysis of arthroscopically assisted PCL reconstructions: 3 to 8 year follow-up. Arthroscopy 1999;15(5):567.

Chapter Twelve

Arthroscopically Assisted Posterior Cruciate Ligament Reconstruction: Tibial Inlay Technique

Richard D. Parker, John A. Bergfeld, David R. McAllister, and Gary J. Calabrese

Posterior cruciate ligament (PCL) injuries of the knee, though less common than anterior cruciate ligament (ACL) injuries, are more prevalent than once believed, representing up to 20% of significant knee ligament injuries.[1] Many of these injuries, especially if isolated, are not diagnosed initially, and surface clinically at a later date or coincidentally during routine physical examination. It has been estimated that every year at the combined National Football League's physical examination of potential draftees, ~2% of the athletes had sustained an isolated PCL injury at some point in their career.[2] Interestingly, these athletes are usually minimally symptomatic and are functioning at an elite level.

Posterior cruciate ligament injuries are divided into two main categories: (1) isolated injury to the PCL, and (2) injury to the PCL combined with injury to other capsuloligamentous structures about the knee. Treatment of PCL injuries is dependent on several factors including degree of injury, associated injuries, occupation, activity level, symptoms, and sports participation. Though most authors agree that operative treatment of combined PCL with other capsuloligamentous injuries is preferable, there is controversy surrounding the treatment of the isolated PCL. Studies documenting the deterioration of clinical outcome with chronic PCL injuries advocate PCL reconstruction.[3,4] Other studies evaluating isolated PCL injuries have shown that the majority of these patients can be treated with rehabilitation with good short-, intermediate-, and long-term functional outcome.[5,6] Shelbourne and Patel[7] presented their attempt at a natural history study of the nonoperatively treated isolated PCL and found that PCL laxity and subjective knee scores did not deteriorate with time. As a result of these studies,[5-7] and the fact that no study has shown that PCL reconstruction will prevent the sequelae of the PCL-deficient knee, we believe nonoperative treatment for the isolated PCL injury is preferred. Surgical treatment is indicated for combined injuries to the PCL with capsuloligamentous structures and chronic symptomatic isolated grade III PCL injuries.

Intraarticular PCL reconstruction varies from surgeon to surgeon with respect to indications, tissue, technique, treatment of associated injuries, and postoperative rehabilitation. This chapter presents the specifics of our physical examination, treatment (nonoperative and operative) indications, our preferred surgical reconstructive technique with scientific rationale, and our postoperative rehabilitation guidelines.

Physical Examination

Physical examination of the PCL-injured knee is critically important in determining the appropriate form of treatment. In addition to the history, the physical examination differentiates between the isolated PCL injury and the combined PCL with associated capsuloligamentous injury. As with any knee condition, a complete physical examination of the knee should be performed. There are specific physical examination tests for PCL and associated capsuloligamentous injuries.

General Appearance of the Knee

Isolated PCL injury to the knee usually presents with mild to moderate effusion and a complaint of anterior and posterior soreness. On the other hand, the combined PCL and capsuloligamentous injured knee will typically present with a moderate to tense effusion, peripheral ecchymosis, peripheral edema, and even lacerations or abrasions. It is important to consider the possibility of a traumatic knee dislocation if the patient presents with all of the above appearance characteristics.

As in every physical examination, the joint above (hip) and below (ankle) should be examined. In addition, a complete neurovascular examination should be performed, especially if the patient presents with a potential knee dislocation. If possible, lower-extremity skeletal alignment is assessed especially in the chronic combined PCL and posterolateral corner injury. In this combination, a varus alignment of the lower extremity must be recognized and treated to ensure a satisfactory outcome. Grossly, an office gait analysis should be performed to assess for a varus thrust.

Posterior Drawer (Neutral/Internal/External Rotation) Test

Maximum posterior translation of the tibia on the femur with PCL deficiency occurs at between 70 and 90 degrees of knee flexion.[8] This biomechanical fact makes the posterior drawer at 90 degrees a very accurate test for the integrity of the PCL.[8] The posterior drawer at 90 degrees is performed with the tibia in neutral, internal, and external rotation. It is important to assess the medial joint line position for normal step-off of the tibia on the femur. If this is not present, then an anterior drawer is performed to reduce the posterior sagging tibia to a more appropriate starting point. Clancy et al.[3,9] postulated that internal rotation tightens the meniscofemoral ligaments acting as a secondary stabilizer to posterior translation of the knee, thus decreasing the amount of posterior translation compared to neutral rotation. However, Ritchie and Bergfeld et al.[2] reported in a selective cutting study that the meniscofemoral ligaments are not responsible for the decrease in posterior translation with internal tibial, but rather it is actually the superficial medial collateral ligament (MCL) that is responsible.

Parolie and Bergfeld[5] have found that clinically, the posterior drawer test at 90 degrees in neutral and internal rotation (Fig. 12.1) helps to differentiate between an isolated PCL and combined PCL and capsuloligamentous injury. An isolated PCL injury will translate up to an additional 10 ($\pm$2) mm (3+) in neutral tibial rotation compared to the contralateral intact knee, whereas when the tibia is internally rotated, the posterior drawer at 90 degrees decreases 4 ($\pm$2) mm (1+/2+). On the other hand, the posterior drawer in neutral rotation will be consistently >10 mm in the combined PCL and capsuloligamentous injury and will not decrease signifi-

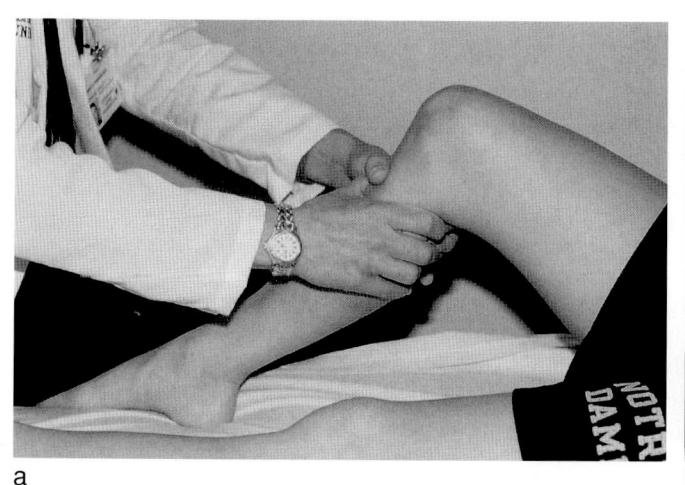

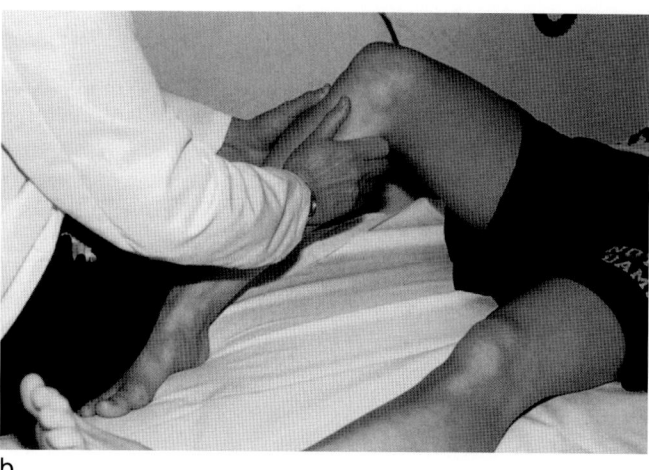

Fig. 12.1. The posterior drawer is performed with the knee in 90 degrees of knee flexion in both (a) neutral and (b) internal rotation.

cantly with internal tibial rotation.[2] Thus, we advocate the posterior drawer being performed at 90 degrees in neutral and internal rotation, and we feel it helps differentiate clinically between an isolated PCL and a combined PCL and capsuloligamentous injury.

The posterior drawer test at 90 degrees in external rotation helps to evaluate the status of not only the PCL but also the posterolateral corner structures. If positive, then a combined PCL and posterolateral corner injury should be anticipated.

Varus/Valgus Tests

The primary components of the capsuloligamentous structures are the collateral ligaments. The integrity of these structures is tested with varus and valgus stress at 0 degrees (full extension) and 30 degrees of flexion. Examining the knee in full extension tests the collateral ligaments as well as the associated posteromedial (valgus) and posterolateral (varus) capsular structures because extension tightens these structures. Examining the knee in 30 degrees of flexion isolates the MCL (valgus) and lateral collateral ligament (varus) because flexion relaxes the posterior structures.

An isolated PCL injury should exhibit no asymmetric varus or valgus patholaxity in 0 or 30 degrees of knee flexion. On the other hand, a combined PCL and medial corner injury will reveal a positive posterior drawer in neutral and internal tibial rotation and valgus patholaxity at 0 and 30 degrees of knee flexion. A combined PCL and posterolateral corner injury will reveal a positive posterior drawer test in neutral and external tibial rotation and varus patholaxity at 0 and 30 degrees of knee flexion.

Additional Tests

The authors rely heavily on the above tests to distinguish between an isolated PCL injury and a combined PCL with associated capsuloligamentous injury. Additional tests include the Godfrey (posterior sag) test, quadriceps active test, posterior Lachman test at 30 degrees, Hughston posterolateral corner sag, reverse pivot test, and prone assessment of external tibial rotation at 30 and 90 degrees knee flexion.[10] These tests are excellent, and they further confirm the presence or absence of structural injury. These tests have been described in detail in Chapter 4 and therefore will not be described here.

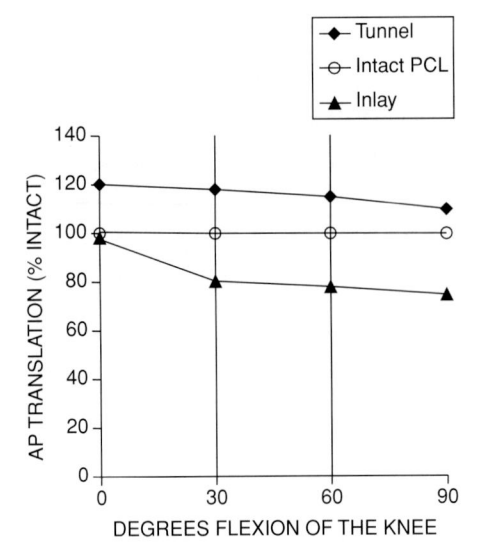

Fig. 12.2. Anteroposterior (AP) translation following posterior cruciate ligament (PCL) reconstruction using the tunnel and inlay reconstruction techniques. The inlay technique is statistically significant in terms of stability ($p < .01$) at 30, 60, and 90 degrees of flexion, as compared to the tunnel technique. When compared to the intact knee, statistical significance is reached at 60 ($p < .02$) and 90 ($p < .01$) degrees of flexion.

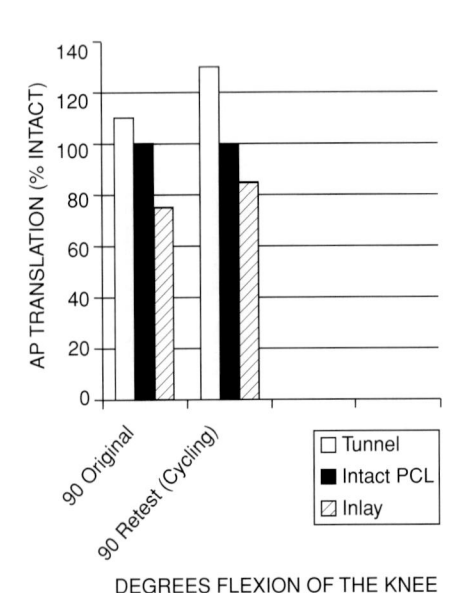

Fig. 12.3. Repetitive loading (cycling) retest data show increased AP translation significantly different from the intact knee for the tunnel reconstruction ($p < .01$), while the inlay technique was similar to the intact knee ($p > .08$).

Treatment Indications

Successful treatment of the isolated PCL injury and combined PCL with capsuloligamentous injury is dependent on an accurate history and physical examination. Based on our clinical experience and review of the literature, our treatment options can be categorized as nonoperative and operative. Nonoperative treatment is recommended for:

1. acute isolated PCL injury,
2. chronic isolated PCL injury newly diagnosed with no history of prior rehabilitation,
3. acute or chronic isolated PCL injury or PCL combined with capsuloligamentous injury in a patient who is noncompliant or incapable of complying with the postoperative rehabilitation.

Operative treatment is recommended for:

1. acute PCL injury combined with additional capsuloligamentous injury,
2. chronic PCL injury combined with additional capsuloligamentous injury,
3. chronic isolated PCL injury that has failed rehabilitation and has symptoms of pain (disability) and instability with activities important to the patient.

Often we utilize a bone scan to evaluate the osseous metabolic status of the knee in the chronic isolated PCL patient who has failed rehabilitation. If increased uptake is present in the medial and/or patellofemoral compartment on the delayed images, surgery is strongly considered.

Preferred Reconstruction Technique: Cleveland Clinic Foundation Arthroscopically Assisted PCL Tibial Inlay Technique

Scientific Rationale

Surgical reconstruction of the PCL can be categorized as arthroscopic, arthroscopically assisted, or open. With respect to the tibial portion of the reconstructive procedure, the procedure can be categorized further as via a tibial tunnel or inlay on the posterior aspect of the tibia at the tibial anatomic attachment site. Our preferred technique for PCL reconstruction is categorized as an arthroscopically assisted tibial inlay technique. This technique has evolved as a direct result of our (R.D.P and J.A.B.) collective dissatisfaction with the arthroscopic and tibial tunnel technique's clinical outcome in terms of stability and patient satisfaction. This technique is not new, but rather modified. Benedetto et al.,[11] Jakob,[12] and Thomann and Gaechter[13] have popularized this procedure in Europe, while Berg[14] has reported his experience in a small series of patients. The senior authors (R.D.P. and J.A.B.) began performing PCL reconstructions in this manner after carrying out preliminary cadaveric reconstructions and biomechanical studies in 1992. Based on early enthusiasm with its clinical outcome, we have continued to study this in our laboratory and performed the procedure on appropriate patients at the Cleveland Clinic Foundation since 1992 and have been even further convinced of its advantages when compared to the tibial tunnel technique.

12. Arthroscopically Assisted PCL Reconstruction: Tibial Inlay Technique

In a study recently completed in our laboratory, McAllister, Bergfeld, Parker, et al.[15] showed that the tibial inlay technique was more stable to posterior translation at all flexion angles than the tibial tunnel technique and the intact knee (Fig. 12.2). This was especially true at higher flexion angles where statistical significance was reached at 30, 60, and 90 degrees ($p < .01$) compared to the tunnel technique, whereas statistical significance was reached at 60 ($p < .02$) and 90 ($p < .01$) degrees compared to the intact knee. In addition, we found with repetitive loading (cycling) of the reconstructed knees that the inlay technique maintained its statistically significant stability in comparison to the tunnel technique and became similar in terms of stability to the intact knee (Fig. 12.3).[15] After cycling the reconstructed knees we examined the grafts macroscopically and there appeared to be thinning and damage just above the bone tendon junction at the exit of the tibial tunnel (at the "killer curve") on the tunnel technique graft (Fig. 12.4a). When both grafts were placed under a bright light, it became quite evident that there was thinning of the tunnel technique graft (Fig. 12.4b).

Patient Education

Extensive preoperative and perioperative education of the patient and family is performed. We feel this reduces hospital stay and postoperative pain.[16] Perioperative education includes preoperative education regarding intraoperative procedures, postoperative pain control, postoperative brace instruction, postoperative rehabilitation expectations, as well as the potential risks, benefits, expectations, alternatives, and possible complications.[16] The patient is also informed that after the procedure is performed, the patient is admitted to the 23-hour stay unit, because at our institution this technique is considered an outpatient procedure.

Patient Positioning and Preparation

The patient is positioned supine on the operating room table and general endotrachial anesthesia is induced. The patient is given a broad-spectrum cephalosporin intravenously prior to the surgical incision. Under anesthesia, an examination of both knees is performed and compared to the preoperative physical examination. Special attention is paid to the secondary

Fig. 12.4. Side-by-side comparison of the tibial tunnel technique (top) and inlay technique (bottom). (a) Macroscopic evaluation reveals apparent thinning and damage (arrow) to the tendon just above the bone tendon junction, which corresponds to the exit site of the tibial tunnel (at the "killer curve") of the graft from the tibial tunnel technique compared to the inlay technique graft. (b) Bright light evaluation clearly reveals the thinning of the tendon bone junction of the tibial tunnel technique compared to the inlay technique.

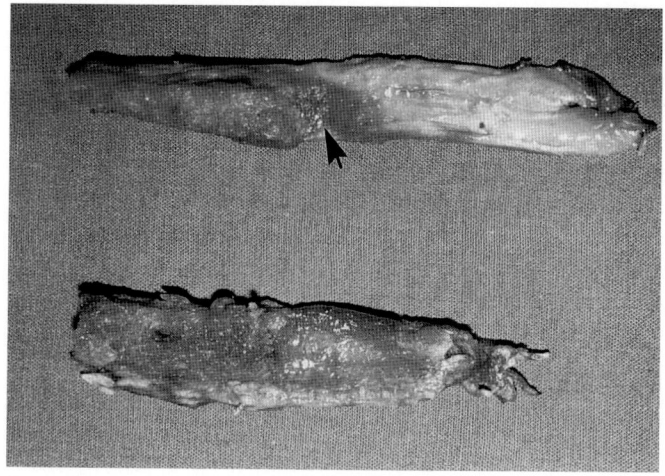

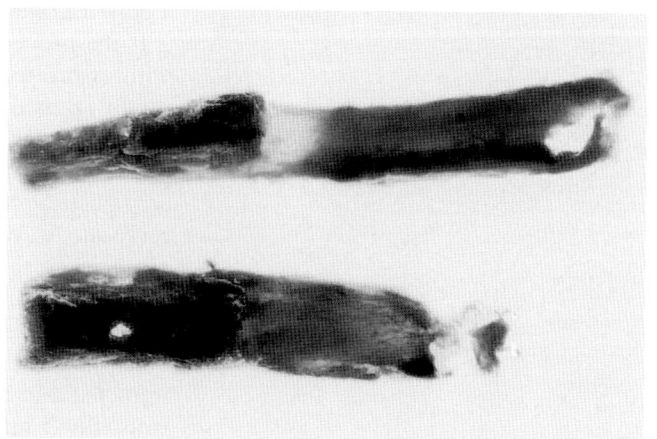

a

b

restraints during the physical examination, to ensure not overlooking any associated capsuloligamentous injuries. Prior to prepping and draping, the knee is injected intraarticularly with 0.25% bupivacaine and 1:400,000 epinephrine and 1 to 3 mg of morphine sulfate, which has been shown to be efficacious in our ACL reconstructions.[17] In addition, the surgical incisions are marked and injected with a mixture of 1% lidocaine and 0.25% bupivacaine and 1:400,000 epinephrine.

Diagnostic/Operative Arthroscopy

We begin the surgical procedure with a diagnostic/operative arthroscopy. This is performed without elevating the tourniquet. The PCL injury is documented and any associated internal derangements such as meniscal tears, chondrosis, and adhesions are addressed. It should be noted that the ACL may appear to be lax in the PCL-deficient knee because the knee is subluxated posteriorly. By placing an anterior drawer, the ACL will regain its normal tension.

Once the diagnostic/operative arthroscopy is completed, removing the PCL remnant/scar begins the arthroscopically assisted PCL reconstruction. This allows for visualization of the femoral anatomic attachment site (FAAS). We preserve the ligaments of Humphry and Wrisberg if present and work around them, as much as possible. We mark our FAAS with a sharp curette. The FAAS is anterior and distal in the femoral notch, usually 8 to 10 mm from the articular cartilage of the medial femoral condyle and at the 11 o'clock position on the left knee and the 1 o'clock position on the right knee. This position of the FAAS may vary from knee to knee based on the size of the patient and appearance of the normal PCL footprint. We only mark the FAAS at this stage, because we recheck it through a small medial arthrotomy at the time of guide pin placement.

If an associated ACL reconstruction is being performed, the ACL FAAS and ACL tibial anatomic attachment site (TAAS) are identified and drilled.

Autograft Bone–Patellar Tendon–Bone or Quadriceps Tendon-Bone Harvest and Graft Preparation

The arthroscopy is stopped and we proceed to harvesting either the central-third bone–patellar tendon–bone (BPTB) or quadriceps tendon-bone (QB) graft. The tourniquet is elevated 100 mm Hg above the systolic blood pressure (~250 mm Hg) after exsanguinating the leg (midthigh to toes) with an Esmarch wrap. If multiple procedures are planned and tourniquet time a concern, we harvest the graft without the tourniquet elevated.

If the procedure being performed is for an isolated PCL injury or PCL with associated capsuloligamentous injury other than the ACL, we prefer the BPTB as our graft of choice. On the other hand, if both the ACL and PCL are being reconstructed along with capsuloligamentous repair/reconstruction, we prefer the BPTB for the ACL and the QB for the PCL reconstruction. Sometimes we consider allograft tissue, especially in multiple revision. A 15-cm curvilinear (hockey-stick) incision is placed proximal posterior to the patella distally along the medial border of the patella and patellar tendon to a point 1 cm distal to the patellar tendon insertion into the tibial tubercle. The subcutaneous skin is incised and elevated, and hemostasis achieved. The paratenon over the patellar tendon is incised and retracted and the central-third BPTB is harvested. The patellar bone plug is usually 9 mm wide by 30 mm long, and the tibial bone plug is 11 mm wide and 20 mm long. The graft is then transported to the back table by the surgeon and handed to an assistant, who performs the final preparations of the graft.

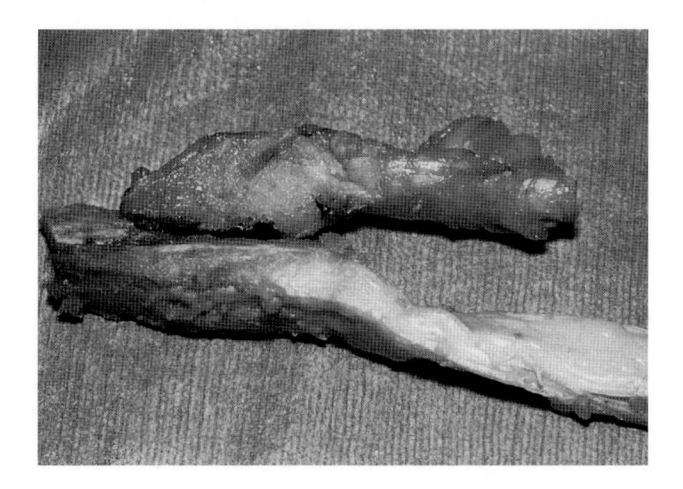

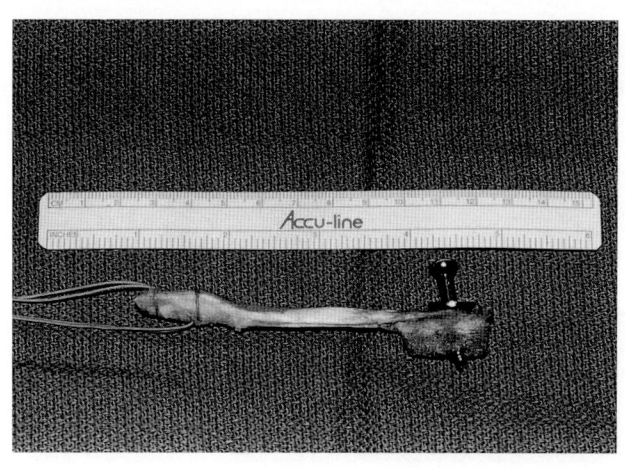

The patellar bone plug is fashioned to fit the femoral tunnel, while the tibial bone plug is prepared for the posterior tibial inlay. We have noted in side-by-side comparisons of the tibial insertion of the normal PCL and the patellar tendon insertion into the tibial tubercle that there is a very similar fiber insertion orientation (Fig. 12.5). We feel this fiber insertion orientation can be replicated by the tibial inlay technique. The femoral (patellar) bone plug is sculptured to be as cylindrical as possible and to fit through a 9-mm sizer. Two corticocancellous 2-mm drill holes are placed 5 mm and 15 mm from the distal tip, respectively, and a no. 5 nonabsorbable suture is placed through each hole to allow for passing and tensioning of the graft at the end of the procedure. The tibial bone plug is fashioned into a corticocancellous rectangle 11 mm wide and 20 mm in length. Through the center of this bone plug, a 3.2-mm pilot drill hole is placed. This hole is drilled from the cancellous side through the cortical side to avoid damaging the soft tissue part of the graft. This 3.2-mm pilot drill hole is overdrilled with a 4.5-mm drill and then tapped with the 6.5-mm tap from the cancellous to cortical side. A 35-mm long, 6.5-mm short threaded cancellous screw with metal washer is placed through the soft tissue from the cortical to cancellous side such that the tip of the screw protrudes 5 mm past the cancellous surface (Fig. 12.6).

As stated previously, we advocate using BPTB for routine isolated PCL reconstruction; however, in multiple ligament injuries or revision PCL reconstruction, the QB bone is our second graft of choice. This is harvested through the same incision. The central anterior portion of the quadriceps tendon, 10 cm in length, is harvested with an 11 mm by 20 mm bone plug from the patella. Two no. 5 nonabsorbable sutures are placed in a Krackow grasping technique in the quadriceps tendon and the patellar bone plug is prepared in exactly the same way as the tibial portion of the BPTB plug (Fig. 12.7).[18]

Fig. 12.5. Side-by-side comparison of the PCL fiber insertion into the tibia (top) is similar to the fiber insertion of the patellar tendon into the tibial tubercle (bottom).

Fig. 12.6. Final bone–patellar tendon graft preparation. The patellar portion has been sculptured to fit into the femoral tunnel. It is cylindrical (9 mm) and two no. 5 nonabsorbable sutures are present to aid in graft placement and tensioning. The tibial portion is 11 mm wide and 20 mm long, and the 6.5-mm screw is placed with its tip 5 mm proud.

Femoral Tunnel Selection

While the assistant is preparing the harvested graft on the back table, the femoral tunnel is selected and prepared via a mini-medial arthrotomy. The authors believe this step enhances the accuracy of femoral tunnel place-

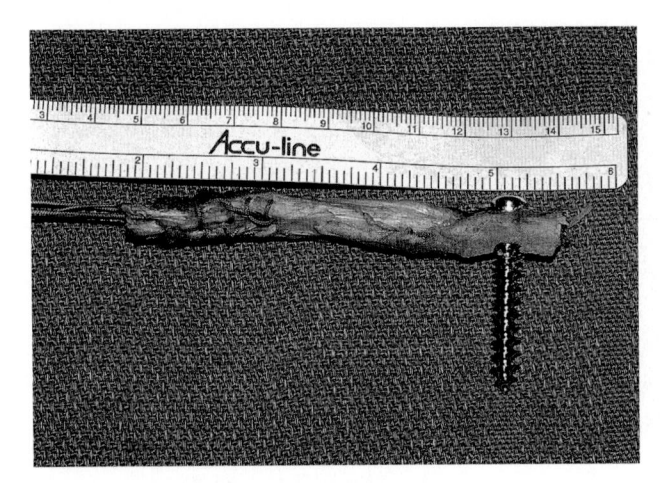

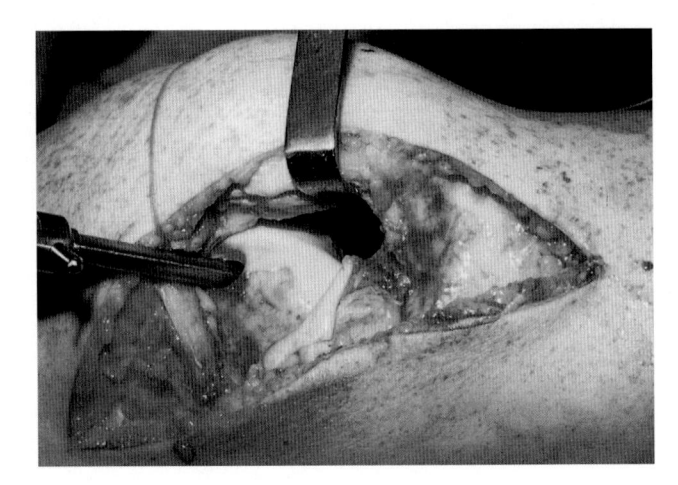

Fig. 12.7. The quadriceps tendon–bone graft is the second choice for reconstruction of the PCL. The quadriceps tendon is secured by two no. 5 nonabsorbable sutures placed in a Krackow technique, which optimizes graft purchase. The patellar bone is prepared in the same manner as the tibial portion of the bone–patellar tendon–bone graft. Note that a fully threaded cancellous screw is being used.

Fig. 12.8. A left knee femoral tunnel selection and preparation. The mini-arthrotomy extends from the proximal tibia above the medial meniscus to the vastus medialis obliquus (VMO) and then posteromedially along the VMO. With gentle retraction, the intercondylar notch and the medial femoral condyle, especially the transition to articular cartilage, are visualized. This allows for accurate femoral tunnel placement and even ensures a femoral tunnel directed proximal and posteriorly, thus decreasing the obliquity of the tunnel as it exits the femur in the intercondylar notch. Note the tunnel's entrance point is 1 to 2 mm proximal to the articular surface.

ment, since PCL reconstruction is an uncommonly performed procedure (<10 times/year). The mini-arthrotomy is established through the same incision as the graft harvest extending from the proximal tibia (just above the medial meniscus) to the insertion of the vastus medialis obliquus (VMO) and then posteromedially along the VMO to allow visualization of the intercondylar notch and distal medial femoral condyle, especially the transition to the articular surface (Fig. 12.8). The intercondylar notch is visualized by retracting the patella and patellar tendon laterally. It is not necessary to dislocate the patella. To enhance visualization, the arthroscope's light can be used and the previously marked FAAS is identified and checked for accuracy. Often fine-tuning of the FAAS is required. A PCL femoral guide is placed with the tip of the guide in the marked FAAS. It should be positioned such that the entrance point on the medial femoral condyle is 1 to 2 mm proximal to the articular surface and directed proximally and posterior. This position and direction of the femoral tunnel is important since this decreases the obliquity of the femoral tunnel and decreases the stress on the PCL autograft as it exits the femoral tunnel.

A guide wire is placed and its position rechecked, and it is then overdrilled with the appropriate cannulated drill. The drill size is determined by the diameter of the femoral graft plus 1 mm, and is usually 10 mm. Once the femoral tunnel is drilled, a rasp is used to smooth the tunnel edges, thus limiting graft abrasion. Then a double-looped malleable 18-gauge wire or commercially available graft passer is placed into the tunnel and directed posteriorly toward the posterior intercondylar notch. This will facilitate graft passage once the graft is secured on the posterior tibia.

Once the femoral tunnel is placed and prepared, the tourniquet is deflated and the wound is loosely approximated and covered with a sterile dressing and circumferential wrap. The entire leg is placed in a sterile bag, such as a commercially available image cover or Mayo stand cover, and wrapped again circumferentially. Another operating room table is brought into the room and positioned next to the patient. The operating team under the direction of the surgeon and anesthesiologist then turns the patient to the prone position. Care is taken to position and protect the important neurovascular structures from positional injury (Fig. 12.9). Once the patient is turned and positioned, the original operating room table is removed from the room, cleaned, and prepared for placing the patient back in the supine position toward the end of the surgical procedure.

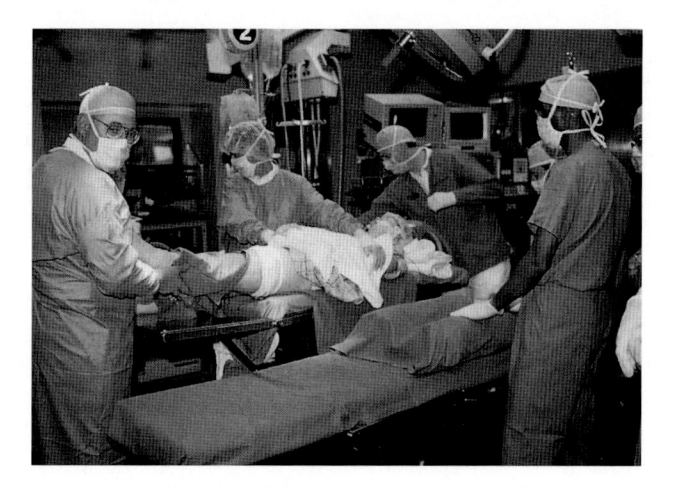

Fig. 12.9. A second operating room table is positioned alongside the patient and the patient is turned under the direction of the surgeon and anesthesiologist.

Posterior Approach to the Tibia

Once the patient is in the prone position, the sterile dressings are removed and the entire leg is reprepped and draped circumferentially from the proximal thigh to the tips of the toes. The authors have tried many different patient body positions for the approach to the posterior tibia, such as turning the patient into the lateral decubitus position and utilizing a posterolateral or posteromedial incision. These approaches are enticing to the surgeon but are fraught with difficulty. Portions of the procedure such as visualizing the anatomic landmarks of the posterior tibia, placing osteotomes perpendicular to the posterior cortex of the tibia, and, most of all, placing the screw and posterior bone block perpendicular to the posterior tibia are difficult and often impossible through these approaches. Therefore, the authors have found positioning the patient in the prone position and utilizing this approach to the posterior tibia make the exposure easier and result in consistent positioning and placement of the tibial inlay portion of this procedure. We have found turning the patient to the prone position and reprepping the leg only requires an additional 10 to 15 minutes of operating room time and is well worth the effort.

The incision is marked and injected with a mixture of 1% lidocaine and 0.25% bupivacaine and 1:400,000 epinephrine. The landmarks for this modified Burks' posterior approach to the posterior tibia are (1) the medial border of the medial head of the gastrocnemius, (2) the posterolateral border of the semitendinosus, (3) the popliteal crease, and (4) the midline of the distal thigh (Fig. 12.10).[19] The authors prefer to perform this portion of the procedure without elevating the tourniquet, thus requiring and ensuring hemostasis. The skin is incised to the fascia overlying the popliteal fossa and elevated subcutaneously. Care should be taken toward the midline, since the medial sural cutaneous nerve usually exits the fascia just distal to the popliteal crease. The fascia is incised at the interval between the medial head of the gastrocnemius and the semitendinosus and carried proximally and toward the midline. The proximal and distal extent of the dissection is determined ultimately by the amount of visualization of the posterior tibia and often requires further extension.

A broad low profile blunt retractor is placed in the interval between the medial head of the gastrocnemius and the semitendinosus and the medial head of the gastrocnemius is retracted laterally. By retracting the medial head of the gastrocnemius laterally, the neurovascular bundle (tibial artery, vein, and nerve) is protected from injury and does not require visualization

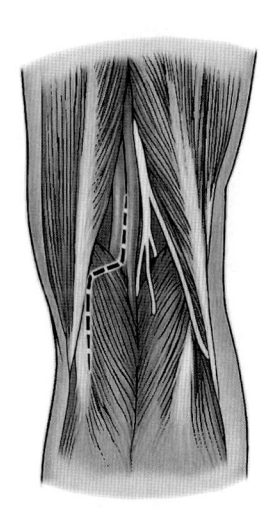

Fig. 12.10. The modified posterior approach to the posterior tibia: the incision landmarks are (1) the medial border of the medial head of the gastrocnemius, (2) posterolateral border of the semitendinosus, (3) popliteal crease, and (4) midline of the distal thigh.

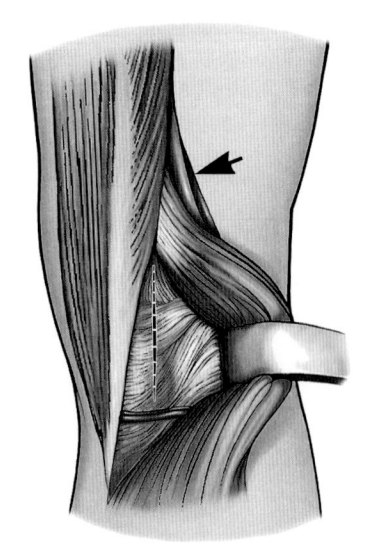

Fig. 12.11. A broad low-profile blunt retractor is placed in the interval between the medial head of the gastrocnemius and the semitendinosus, and the medial head of the gastrocnemius is retracted laterally. By retracting the medial head of the gastrocnemius laterally, the neurovascular bundle (tibial artery, vein, and nerve) (arrow) is protected from injury and does not require visualization.

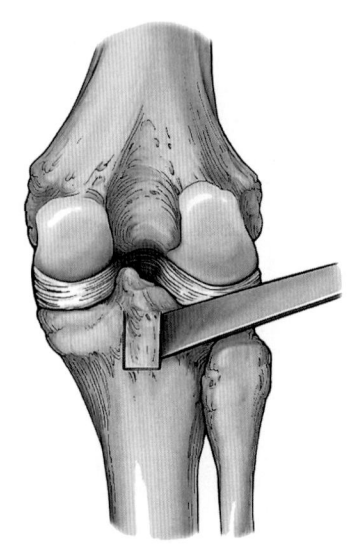

Fig. 12.12. Once the tibial attachment site is determined, the knee is extended and a cortical window is made with an osteotome. Its dimensions are determined by the size of the tibial bone graft previously harvested and prepared, and is usually 11 mm wide and 20 mm long.

(Fig. 12.11). If the medial head of the gastrocnemius is large, its origin can be partially detached to enhance visualization of the posterior tibia. The small vessels lying against the posterior capsule and posterior tibia, such as the middle and medial geniculate arteries and veins, are ligated and transected prior to incising the posterior capsule and exposing the proximal posterior tibia.

Once incised, the posterior capsule is reflected medially and laterally and remaining scar tissue and remnants of the PCL in the intercondylar notch are excised visualizing the femoral tunnel and graft passer previously placed. Care is taken to preserve the ligaments of Wrisberg and Humphry if present. The insertion of the PCL into the posterior tibia is identified and marked. The fascia over the proximal popliteus muscle near the distal portion of the insertion of the PCL is incised to allow for distal retraction of the popliteus muscle if necessary. The BPTB or QB graft is brought to the operative field and the femoral portion is passed into the femoral tunnel with the previously placed graft passer. If a BPTB is used, the tibial bone portion is positioned at the tibial insertion of the PCL and the knee is flexed to 90 degrees. The slack is removed from the tendinous portion of the graft by placing tension on the femoral bone plug via the no. 5 nonabsorbable sutures. The femoral bone plug's position is assessed. If 25 to 30 mm of the femoral bone plug is in the femoral bone tunnel, the tibial attachment site is accepted. On the other hand, if <25 mm of the femoral bone plug is in the bone tunnel, the tibial attachment site is adjusted distally. We have found clinically we can distalize the tibial attachment of the PCL graft by as much as 10 mm without affecting the kinematics of the reconstruction. If it is deemed that more than 10 mm of distalization is necessary to have enough bone in the femoral tunnel, the authors will not distalize the tibial attachment and will ultimately tie the femoral graft's sutures over a post or staple the femoral graft to the medial femoral condyle. If the QB graft is used, the tibial graft is placed anatomically because the femoral tunnel is filled with soft tissue.

Once the tibial attachment site is determined, the knee is extended and a cortical window is made with an osteotome. Its dimensions are determined by the size of the tibial bone graft previously harvested and prepared and is usually 11 mm wide and 20 mm in length (Fig. 12.12). Internal rotation of the tibia aids in this exposure. Once osteotomized, the cortex is removed, leaving a cancellous bone window. The tibial bone graft with its 6.5-mm screw and washer already in place is keyed into the cortical window and the screw is advanced. We have not found it necessary to predrill the cancellous bone. Often minor adjustments of graft position are necessary before the screw can be fully tightened and the graft fully seated. If an ACL reconstruction is being performed in conjunction with the PCL reconstruction, the screw is directed a bit laterally to avoid the ACL's tibial tunnel. Tension is placed on the femoral plug's sutures to ensure adequate tibial fixation has been achieved. Intraoperative radiographs (anteroposterior and lateral) are obtained at this time to check screw placement and direction.

Once the surgeon is convinced that tibial graft fixation and position are appropriate, the posterior wound is irrigated with normal saline, and the retractor removed. A few moments are allowed to pass to ensure no significant bleeding. The posterior capsule is closed or reapproximated with a no. 1 absorbable suture. The fascia is gently approximated with no. 1 absorbable suture, and the subcutaneous skin closed with no. 2-0 absorbable subcuticular technique. The posterior and anterior wounds are dressed and covered circumferentially and once again a sterile commercially available image cover bag or Mayo stand cover is applied and circumferentially covered. The original operating room table is positioned next to the patient and the patient is turned back to the supine position under direction of the surgeon and anesthesiologist.

Graft Tensioning and Femoral Fixation

The leg is once again reprepped and redraped for final tensioning of the PCL graft, fixation of the femoral plug, and anterior wound closure. If more than 2 hours of operative time has passed, the patient is given another dose of the broad-spectrum cephalosporin given at the beginning of the procedure. The anterior wounds are irrigated and hemostasis achieved. If associated capsoligamentous repair and/or reconstruction is necessary, it is performed and secured at this time.

The knee is flexed to 90 degrees with the tibia in neutral rotation. A posterior-anterior force (anterior drawer) is placed on the proximal tibia and the femoral bone plug's sutures (if BPTB) or femoral quad tendon's sutures (if QB) are tensioned with approximately 10 to 20 lb of force and fixed. If BPTB is used and 20 to 25 mm of bone is within the femoral tunnel, we prefer to use interference screw fixation with a 9- $\times$ 20-mm interference screw placed proximally to secure the bone plug in the distal aspect of the bone tunnel (toward the articular surface). If insufficient bone is present in the femoral tunnel (<20 mm of bone), a trough is made in the medial femoral condyle and fixation achieved with either two staples, or tying the no. 5 nonabsorbable sutures to a 6.5-mm (with washer) post in the medial femoral condyle. If QB is used, the no. 5 nonabsorbable sutures are tied to a 6.5-mm (with washer) post in the medial femoral condyle, or if enough tendon is present outside of the femoral tunnel, staples or soft tissue screw and washer fixation can be considered. Soft tissue interference screw fixation can be considered if QB is used or if insufficient bone (BPTB) is present in the femoral tunnel. The authors have no experience with this form of fixation. Once femoral fixation is achieved, the knee is flexed and extended to ensure a full range of motion and then flexed to 90 degrees and a gentle posterior drawer tested. If the posterior drawer is not negative or grade I, fixation should be checked or the graft retensioned.

If an associated ACL reconstruction has been performed, the ACL graft is passed prior to PCL tensioning and fixation. The cruciate ligaments are balanced by fixating the tibial side of both the ACL and PCL grafts and then tensioning and fixating the femoral sides together. Once the surgeon is satisfied with the knee stability, the wounds are irrigated and hemostasis checked. A decision is made regarding an intraarticular drain placement and is dependent on the amount of bone bleeding. The wounds are closed in a layered fashion with cosmesis in mind, and a bulky sterile dressing is applied. A thigh-high compressive stocking is applied with a cryotherapy pad incorporated into the dressing. A postoperative knee brace or knee immobilizer is applied and locked at 0 degrees extension.

Postoperative Care

The patient is admitted for a 23-hour stay, because in our institution this technique is considered an outpatient procedure. The patient is allowed toe-touch weight bearing on the operative leg with crutches, which is secured in a postoperative knee brace locked in 0 degrees extension. The cryotherapy unit is used continuously for the first 24 hours, and the patient is treated proactively with narcotics (intramuscularly or orally) for

Fig. 12.13. Artist's drawings (a and c) and radiographs (b and d) of a bone–patellar tendon–bone arthroscopically assisted PCL reconstruction: tibial inlay technique. Anteroposterior drawing (a) and radiograph (b). Lateral drawing (c) and radiograph (d).

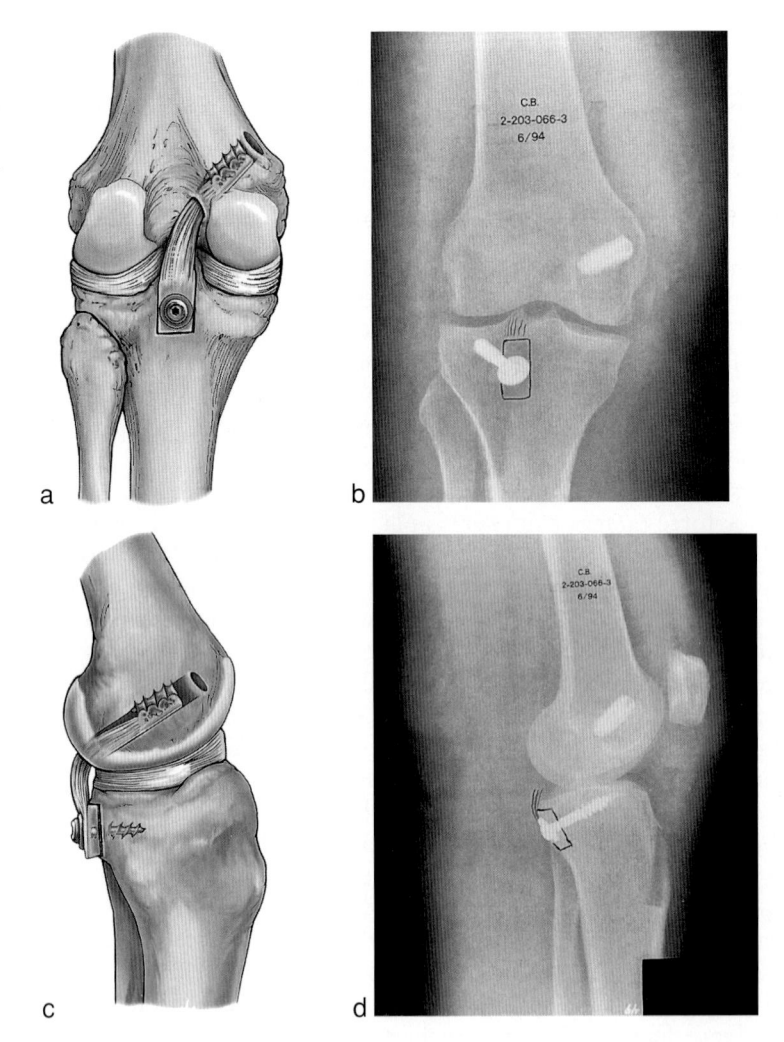

pain. The morning after surgery, the drain is removed (if present) and the patient is discharged. The brace is left locked at 0 degrees extension until the first postoperative rehabilitation visit, which is within 1 week of surgery. At the first postoperative visit, the dressing is removed, wounds cleaned, and a less bulky dressing applied. The first rehabilitation visit occurs the day of the first postoperative visit. The patient will return for suture removal and wound check between the 10th and 14th postoperative day. At some point during the initial postoperative period, radiographs are obtained (Fig. 12.13).

Postoperative Posterior Cruciate Reconstruction Rehabilitation

Ligamentous disruption equates to loss of normal biomechanical function of the knee. Stresses to the secondary static and dynamic stabilizers, namely the articular surface and patellofemoral articulation, increase dramatically with disruption of the normal joint kinematics. The rehabilitation program must take into account the biomechanical stresses placed on the knee joint and patellofemoral articulation while incorporating appropriate open kinetic chain (OKC) and closed kinetic chain (CKC) exercises. The effect of OKC and CKC exercises on knee function remains an area of continued research. During OKC exercises, the distal segment of the limb is free to move in space, whereas CKC has the distal segment fixed on a movable or immovable object. OKC results in isolated muscular contractions while CKC contractions allow for coordinated co-contraction of many muscle groups in a predictable manner. OKC and CKC exercises produce different effects on the tibiofemoral and patellofemoral joint.

Lutz et al.[20] investigated the shear force development in the tibiofemoral joint with OKC isometric contractions in 30, 60, and 90 degrees of knee flexion and extension to CKC exercises. Their results indicated that significantly less posterior shear force developed during the CKC exercises as compared with OKC knee flexion at 60 and 90 degrees. A maximum posterior shear force of 1,780 N was reported at 90 degrees of knee flexion. These tremendous shear forces indicate that OKC knee flexion should be avoided so as not to induce posterior tibiofemoral shear. OKC resisted knee extension produces an anterior tibial shear force and is included early in the rehabilitation of the PCL-deficient knee program. The rehabilitation program must balance the intensity of the exercises to avoid irritation of the patellofemoral articulation while performing OKC knee extension exercises.

Closed kinetic chain exercises are not benign in relation to PCL shear stress. Dahlkuist et al.[21] demonstrated an increase in posterior shear up to three times body weight during squatting exercises. Others have indicated that the posterior shear force increases proportionately with increased knee flexion during front knee squats and lunges.[22-24] CKC exercises reproduce functional positioning and co-contraction muscle firing patterns, which OKC exercises cannot, that are necessary for functional activity. These activities can be utilized to reduce the patellofemoral joint compressive forces experienced with OKC exercises.

The research and data presented indicate that it is important to include both OKC knee extension and CKC exercises in a comprehensive PCL rehabilitation program. Both OKC extension from 60 to 0 degrees and CKC exercises from 0 to 60 degrees appear to have a beneficial effect of decreasing posterior shear force and thus are safe to perform with PCL-deficient patients.

Rehabilitation following PCL reconstructive surgery has not been given adequate emphasis in the published literature and has characteristically been referred to as "the ACL turned around," a perplexing situation and a clinical challenge.[25] Hughston et al.[26] further stated, "In my estimation, rehabilitation accounts for 50 percent of a successful result following injury or operation." Reconstructive PCL procedures and rehabilitation are performed to allow functional return to previous activity and to prevent further articular cartilage degeneration. The lack of research data on the graft maturation sequence, the forces imposed on the healing graft during rehabilitation, and prospectively randomized research on isolated PCL reconstructed knees create a significant challenge for the clinician. Using a functional progression protocol, the frequency of clinical visits for rehabilitation or reevaluation is determined by the patient's level of function. The goal of rehabilitation is to return the athlete to previous levels of function within 9 to 12 months. The quantification of the patient's functional status is determined by a combination of subjective patient assessment and objective evaluation. Objective evaluation includes postoperative range of motion (ROM), swelling (effusion), gait abnormalities, incision healing, patellar mobility, balance/proprioception deficits, muscular strength, patellofemoral pathology, and joint stability. These factors aid the clinician in determining the level of progression and can be useful in the recognition and management of potential complications early in the rehabilitation process.

General rehabilitation guidelines are designed to do the following:

1. Minimize postoperative immobilization,
2. Allow for tissue healing and remodeling to occur in a non-stressed environment,
3. Identify potential complications from the start,
4. Base rehabilitation progression and treatments on available basic science and applied clinical research,
5. Utilize a functional progression based on critical milestones and reevaluation,
6. Individualize the rehabilitation with the patient's goals as the functional targets, and
7. Utilize a team approach—physician, physical therapist, athletic trainer, and coaching staff—to care for the patient and optimize results of the surgery and rehabilitation.

Rehabilitation after PCL reconstruction is divided into overlapping phases of acute immediate postoperative, acute, progressive ROM/strengthening, and functional return to activity (Table 12.1). In the acute immediate postoperative period, the goals of rehabilitation are to control pain and decrease joint effusion, progress ROM from active assisted to active, decrease the incidence of fibrosis, initiate quadriceps strengthening exercises, and educate the patient on positions that counteract gravitational stresses on the PCL graft, such as lying/sleeping prone rather than supine, and the avoidance of aggressive hamstring contractions. In this phase, the patient is placed into a hinged brace locked at 0 degrees extension. The patient is allowed to ambulate with crutches with partial weight bearing (25% to 50%), and a normal heel-toe pattern is encouraged. Cryotherapy application with a commercially available continuous cold system is utilized in conjunction with elevation, and ankle pumps. The brace is unlocked each hour for patient-assisted tibial lifts into flexion, limited to 0 to 60 degrees of knee motion. Strengthening exercises for the quadriceps are initiated utilizing functional electrical stimulation for quadriceps sets, straight leg raises in all directions, knee extension 0 to 40 degrees and progressed to 0 to 60 degrees by postoperative week 4. Functional electrical stimulation of the quadriceps in the immediate postoperative period has been reported in the literature to improve the neuromuscular control of the quadriceps.[27,28] This should be initiated if the patient fails to obtain a good quadriceps contraction in the first week.

The acute-phase (weeks 1–3) goals include progressive weight bearing as tolerated with crutches, emphasizing a normalized gait pattern. ROM is increased to 0 to 90 degrees of knee flexion both actively and assisted. Patellar mobilization is guided by the degree of knee joint effusion. Aggressive patellar mobilization is avoided if marked effusion is present due to the potential for exacerbating patellofemoral fibrosis that may occur with prolonged effusion and limited ROM. Continuation of isometric quadriceps exercises includes multiangle extensions from 60 to 0 degrees, straight leg raises with proximal resistance transferred progressively to the ankle if quadriceps control allows no quadriceps lag, and knee extension from 60 to 0 degrees without resistance. Initiation of CKC exercises includes minisquats (after three weeks postoperatively and acute range of motion is at least 0 to 100 degrees) for proprioceptive training, reciprocal bike and weight shifts starting earlier for ROM and contralateral cycling, and standing weights shifts. The patient is fitted with a functional PCL brace at between 4 and 6 weeks postoperative.

Table 12.1. Rehabilitation guidelines after PCL reconstruction

Acute immediate postoperative phase (early protection phase) (postoperative weeks 0–1)
  Bracing: postoperative hinged brace or knee immobilizer locked at 0 degrees
  Range of motion (ROM): passive range of motion (PROM) with patient-assisted tibial lift into flexion 0–60 degrees
  Exercises: quadricep isometrics, straight leg raises in hip adduction, abduction; knee extension 60–0 degrees, and electrical stimulation to quadriceps

Acute phase (maximal protection phase) (postoperative weeks 2–6)
  Goals
    Minimize external forces to protect graft
    Prevention of quadriceps atrophy
    Control postsurgical effusion
  Weight bearing: as tolerated with assistive device such as crutches
  ROM: as tolerated to 90 degrees
  Exercises
    Continue isometric quadriceps strengthening
    Closed kinetic chain mini-squats, shuttle, bike
    Open kinetic chain knee extension (60–0 degrees)
    Proprioception training
    Weight shifts
  Brace: fit with functional brace at 4–6 weeks postoperative

Progressive range of motion/strengthening phase (postoperative weeks 7–24)
  Weight bearing: as tolerated without assistive device
  Range of motion: as tolerated to 125 degrees flexion
  Exercises
    Continue isometric quadriceps strengthening
    Begin isotonic quadriceps strengthening
    Leg press (0–60 degrees)
    Step-ups
    Sport-cord progression program
    Rowing, Nordic Track
    Initiate closed kinetic chain terminal knee extension

Functional activity phase (postoperative weeks 25+)
  Few scientific data support a progression of the rehabilitation program into functional stages; therefore, progression should be based on the patient's tolerance to exercise and level of function; overall functional tests for power and endurance, popularized during anterior cruciate ligament (ACL) programs, should theoretically measure total length, strength, and endurance for the PCL-reconstructed knee
  Anticipated return to activity following PCL reconstruction is between 9–12 months

The progressive ROM and strengthening phase is characterized by increasing the ROM to full motion, and full–weight-bearing ambulation without assistive device. Exercise continues to focus on total leg strengthening, especially quadriceps activity with isotonic strengthening, leg press (0–60 degrees), concentric and eccentric step drills progressing from 2 to 8 inches, resisted forward/backward walking, seated Versa-climber, and biking. Proximal and distal (hip and ankle) strengthening is addressed as the patient achieves full weight bearing. Little scientific research exists to support a progression of the rehabilitation program into truly functional stages. Therefore, progression should be based on the patient's tolerance of exercises geared to individual activities and level of function. Functional tests that estimate power and endurance, popularized in ACL rehabilitation, should give the clinician a standard measure of total lower limb functional ability and can be utilized for the PCL-reconstructed knee.

The final stage, the functional activity phase, is based on the patient's tolerance of exercise and level of function. Overall functional tests for power and endurance, popularized during ACL programs, should theoretically measure total length strength and endurance for the PCL-reconstructed knee. The anticipated return to full, unrestricted activity following PCL reconstruction is between 9 and 12 months.

References

1. Clendenin MB, DeLee JC, Heckman JD. Interstitial tears of the posterior cruciate ligament of the knee. Orthopedics 1980;3:764–772.
2. Ritchie JR, Bergfeld JA, Kambic H, et al. Isolated sectioning of the medial and posteromedial capsular ligaments in the posterior cruciate ligament–deficient knee. Influence on posterior tibial translation. Am J Sports Med 1998;26:389–394.
3. Clancy WG Jr, Shelbourne KD, Zoellner GB, et al. Treatment of the knee joint instability secondary to rupture of the posterior cruciate ligament: report of a new procedure. J Bone Joint Surg 1983;65A:310–322.
4. Cross MJ, Powell JF. Long-term follow-up of posterior cruciate ligament rupture: a study of 116 cases. Am J Sports Med 1984;12:292–297.
5. Parolie JM, Bergfeld JA. Long-term results of nonoperative treatment of isolated posterior cruciate ligament injuries in the athlete. Am J Sports Med 1986;14:35–38.
6. Dandy DJ, Pusey RJ. The long-term results of unrepaired tears of the posterior cruciate ligment. J Bone Joint Surg 1982;64B:92–94.
7. Shelbourne KD, Patel DV. The natural history of acute isolated nonoperatively treated posterior cruciate ligament of the knee: a prospective study. Am J Sports Med 1999;27:276–283.
8. Noyes FR, Stowers SF, Grood ES, et al. Posterior subluxation of the medial and lateral tibiofemoral compartments: an in vitro ligament sectioning study in cadaveric knees. Am J Sports Med 1993;21:407–414.
9. Clancy WG Jr. Repair and reconstruction of the posterior cruciate ligament. In: Chapman MW, ed. Operative Orthopaedics, 2nd ed. Philadelphia: JB Lippincott, 1993:2093–2107.
10. Ritchie JR, Miller MD, Harner CD. History and physical evaluation. In: Fu FH, Harner CD, Vince KG, eds. Knee surgery, vol 1. Baltimore: Williams & Wilkins, 1994:253–273.
11. Benedetto KP, Hackl W, Fink C. Mittelfristige Ergebnisse der hinteren Kreuzbandrekonstruktion mit dem LAD-augmentiereten Lig. Patellae. Arthroskopic 1995;8:95–99.
12. Jakob RP. PCL and posterolateral instability. Paper presented at the First World Congress of Sports Trauma, Palma de Mallorca, May 26, 1992.
13. Thomann YR, Gaechter A. Dorsal approach for reconstruction of the PCL. Arch Orthop Trauma Surg 1994;113:142–148.
14. Berg EE. Posterior cruciate ligament tibial inlay reconstruction. J Arthrosc Rel Res 1995;11(1):69–76.
15. Bergfeld JA, McAllister D, Kambic H, Parker RD, Valdevit A. A biomechanical comparison of PCL reconstructive techniques. Am J Sports Med (submitted).
16. Calabrese GJ, Passerallo AJ, Duriak K. Perioperative education and rehabilitation and its effects on pain. Sports Med Arthrosc Rev 1998;6(3):216–222.
17. Gatt CJ, Parker RD, Tezlaff JE, et al. Preemptive analgesia: its role and efficacy in anterior cruciate ligament reconstruction. Am J Sports Med 1998;26(4):524–529.
18. Krackow KA, Thomas SC, Jones LC. A new stitch for ligament tendon fixation. J Bone Joint Surg 1986;68A:764–765.
19. Burks RT, Schaffer JJ. A simplified approach to the tibial attachment of the posterior cruciate ligament. Clin Orthop Rel Res 1990;254:216–219.
20. Lutz GE, Palmitier RA, An KN, et al. Comparison of tibiofemoral joint forces during open-kinetic-chain and closed-kinetic-chain exercises. J Bone Joint Surg 1993;75A:732–739.
21. Dahlkuits NJ, Mago P, Seedholm BB. Forces during squatting and rising from a deep squat. Eng Med 1982;11:69–76.
22. Morrison JB. Function of the knee joint in various activities. Biomech Eng 1969;4:573–580.
23. Smidt GL. Biomechanical analysis of knee extension and flexion. J Biomech 1973;6:79–92.
24. Ohkoshi Y, Yasuda K, Kaneda K, et al. Biomechanical analysis of rehabilitation in the standing position. Am J Sports Med 1991;19:605–611.

25. Wilk KE. Rehabilitation of isolated and combined posterior cruciate ligament injuries. Clin Sports Med 1994;13(3):649–677.
26. Hughston JC, Bowden JA, Andrews JR, et al. Acute tears of the posterior cruciate ligament. J Bone Joint Surg 1980;62:438–450.
27. Morrissey MC, Brewster CE, Shields CL, et al. The effects of electrical stimulation on the quadriceps during post-operative knee immobilization. Am J Sports Med 1985;13:40–45.
28. Lossing I, Gremby G, Johnson T. Effects of electrical muscle stimulation combined with voluntary contractions after knee ligament surgery. Med Sci Sports Exer 1988;20:93–98.

Chapter Thirteen

Arthroscopically Assisted Posterior Cruciate Ligament Reconstruction: Double Femoral Tunnel Technique

Michael H. Metcalf and Roger V. Larson

Posterior cruciate ligament (PCL) injuries account for 3% to 20% of all knee ligament injuries.[1] This wide range in the reports of the incidence of PCL injuries reflects differences in study populations as well as the difficulty in diagnosing acute PCL injuries.[2,3]

Isolated PCL tears have traditionally been treated with nonoperative care, and the initial short-term reports indicated that the patients functioned relatively well.[4-7] However, long-term follow-up studies have shown a higher incidence of osteoarthritis and poorer patient function.[2,8] The reasons for failures of nonoperative treatment are multifactorial, but include increased compartment pressures and progressive degenerative changes.[3,7,9] The medial and patellofemoral compartments appear to be the most susceptible to these changes.[9] It has also been observed that in some cases progressive laxity of the knee joint develops due to stretching out of the secondary stabilizers.[3,4,10-12] The issues of natural history have been addressed in detail in preceding chapters.

Prior Reconstruction Techniques

With the exception of primary repair of avulsion fractures, reconstructions of the PCL have not fared as well as ACL reconstructions. There have been many techniques that have been developed for PCL reconstruction. Early procedures focused primarily on the type of graft used for reconstruction, including bone–patellar tendon–bone graft,[13-16] semimembranosus,[17] semitendinosus,[18-22] medial gastrocnemius,[3,20,23-27] meniscus,[28-31] fascia lata,[25,32,33] popliteal tendon,[34,35] and allograft.[36,37] Each of these procedures relied upon a single band of tissue to re-create the function of the PCL. Longer follow-up studies have revealed a significant degree of premature arthritis and increased joint laxity despite operative intervention.[38,39] Whether any of these procedures produce better results in isolated PCL injuries than nonoperative treatment has not been demonstrated.

Physical Examination for PCL Insufficiency

An isolated injury to the PCL is best demonstrated by the posterior drawer test with the knee flexed to 90 degrees. If the injury is truly an isolated PCL injury, it is unlikely that more posterior laxity will be present than what is necessary to create a neutral tibial step-off. This is a situation

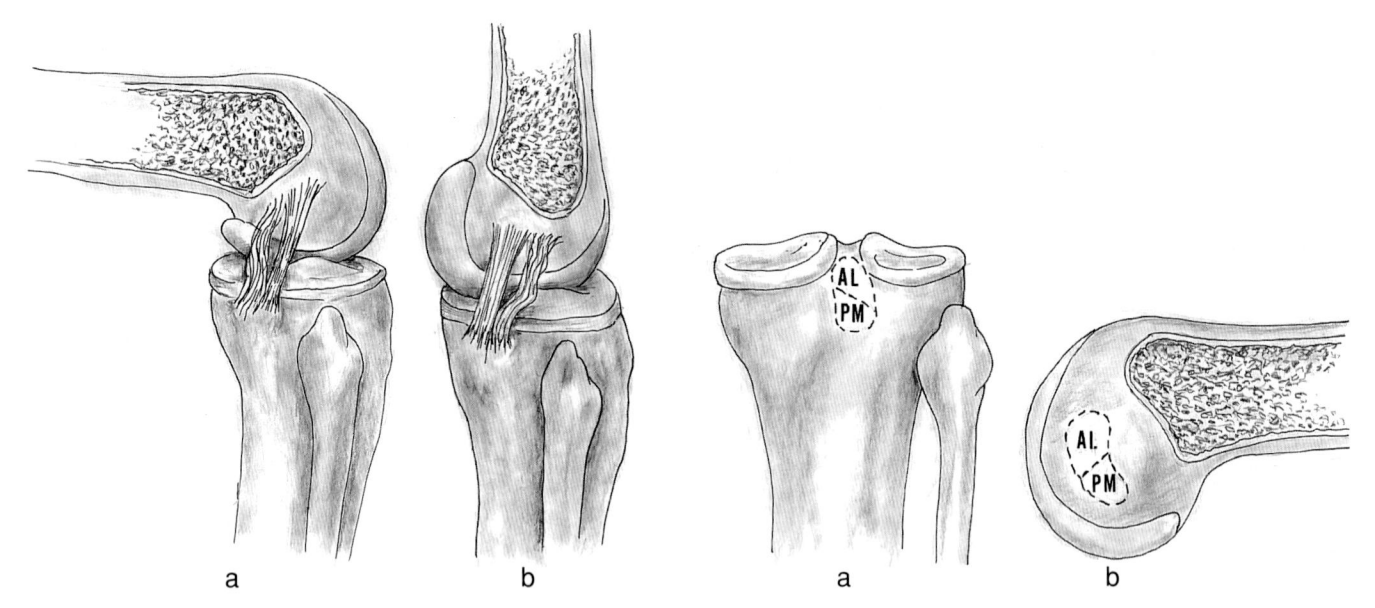

Fig. 13.1 The posterior cruciate ligament (PCL) is not an isometric ligament as it functions as at least two separate ligaments: an anterolateral band (AL) that is taut and supportive when the knee is in flexion and the posteromedial band (PM) that is taut and supportive when the knee is in extension.

Fig. 13.2. The regions within the anatomic PCL footprints on the tibia and femur that represent the attachment site of the anterolateral (AL) and posteromedial (PM) bands of the PCL.

where the anterior tibia with posterior displacement sits flush with, or slightly anterior to, the distal femur. If there is further posterior displacement of the tibia and a "negative" step-off or grade III laxity, then it is likely that there are more structures injured than the PCL. These structures may include the posterior capsule, the posterolateral corner, the posterior oblique complex, or the medial collateral ligament. If both the anterolateral and posteromedial bands of the PCL have been disrupted, there will also be increased translation posteriorly with Lachman testing at 20 to 30 degrees of knee flexion. If there is increased varus or valgus laxity or increased external tibial rotation, then there have been injuries to the collateral ligament complexes or the posterolateral corner. Further details of physical examination are discussed in Chapter 4.

Anatomy

Anatomic studies reveal that the PCL functions as multiple bands, with two major bands: the anterolateral band, which tightens in flexion and relaxes in extension, and the posteromedial band, which tightens in extension and relaxes in flexion (Fig. 13.1).[3,40–42] If a single graft is used to reconstruct the PCL, a decision must be made as to which of these two bands to reconstruct. A single graft cannot create normal PCL laxity through a full range of motion. The anatomic attachment points of the two bands are shown in Fig. 13.2.

Several authors have shown that the anterolateral band has greater linear stiffness and ultimate load strength and failure energy than either the posteromedial band or the meniscofemoral ligaments.[43,44] Based on this finding, most authors advocate reconstruction of the anterolateral band, realizing that this type of reconstruction, if appropriately tensioned, will provide resistance to posterior tibial translation when the knee is in greater flexion angles. It is accepted that a graft so positioned will become somewhat lax as the knee approaches full extension. If the posterior capsule and posterolateral corner are intact, however, these structures will serve as important secondary restraints to posterior tibial translation near full extension when the intraarticular graft is less supportive. The posteromedial band, if it were to be reconstructed and appropriately tensioned, would provide resistance to posterior tibial translation near full extension, but would become somewhat lax and nonsupportive when the knee was in greater flexion angles where there is little help in preventing posterior tibial displacement.

Isometry

Unlike the anterior cruciate ligament (ACL), the PCL does not have an area of isometry that is large enough to accept a finite-sized graft.[45–47] The most isometric force on the femur relative to the anatomic tibial attachment site is located within the anterior segment of the posterior fibers of the PCL footprint (Fig. 13.3).[48] The area is so small, however, that it will not allow the isometric placement of a finite-sized graft. The femoral attachment site of the PCL graft appears to be the primary determinant to allow isometric movement through a defined range of knee motion.[45,46,48–51] Length changes are relatively insensitive to the tibial attachment site within the confines of the anatomic footprint. Recommendations for the placement of the femoral site include both anterior along the proximal edge of the PCL[45,48] and more posterior, near the insertion of the PCL oblique band.[49,51] Although both of these placements are nonisometric, Galloway et al.[52] found that the stability of the knee after reconstruction of cadaveric knees with a single autograft was strongly dependent on the femoral attachment location. They further suggested that the femoral position of the graft that most closely reproduced the motion of the intact knee was not centered at the point where a single fiber was most isometric. Instead, they showed that when utilizing a single graft to reconstruct the ACL, a femoral attachment site high and forward in the notch is the most ideal position.[52]

It has been postulated that a reconstruction that uses two separate bands of tissue to reconstruct the PCL may allow for a more supportive reconstruction of the PCL throughout a full range of motion. If a traditional anterolateral band reconstruction is augmented by a posteromedial band reconstruction, it can be theorized that the anterolateral band will prevent posterior tibial displacement at greater flexion angles, and as it becomes lax near full extension, the posteromedial band will come under tension and resist posterior tibial translation. Race and Amis[53] demonstrated that reconstruction in this manner more closely restored normal knee laxity across the full range of flexion.

Similarly, Metcalf et al.[54] compared motion patterns of the intact knee with motion patterns following anterolateral band PCL reconstruction as well as anterolateral band reconstruction augmented by a posteromedial band reconstruction. The results of these studies are graphically demonstrated in Fig. 13.4. The study demonstrates that preinjury motion patterns can be more closely reproduced following a two-band PCL reconstruction than they can be with reconstruction of the anterolateral band alone.

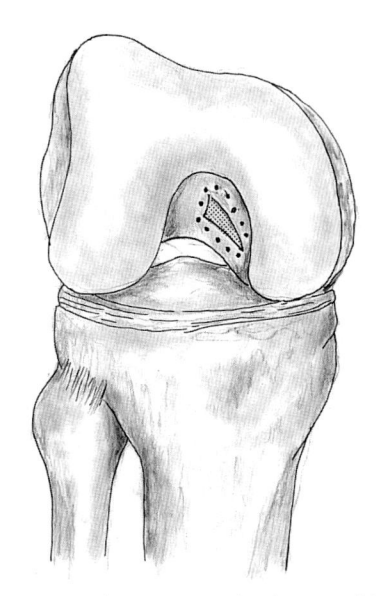

Fig. 13.3. The area on the femur of least strain for a single fiber relative to an anatomic tibial attachment site. The area is centered in the distalmost part of the proximal fibers. The cross-hatched area strains less than 10%. The bulk of the PCL attachment on the femur is not isometric and strains more than 10% through a full range of motion.

Operative Principles of Double Femoral Tunnel PCL Reconstruction

Graft Choices

There are several graft options for double femoral tunnel PCL reconstruction. It is important that tissue be utilized that is strong enough to func-

Two Band PCL Reconstruction

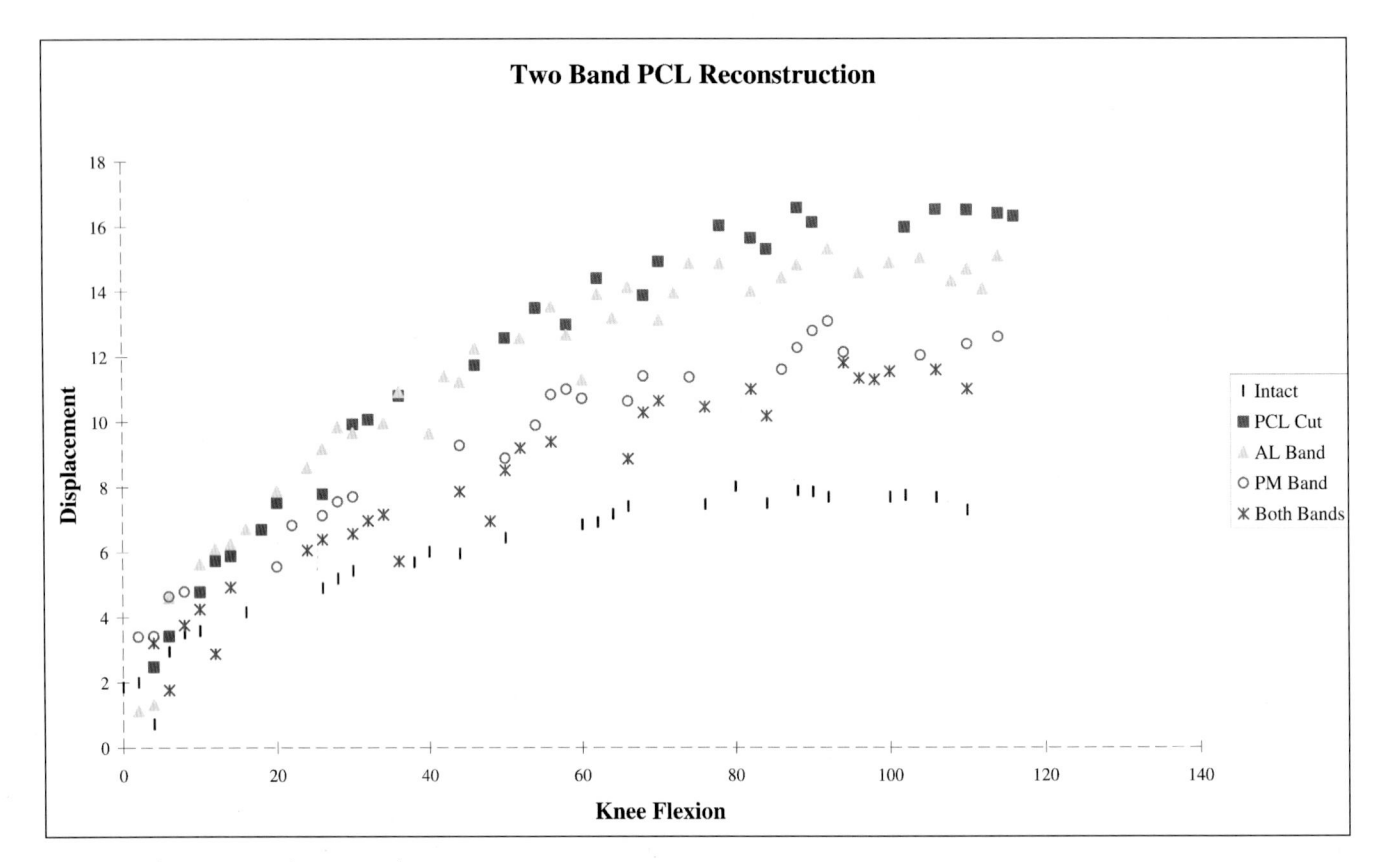

[Scatter plot showing Displacement (y-axis, 0-18) vs Knee Flexion (x-axis, 0-140) with five data series: Intact, PCL Cut, AL Band, PM Band, Both Bands]

Fig. 13.4. The amount of posterior laxity in the knee as a function of flexion angle with a constant 100-N posterior force. With reconstruction of both the posteromedial band and the anterolateral band, normal laxity is more closely approximated through a full range of motion.

tion in this location, that can be adequately fixed, and that can be preferentially tensioned. The options for tissue include autogenous hamstring tendons, preferably a semitendinosus from each lower extremity, two bone–patellar tendon–bone grafts, a split quadriceps tendon graft, and allograft tissue for either one or both grafts.

The use of hamstring tendons is particularly convenient for a double femoral tunnel PCL reconstruction technique. To obtain a large and long tendon for both grafts, the use of a semitendinosus tendon from each lower extremity is optimal. Utilizing two semitendinosus tendons allows for fixing the looped end of each graft into the femur, facilitating separate tensioning and direct fixation of each graft on the tibia. In some instances it is reasonable to utilize the gracilis tendon or an allograft hamstring tendon for the posteromedial band.

An allograft Achilles tendon or autogenous quadriceps tendon also provides adequate graft tissue for a double femoral tunnel PCL reconstruction technique. These grafts offer the advantage of an optional tibial onlay procedure posterior to avoid a transtibial tunnel and the associated posterior bending angle. This procedure however requires tensioning on the femur, which is somewhat more challenging than tensioning of the tibial ends of the grafts.

It is unlikely that a 10-mm-wide patellar tendon autograft is large enough to reconstruct both bands of the PCL. If utilizing this autogenous tissue, it is likely necessary to go to the opposite knee to obtain a second 10-mm-wide graft to provide the second graft. It may also be appropriate to utilize a bone–patellar tendon allograft for one graft or an autogenous semitendinosus from the same extremity for a second graft. See Chapter 10 for a more detailed discussion of graft choices.

Optimizing Tunnel Placement

Tibial Tunnel Placement

The appropriate location of the tibial tunnel for an anterolateral band PCL reconstruction is a point in the distal and lateral portion of the tibial anatomic footprint (Fig. 13.5). The most common error in creating the tibial tunnel is to have the posterior exit point too proximal. This location can lead to insufficient bony integrity of the superior roof of the tunnel and a tendency for the graft to cut through and migrate anteriorly on the tibia. It is therefore advisable to have a guide pin exit the anatomic attachment site near the lateral aspect of the most distal fibers.

The isometry of both the anterolateral and posteromedial bands of the PCL is quite insensitive to the tibial attachment site within the anatomic footprint.[46] It is therefore not thought to be necessary to create two separate tibial tunnels when performing a double graft technique. The position described above for the anterolateral band will be adequate for the posteromedial band as well. If it is desirable for tensioning or fixation purposes to create a second tibial tunnel, it should be created parallel to the first slightly more medially and proximally within the footprint.

To reduce bending angles and to facilitate graft passage, the start point on the anterior tibia should be quite distal. The start point is usually approximately 2 inches distal to the joint line. This provides a tunnel path with a more favorable passing angle as it exits the posterior tibia. A more proximal start point would create a tunnel with a sharper bending angle and less bony integrity (Fig. 13.6). The start point can be on either the medial or lateral side of the tibial crest. It is easier to access the medial side, but a lateral start point provides a straighter graft course to the femoral attachment site (Fig. 13.7), and it also allows placement of anterior compartment musculature over the fixation device on the tibia. To shorten the graft construct and to avoid the sharp bending angle posteriorly, a posterior approach to the tibial attachment site can be undertaken and a bone block directly screwed to the posterior tibia (Fig. 13.8). If this approach is undertaken, the bone block can be placed directly at the PCL attachment location, avoiding the bending angle necessitated by a tibial tunnel.

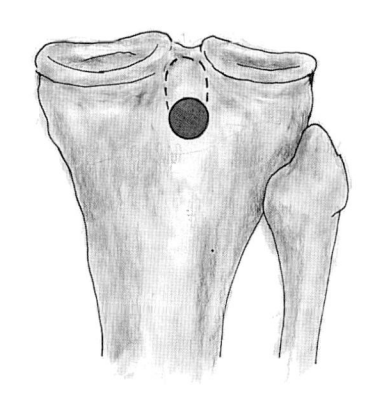

Fig. 13.5. The location of the tibial tunnel exit point should include the distalmost and lateralmost fibers of the PCL anatomic footprint.

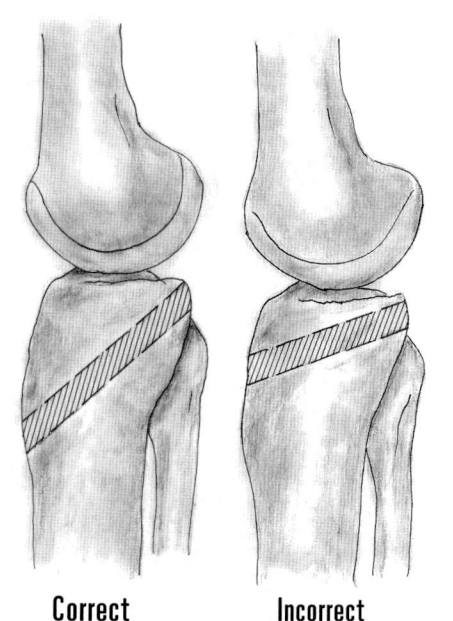

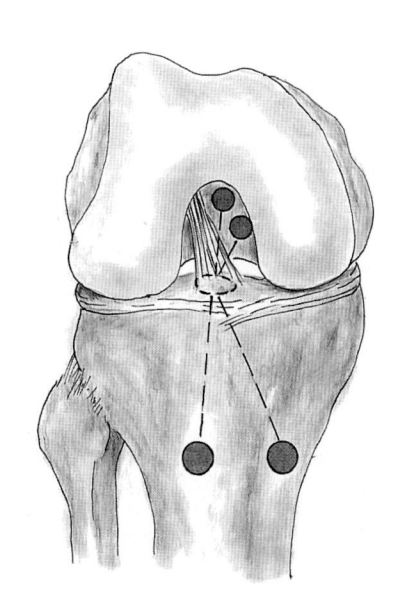

Fig. 13.6. Correct (left) and incorrect (right) angle of the tibial tunnel.

Fig. 13.7. The start point for the tibial tunnel can be on either the medial or lateral aspect of the tibia. A start point on the lateral side provides a straighter pathway for graft passage.

Correct   **Incorrect**

Fig. 13.8. To avoid a tibial drill hole and the sharp posterior bending angle, a tibial onlay procedure can be utilized to fix the graft to the tibia. With this technique, graft tensioning is done on the femoral side.

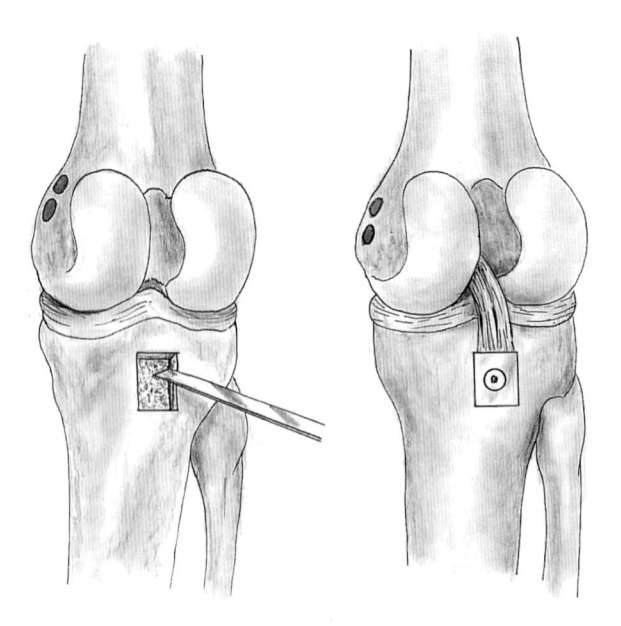

Femoral Tunnel Placement

Optimal graft attachment to the femur is critical to obtain the desired function from a PCL graft. For reconstruction of the anterolateral portion of the PCL, the attachment needs to be high and "forward" in the notch as viewed arthroscopically. Bony landmarks are used to define this position. In general an acceptable position can be obtained if one centers a drill hole at either the 11 o'clock or 1 o'clock position and places the guide pin 7 to 10 mm from the articular cartilage margin at the front of the notch (Fig. 13.9). The exact position can be facilitated by observing the PCL footprint on the femur. The tunnel should encompass the anterior and distal most fibers of the footprint (Fig. 13.10). If drilled from outside in, the start point for the femoral tunnel should be approximately 1 cm from the articular cartilage margin and midway between the epicondyle and the trochlea. This minimizes the bending angle at the femoral tunnel and facilitates graft passage. A tunnel in this position is very close to the articular cartilage of the femur, however, and considerable care must be exercised to avoid damaging it.

Fig. 13.9. The tibial guide pin should enter the joint at approximately the 11 o'clock or 1 o'clock position and 7 to 10 mm from the articular cartilage margin. This will allow placement of the anterolateral band tunnel in a position that will include the anteriormost and distalmost fibers of the PCL footprint.

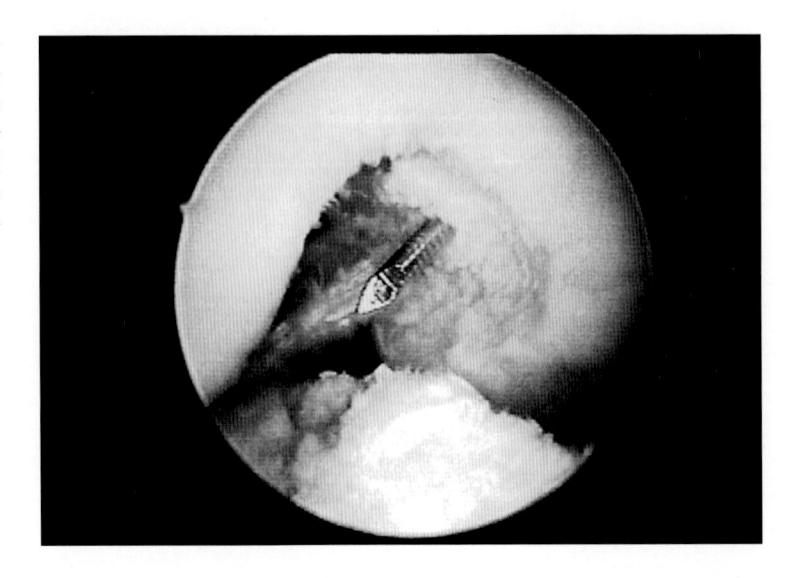

To create the posteromedial band of the PCL, a second femoral tunnel is created within the PCL footprint more inferior and slightly deeper in the intercondylar notch as viewed arthroscopically (Fig. 13.11). This corresponds to the posteromedial band as previously described. If utilizing outside-in drilling for the posteromedial band, the start point should be approximately 2 cm from the start point of the anterolateral band to allow for a bone bridge, which will help with fixation to the femur.

Tensioning of Grafts

The science of tensioning ligament grafts is still an emerging one; however, recent studies can provide some guidelines. It is important to understand the factors that influence appropriate tensioning. Proper tensioning is dictated by the laxity of the patient's normal knee and is influenced by the graft tissue being used, the isometry of the attachment sites, and the biomechanical characteristics of the graft fixation method utilized.[55] The purpose of tensioning is to allow near-normal stability through a complete range of motion. Grafts that are too tight may prevent a fluid motion and result in an overconstraint of the joint. While potential overconstraint is of more concern with ACL reconstruction, it is likely true for PCL reconstructions as well. An overtensioned graft might prevent normal anteroposterior (AP) laxity and cause increased stress on the graft as well as the articular cartilage. This may lead to premature graft failure or premature degenerative arthritis. The tissue being utilized for PCL reconstruction affects how tensioning is performed, since tensioning characteristics are tissue specific. The hamstring tendon graft is not as stiff a construct as a bone-tendon-bone graft with rigidly fixed bone blocks.[56] Therefore, more leeway is possible with tensioning a hamstring tendon graft than with the bone–patellar tendon–bone graft and the risk of overconstraint is less. Fixation devices that rigidly fix the graft at the point of entrance into the joint provide a graft whose properties are more of the graft tissue itself than of the fixation device. With less rigid fixation, the behavior of the graft becomes not only that of the tissue itself, but of the entire construct including the fixation device.

The isometry of the attachment points or the deviations from isometricity are also important factors in how grafts are tensioned. Grafts placed at

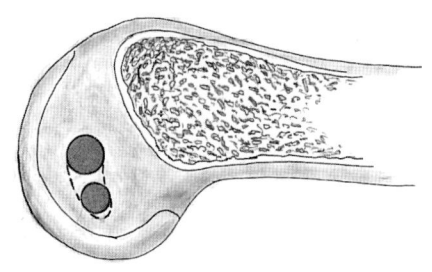

Fig. 13.10. The two femoral tunnels are depicted on the medial intercondylar notch. The anterolateral band tunnel involves that area of the PCL footprint that is most distal and most anterior. The posteromedial band tunnel includes those fibers of the footprint that are more posterior and proximal.

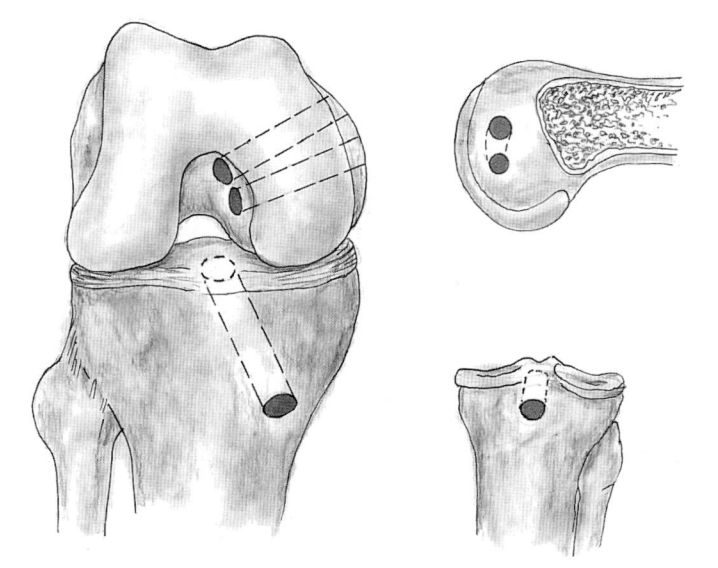

Fig. 13.11. Arthroscopic view of the anterolateral and posteromedial band tunnels.

a truly isometric attachment site can be tensioned to the desired amount at any flexion angle. This is not the case, however, in a double femoral tunnel PCL reconstruction. Since these grafts are not isometric, it is important to understand through which range of motion the graft must be supportive and to tension the graft in that range. It is also essential that in other ranges of motion the graft becomes more lax and not tighter.

To tension the anterolateral band of the PCL, it has been shown that the knee should be flexed to approximately 70 to 90 degrees. The tibia should then be pulled taut against an intact ACL prior to fixing the second end of the graft. A graft so tensioned will provide resistance to posterior tibial translation at greater flexion angles, but will become somewhat lax as the knee passes into full extension. The posteromedial band, therefore, needs to be tensioned at approximately 20 degrees of knee flexion again while pulling an anterior drawer force against an intact ACL. This graft will then be supportive near full extension, but will become somewhat lax as the knee passes into greater flexion angles where the anterolateral band will become supportive.

Surgical Technique Utilizing Hamstring Tendon Autografts

Setup

To obtain adequate tissue for double hamstring PCL reconstruction, it is usually necessary to obtain the semitendinosus tendon from both lower extremities. Therefore, both lower extremities need to be prepped appropriately. The PCL-deficient knee also should have a proximal tourniquet placed and a thigh-holding device. Care should be taken to ascertain that the knee can be flexed 90 degrees and that there is access to both the medial and lateral sides of the knee for instrumentation.

Graft Harvest and Preparation

A skin incision is made vertically overlying the insertion of the pes anserines of the involved extremity. Utilizing both blunt and sharp dissection, the incision is carried down to the semitendinosus tendon, which is isolated. It is released from the tibia, and care is taken to release all extraneous bands that may run from the inferior surface to the medial head of the gastrocnemius. The tendon is then harvested with a tendon stripper. Muscle tissue is removed from the graft and as much length as is possible is maintained. Through a small incision overlying the pes anserines of the opposite knee, the semitendinosus can be similarly harvested and the graft prepared. Both ends of both grafts are then sutured with a Bunnell stitch of heavy nonabsorbable suture. Each graft is then looped and passed through a sizing tube to determine the size of the femoral tunnels to be drilled. The combined double loops of both tendons are then passed through a separate sizing tube to determine the size of the tibial tunnel to be created. It is usual that the femoral tunnels will be 6 to 7 mm in diameter and the tibial tunnel 9 to 10 mm in diameter (Fig. 13.12).

Diagnostic Arthroscopy and Notch Preparation

The involved limb is next exsanguinated and the tourniquet inflated. Diagnostic arthroscopy is carried out and all additional procedures that are

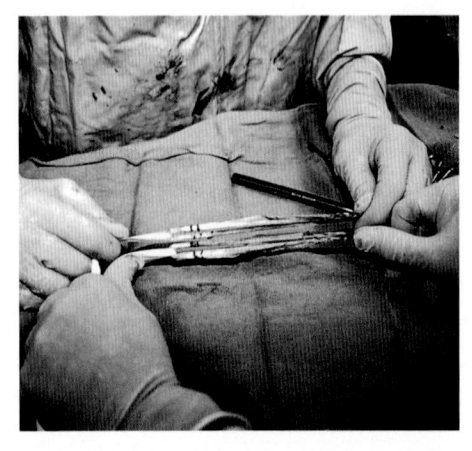

Fig. 13.12. The two semitendinosus tendons have been prepared for double femoral tunnel PCL reconstruction. The looped end of each graft will be pulled into femoral sockets by the attached passing tapes. The free ends of the grafts have been sutured with a Bunnell stitch of heavy nonabsorbable suture. Marks have been placed to gauge the depth of penetration of the looped ends into the femoral sockets.

necessary are completed. Notch preparation is next undertaken. It is usually unnecessary to perform a bony notchplasty. A debridement of the PCL stump, however, is mandatory and can be accomplished utilizing basket forceps and motorized shaver. It is advisable to maintain the footprint of the PCL on both the tibia and femur.

To adequately visualize the tibial attachment of the PCL and to adequately debride the area, it is usually necessary to create an accessory posteromedial arthroscopy portal. Through the posteromedial portal a motorized shaver can be passed directly to the PCL attachment site, which can be viewed through the intercondylar notch with either a 30- or 70-degree arthroscope. If the tibial stump is not adequately debrided, it can interfere significantly with graft passage.

Creation of the Tibial Tunnel

Any of several available drill guides can be utilized to place a guide pin for creation of the tibial tunnel. The start point for the tibial tunnel should be well distal to the joint line and approximately at the location where the semitendinosus tendon has been harvested from the tibia. The guide pin should exit the posterior tibia within the PCL footprint near the distalmost and lateralmost fibers. This position can be verified by viewing from the posteromedial portal and it may also be advisable to place this pin under fluoroscopic control to be certain that it exits appropriately and that it does not extend further posterior into the vulnerable neurovascular bundle. It is oftentimes helpful to place a fingertip at the PCL attachment site while drilling to be sure that instruments do not extend too far posteriorly. An opening for a fingertip can be made at the posterolateral corner of the knee to allow palpation of the tibial attachment site and additional protection from overadvancement. Once the tibial guide pin position has been verified, it can be overdrilled with the appropriately sized drill with due caution being taken to protect the neurovascular structures. Debris from the drilling should then be cleared utilizing the motorized shaver either through the tibial tunnel or from the posteromedial portal. The tibial tunnel is then plugged.

Creation of Femoral Tunnels

A skin incision is next made within Langer lines near the distal border of the vastus medialis obliques (VMO). This incision is then taken down to the medial femoral condyle and the VMO is retracted proximally. A femoral guide is then utilized to pass a guide pin for the anterolateral band tunnel. This guide pin should start at least 1 cm from the articular margin and approximately midway between the trochlea and epicondyle. The entry point into the joint should be 7 to 10 mm from the articular cartilage margin at the 11 o'clock or 1 o'clock position. The completed tunnel should encompass the distalmost and anteriormost fibers of the anatomic footprint, and therefore the exact placement of the guide pin is facilitated by an observable footprint. Once the first guide pin has been placed, a second guide pin should be placed, entering the joint lower in the intercondylar notch and usually slightly deeper in the intercondylar notch. The second guide pin is usually approximately 8 mm from the first and should encompass that portion of the footprint that is lowest and deepest within the notch corresponding to the posteromedial band. The start point for the second guide pin should be approximately 2 cm from the first to allow for the creation of a bone bridge on the medial condyle.

Graft Passage

A band of heavy polyester tape is next attached to the midportion of each of the two hamstring tendons. Two wire loops are next passed from anterior to posterior through the tibial tunnel and are grasped and pulled into the joint near the entrance of the femoral tunnels. Each wire loop is then grabbed through a femoral tunnel and delivered to the medial condyle. A second wire loop is then attached to each of the first wire loops, which allows pulling the looped end of the second wires across the joint into posterior aspect of the tibia and out to the anterior tibia. The polyester loop of each graft is then placed into the appropriate wire loop and is delivered to the medial condyle. The tapes are used to pull the looped end of each graft into the joint and into the appropriate femoral tunnel. Prior to pulling the grafts into the joint, a mark is placed on each graft at a point 20 mm from the looped end. This will serve as a gauge to allow pulling each graft into the appropriate femoral tunnel to a depth of 20 mm. The four free ends of the two tendons remain at the anterior tibia.

Femoral Fixation

The grafts are fixed to the femur by tying the polyester tapes together over the bone bridge. This provides adequate fixation of the grafts; however, if additional fixation at the entry point into the joint is desired, a supplemental bioscrew can be utilized at that location.

Tensioning and Tibial Fixation

The anterolateral band is generally tensioned and fixed first. This is done by placing a titanium screw and washer near the distal ends of the grafts. Around the screw the sutures are tied tightly while the knee is flexed 90 degrees and an anterior drawer is applied pulling the tibia forward against an intact ACL. Once the sutures have been tied, the graft is further tensioned while direct fixation with a soft tissue washer is accomplished.

The posteromedial band is tensioned and fixed in a similar fashion. The grafts, however, may need to be shortened slightly to allow placement of a second washer a slight distance from the first. The fixation sutures for the second graft are tied tightly around the screw post while the knee is flexed to 20 degrees and an anterior drawer is being applied on the tibia to pull it tightly against the intact ACL. After further tensioning the second graft is fixed directly with a second soft tissue washer. The final position of the grafts and the fixation devices are shown in Figs. 13.13 and 13.14.

Alternate Inside-Out Fixation on the Femur

It is possible to create femoral sockets by an inside-out technique. In this situation guide pins are passed from a low and somewhat more laterally placed anterolateral portal. The guide pins can be passed from inside out in the locations previously described. It is generally unnecessary to utilize a drill guide when creating the tunnels from inside out. Sockets can be created to the desired depths and passing channels created to allow for pulling the grafts into the sockets.

When passing grafts by the all-inside femoral technique, it is necessary to pull the graft and the leading sutures into the joint and then to attach them to passing pins, which will deliver the pulling sutures into the femoral tunnels and to the medial side of the knee. The grafts can then be pulled into the created sockets and fixed with interference-fit screws (Fig. 13.15). The passing sutures can then be removed. Tibial fixation would then be accomplished as previously described.

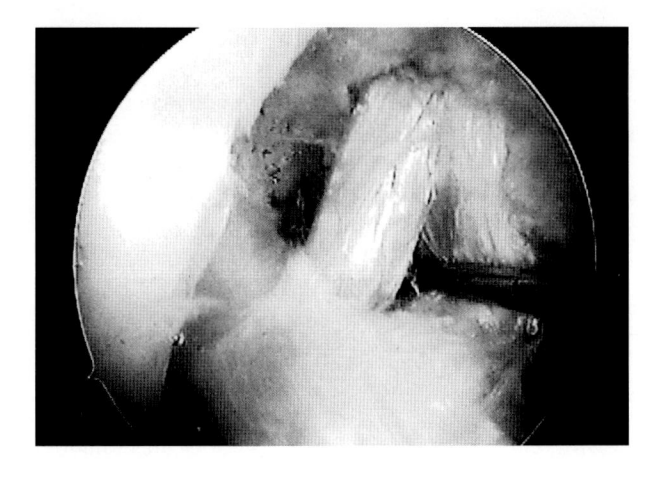

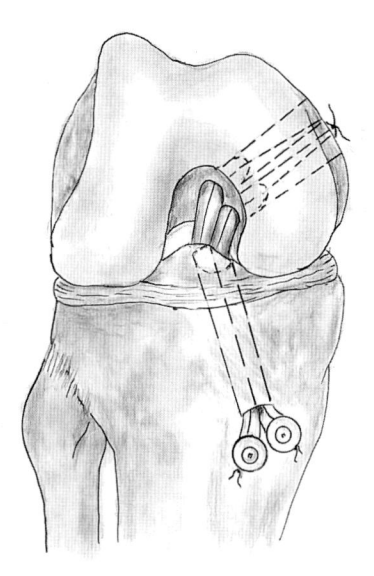

Results of Double Femoral Tunnel PCL Reconstruction

To date there are no published studies comparing the results of PCL reconstruction with single-tunnel versus double-tunnel techniques. There is ample anecdotal information that motion patterns are better restored by these techniques, which supports biomechanical studies. There are no studies as yet, however, that evaluate subjective results or that document any improvement in subsequent arthritic changes. The double femoral tunnel technique of PCL reconstruction is still in its early stages and well-controlled prospective studies will be needed to determine whether clinical results can validate the theoretical advantages of the technique.

Discussion

The PCL is not an isometric ligament but instead functions as at least two distinct ligaments—an anterolateral and posteromedial band. The anterolateral band is positioned to resist posterior tibial displacement when the knee is in flexion, and the posteromedial band is positioned to resist posterior tibial translation when the knee is in extension. Since these ligaments both undergo extensive length changes through a full range of knee motion, it is easy to understand why a single-band PCL reconstruction cannot restore the normal motion pattern to a PCL-deficient knee. The fact that normal motion patterns cannot be restored to a PCL-deficient knee may be an important reason why PCL-deficient knees tend to develop premature arthrosis particularly of the patellofemoral and medial compartment articulations.

In an attempt to better reproduce the motion pattern of a PCL-intact knee, techniques have been emerging to reconstruct both the anterolateral and posteromedial bands of the PCL and to tension them so that one or the other portion of the double graft will resist posterior tibial translations

Fig. 13.13. The arthroscopic view shows the final position of the double femoral tunnel PCL hamstring grafts. The posteromedial band is being displaced by the probe.

Fig. 13.14. Final position of the two grafts for the double femoral tunnel PCL reconstruction technique. The grafts are fixed to the femur by tying the polyester passing tapes together over a bone bridge. Each graft is separately tensioned and fixed on the tibia. Tibial fixation is by sutures and a separate screw and soft tissue washer for each graft.

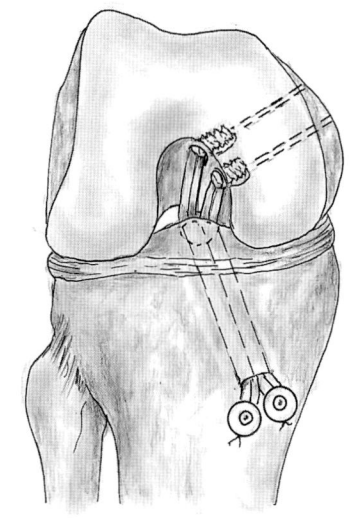

Fig. 13.15. With the inside-out drilling of femoral tunnels, the looped ends of the hamstring tendon grafts can be pulled into the femoral sockets with passing sutures fixed with interference-fit screws.

through a full range of knee motion. Biomechanical studies suggest that this is in fact the case.

The early operative experience is that two-graft PCL reconstructions appear to provide better resistance to posterior tibial translations through a full range of motion. Whether these grafts will continue to function over time and more importantly whether this technique will delay the arthritic changes associated with PCL insufficiency better than a single-graft technique has yet to be determined.

References

1. Miyasaka KC, Daniel DM, Stone ML. The incidence of knee ligament injuries in the general population. Am J Knee Surg 1991;4:3–8.
2. Dandy DJ, Pusey RJ. The long-term results of unrepaired tears of the posterior cruciate ligament. J Bone Joint Surg 1982;64B:92–94.
3. Hughston JC, Bowden JA, Andrews JR, Norwood LA. Acute tears of the posterior cruciate ligament. Results of operative treatment. J Bone Joint Surg 1980;62A:438–450.
4. Torg JS, Barton TM, Pavlov H, Stine R. Natural history of the posterior cruciate ligament-deficient knee. Clin Orthop 1989;246:208–216.
5. Satku K, Chew CN, Seow H. Posterior cruciate ligament injuries. Acta Orthop Scand 1984;55:26–29.
6. Fowler PJ, Messieh SS. Isolated posterior cruciate ligament injuries in athletes. Am J Sports Med 1987;15:553–557.
7. Parolie JM, Bergfeld JA. Long-term results of nonoperative treatment of isolated posterior cruciate ligament injuries in the athlete. Am J Sports Med 1986;14:35–38.
8. Keller PM, Shelbourne D, McCarroll JR, Rettig AC. Non-operatively treated isolated posterior cruciate ligament injuries. Am J Sports Med 1993;21:132–136.
9. Skyhar MJ, Warren RF, Ortiz GJ. The effects of sectioning of the posterior cruciate ligament and the posterolateral complex on the articular contact pressures within the knee. J Bone Joint Surg 1993;75A:695–699.
10. Hughston JC, Degenhardt TC. Reconstruction of the posterior cruciate ligament. Clin Orthop 1982;164:59–77.
11. Hughston JC, Norwood LA. The posterolateral drawer test and external rotational recurvatum test for posterolateral rotatory instability of the knee. Clin Orthop 1980;147:82–87.
12. Brantigan OC, Voshell AF. The mechanics of ligaments and menisci of the knee joint. J Bone Joint Surg 1941;23:44–66.
13. Clancy WCJ, Shelbourne D, Zoellner GB, Keene JS, Reider B, Rosenberg TD. Treatment of knee joint instability secondary to rupture of the posterior cruciate ligament. J Bone Joint Surg 1983;65A:310–322.
14. Clancy WGJ. Repair and reconstruction of the posterior cruciate ligament. In: Chapman MW, ed. Operative Orthopedics, vol 3. Philadelphia: JP Lippincott, 1988:1651-1655.
15. Eriksson E, Haggmark R, Johnson RJ. Reconstruction of the posterior cruciate ligament. Orthopedics 1986;9:217–220.
16. Verdonk R, Vandendriessche G, De Smet L, Overbeke JV, Vandekerckhove B, Claessens H. Free patellar tendon graft for reconstruction of old posterior cruciate ligament ruptures. Acta Orthop Belg 1984;52:554–560.
17. Southmayd WW, Rubin BD. Reconstruction of the posterior cruciate ligament using the semimembranosus tendon. Clin Orthop 1980;150:196–197.
18. Barrett GW, Savoie FH. Operative management of acute PCL injuries with associated pathology: long-term results. Orthopedics 1991;14:687–692.
19. Bianchi M. Acute tears of the posterior cruciate ligament: clinical study and results of operative treatment in 27 cases. Am J Sports Med 1983;11:308–314.
20. Fleming REJ, Blatz DJ, McCarroll JR. Posterior problems in the knee: posterior cruciate insufficiency and posterolateral rotatory insufficiency. Am J Sports Med 1981;9:107–113.

21. Lipscomb AB, Johnston RK, Snyder RB. The technique of cruciate ligament reconstruction. Am J Sports Med 1981;9:77–81.
22. Wirth CJ, Jager M. Dynamic double tendon replacement of the posterior cruciate ligament. Am J Sports Med 1984;12:39–43.
23. Insall JN, Hood RW. Bone-block transfer of the medial head of the gastrocnemius for posterior cruciate insufficiency. J Bone Joint Surg 1982;64A:691–699.
24. Kennedy JC, Galpin RD. The use of the medial head of the gastrocnemius muscle in the posterior cruciate-deficient knee. Am J Sports Med 1982;10:63–74.
25. Clendenin MB, DeLee JC, Heckman JD. Interstitial tears of the posterior cruciate ligament of the knee. Orthopedics 1980;3:764–772.
26. Roth JH, Bray RC, Best TM, Cunning LA, Jacobson RP. Posterior cruciate ligament reconstruction by transfer of the medial gastrocnemius tendon. Am J Sports Med 1988;16:21–28.
27. Tibone JE, Antich TJ, Perry J. Functional analysis of untreated and reconstructed posterior cruciate ligament injuries. Am J Sports Med 1988;16:217–223.
28. Tillberg G. The late repair of torn cruciate ligaments using menisci. J Bone Joint Surg 1977;59B:15–19.
29. Hughston JC. The posterior cruciate ligament in knee-joint stability. J Bone Joint Surg 1969;51A:1045–1046.
30. Coleman HM. Cruciate ligament repair using meniscus. J Bone Joint Surg 1956;38B:778.
31. Lindstrom H. Cruciate ligament plastics with meniscus. Acta Orthop Scand 1959;29:150–151.
32. Cubbins WR, Conley AH, Callahan JJ, Scuderi CS. A new method of operating for the repair of ruptured cruciate ligaments of the knee. Surg Gynecol Obstet 1932;54:299–306.
33. Bosworth DM, Bosworth BM. Use of fascia lata to stabilize the knee in cases of ruptured crucial ligaments. J Bone Joint Surg 1936;18:178–179.
34. Barfod B. Posterior cruciate ligament reconstruction by transposition of the popliteal tendon. Acta Orthop Scand 1971;42:438.
35. McCormick WC, Bagg RJ, Kennedy CW. Reconstruction of the posterior cruciate ligament. Clin Orthop 1976;118:30–34.
36. Noyes FR, Barber SD. Allograft reconstruction of the anterior and posterior cruciate ligaments: report of 10-year experience and results. Instr Course Lect 1993;42:381–396.
37. Noyes FR, Barber Westin SD. Posterior cruciate ligament allograft reconstruction with and without a ligament augmentation device. Arthroscopy 1994;10:371–382.
38. Mariani PP, Adriani E, Santori N, Maresca G. Arthroscopic posterior cruciate ligament reconstruction with bone-tendon-bone patellar graft. Knee Surg Sports Traumatol Arthrosc 1997;5:239–244.
39. Lipscomb AB Jr, Anderson AF, Norwig ED, Hovis WD, Brown DL. Isolated posterior cruciate ligament reconstruction. Long-term results. Am J Sports Med 1993;21:490–496.
40. Van Dommelen BA, Fowler PJ. Anatomy of the posterior cruciate ligament. A review. Am J Sports Med 1989;17:24–29.
41. Girgis FG, Marshall JL, Monajem A. The cruciate ligaments of the knee joint. Anatomical, functional and experimental analysis. Clin Orthop 1975;106:216–231.
42. Johnson CJ, Bach BR. Current concepts review. Posterior cruciate ligament. Am J Knee Surg 1990;3:143–153.
43. Harner CD, Xerogeanes JW, Livesay GA, et al. The human posterior cruciate ligament complex: an interdisciplinary study. Am J Sports Med 1995;23:736–745.
44. Race A, Amis AA. The mechanical properties of the two bundles of the human posterior cruciate ligament. J Biomech 1994;27:13–24.
45. Grood ES, Hefzy MS, Lindenfeld TN. Factors affecting the region of most isometric femoral attachments. Am J Sports Med 1989;17:197–207.
46. Sidles JA, Larson RV, Garbini JL, Downey DL, Matsen FAI. Ligament length relationships in the moving knee. J Orthop Res 1988;6:593–610.
47. Covey DC, Sapega AA, Sherman GM. Testing for isometry during reconstruction of the posterior cruciate ligament. Anatomic and biomechanical considerations. Am J Sports Med 1996;24:740–746.

48. Ogata K, McCarthy JA. Measurement of length and tension patterns during reconstruction of the posterior cruciate ligament. Am J Sports Med 1992;20: 351–355.
49. Trus P, Petermann J, Gotzen L. Posterior cruciate ligament reconstruction—an in vitro study of isometry. Part I: test using a string linkage model. Knee Surg Sports Traumatol Arthrosc 1994;2:100–103.
50. Daluga DJ, Bach BRJ, Mikosz R. Force displacement characteristics of the posterior cruciate ligament with alterations of the femoral and tibial insertion sites. Am J Sports Med 1990;18:551.
51. Petermann J, Gotzen L, Trus P. Posterior cruciate ligament reconstruction—an in-vitro study of isometry. Part II: tests using an experimental PCL graft model. Knee Surg Sports Traumatol Arthrosc 1994;2:104–106.
52. Galloway MT, Grood ES, Mehalik JN, Levy M, Saddler M, Noyes FR. Posterior cruciate ligament reconstruction: an in vitro study of femoral and tibial graft placement. Am J Sports Med 1996;24:415–426.
53. Race A, Amis AA. PCL reconstruction. In vitro biomechanical comparison of "isometric" versus single and double-bundled "anatomic" grafts. J Bone Joint Surg 1998;80B:173–179.
54. Metcalf MH, Larson RV, Harrington RM. Two tunnel vs. single tunnel PCL reconstruction: a biomechanical assessment. Unpublished data, 1998.
55. Corsetti JR, Jackson DW. Failure of anterior cruciate ligament reconstruction: the biologic basis. Clin Orthop 1996;325:42–49.
56. Noyes FR, Butler DL, Grood ES, Zernicke RF, Hefzy MS. Biomechanical analysis of human ligament grafts used in knee ligament repairs and reconstructions. J Bone Joint Surg 1984;66A:344–352.

Chapter Fourteen

Cruciate Ligament Reconstruction with Synthetics

Don Johnson and J.P. Laboureau

I. History and Current Indications
Don Johnson

The use of a synthetic replacement for the cruciate ligament has had a long and fascinating history.[1-3] Hey Groves recorded the original replacement of the anterior cruciate ligament (ACL) with a piece of stainless steel wire in 1917. The quest for the ideal synthetic continued into the modern era of orthopedics; Dr. Jack Kennedy[4] in 1975 use Polyflex, a polyethylene ligament substitute.[5] This was firmly anchored to the bone with a metal collar and bone cement. My only experience with this device was in attempting to remove a failed ligament, and I was left with a large hole, requiring bone grafting.

Marshall first studied the use of a Dacron mesh in dogs in 1975. The clinical use of Dacron continued for the next 10 years.[6,7]

Carbon fiber came into use to replace the ACL in 1975.[8-22] The original ligament was coated with polylactic acid polymer to prevent particulate migration. This still did not stop the carbon fibers from staining the synovium and being picked up by the regional lymph nodes. The theoretical idea of the carbon fiber was that it would act as a stent for the proliferating fibroblasts to lay collagen onto. Unfortunately, the carbon fiber was very brittle and would break up around the corners of tunnels.

The ABC ligament was developed by Mr. Angus Strover in the United Kingdom, about the same time as Gore-Tex was being developed in North America. The ABC ligament evolved from the carbon fiber ligament of the 1970s. The carbon fiber was woven into a polyester mesh, and a composite ligament was fabricated. The idea was that the carbon fiber would stimulate ingrowth of fibroblasts into the polyester and a neoligament would be formed. The results were inconsistent.

Kennedy in 1979 developed a ligament augmentation device (LAD).[23-25] It was made from polypropylene, and was designed to be sutured to a weak tendon to augment its strength.

Trevira (Austin and Associates, Fallston, MD), developed in Germany, was another device that was very similar to the Kennedy LAD—a flat woven polyester graft.[26] It initially was designed as a stent to augment ligament repairs, but gradually evolved into a stand-alone replacement for cruciate ligaments. The device was rolled for use in the tunnels and the intraarticular portion, but was allowed to lie flat over the edge of the tunnel to prevent abrasion. The Trevira ligament is available in different widths and it is still used in Germany.

The Leeds-Keio ligament was a polyester open weave ligament that was developed as a cooperative effort between Japan and England.[27,28] It be-

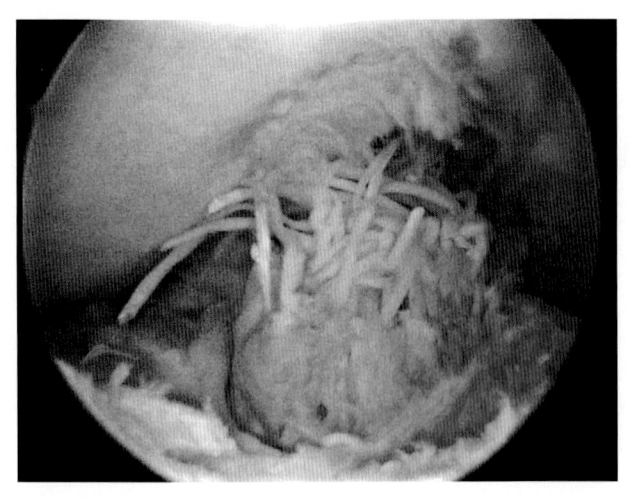

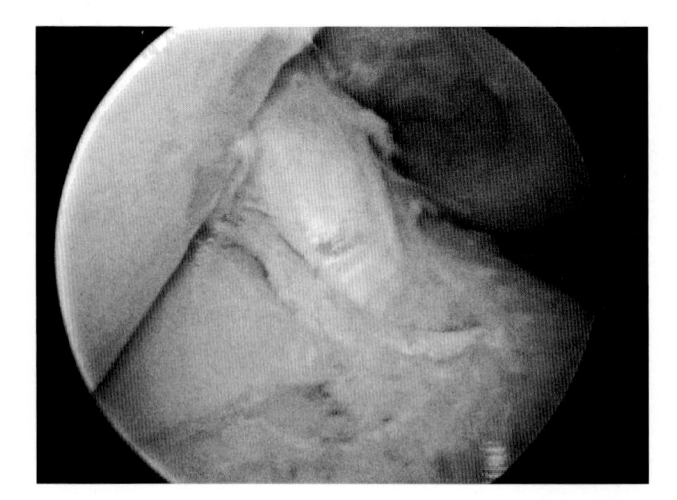

Fig. 14.1. The arthroscopic view of a failed original Gore-Tex ligament. The braided portion has failed intraarticularly. The tibial tunnel is also somewhat anterior, contributing to the failure by anterior notch impingement.

Fig. 14.2. The arthroscopic appearance of the compact diameter Gore-Tex graft 4 years after implantation. The intraarticular portion was intact, and sometimes covered in fibrous tissue and synovium.

came very popular in Europe. It initially had very favorable short-term results. One of the advantages was that this ligament was anchored in the tunnels with a bone plug taken with a coring reamer.

The original Gore-Tex fiber ligament was designed as a braided ligament.[29–31] It was first implanted in patients in 1982. The second-generation ligament was released in 1989, after the intraarticular failure of the original was noted. As with many of the previous synthetics, the initial optimism gave way to reality when the ligament was noted to fail over the edge of the femoral tunnel. By the time this was noticed, and the graft placed in the over-the-top position, with better results, the company withdrew the device from the market. Once again, one of the best features of this ligament was the firm anchorage to bone, with bicortical screws placed through an eyelet in the end of the ligament.

After several years of use of the synthetics to reconstruct the ACL,[1–3] there were numerous studies to suggest that the failure rate was unacceptably high. There were very few published papers on the posterior cruciate ligament (PCL) reconstruction, but the long-term results were similar.

At the present time there are no synthetic ligaments approved by the Food and Drug Administration (FDA) for use in the United States.

During the 1980s, my personal experience was with the use of the Gore-Tex ligament device (Fig. 14.1). The intraarticular portion of the compact diameter second-generation graft was wrapped with Gore-Tex fiber to eliminate the fraying of the graft. This graft was available from 1989 to 1994. The technique of making an extraarticular femoral drill hole seemed to be a good idea, but eventually the graft failed at the femoral tunnel edge, the return of the "killer tunnel angle" (Fig. 14.2).

Problems with Synthetics

The killer tunnel angle is what dooms all synthetics. They have excellent longitudinal strain characteristics, but poor cyclical bending strength. The common site of failure is wear over the edge of the tunnel exit—the killer tunnel angle (Fig. 14.3).

The Gore-Tex compact diameter graft failed where the graft was cycled over the edge of the bony femoral tunnel. When we stopped making a femoral tunnel, and simply put the graft over the top, the results improved.

The LARS Synthetic Ligament

Dr. J.P. Laboureau, Dijon, France, developed the Ligament Advanced Reinforcement System (LARS) synthetic ligament in an attempt to solve some of these problems encountered with the synthetic ligament over the past 20 years (Fig. 14.4). Laboureau improved the surgical technique by:

- Designing guides to place the tunnels in correct anatomic positions,
- Using the two-bundle technique to reconstruct the PCL,
- Reducing the femoral tunnel angle.

The design of the material was also improved by:

- Using a polyester ligament that is well tolerated,
- Using an open weave that at the tunnel exit lies flat,
- Using an open weave to bury the synthetic ligament into the residual ligament, and encourage incorporation of the fibrous tissue.

What Have We Learned from Synthetic Reconstruction of Cruciate Ligaments

Have synthetics failed us, or have we failed synthetics? After 15 years of use of various synthetics, I think that in many cases we did the synthetic ligament a disservice by poor engineering design and improperly implanting the device. However, even with the ideal surgical technique, often the graft fails because of excessive mechanical stresses. The first choice for cruciate ligament reconstruction still should be an autogenous graft.

Laboureau's LARS technique was an open arthrotomy, with the PCL peeled off the femur and pushed to the back of the joint. The arthroscopic modification was to excise all the PCL, as we had done for patellar tendon PCL reconstructions (to make the graft passage easier). I think that leaving the entire residual ligament may allow for more tissue in-growth, and improve the final results. The use of the LARS synthetic ligament is described in detail on pages 192 through 212. In my review of 30 patients at 2-year follow-up, only 40% had good results. Many failures occurred early, within 6 months, which points to a surgical technical error. The major technical problem that I can identify is the complete removal of the ligament in order to do the operation arthroscopically. The residual ligament should be left to allow incorporation into the polyester synthetic ligament.

Current Indications

Synthetics could be used in the older patient with multiple ligament injuries, who needs an internal splint while the capsular ligaments heal. The splint holds the tibia in a reduced position and allows the injured PCL to heal as quickly as possible. This prevents the tibia from healing in a posterior subluxed position. Synthetics could also be used when there is no other graft source available, i.e., a revision situation.

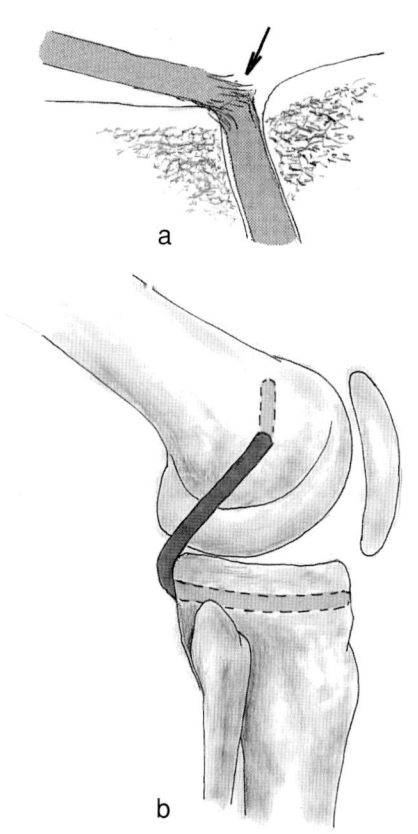

Fig. 14.3. (a and b) The edge of the tunnel produces a bending angle that grafts, and especially synthetics, do not tolerate well. This has been called the "killer tunnel angle."

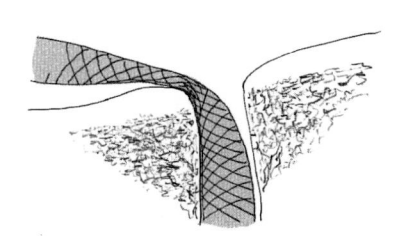

Fig. 14.4. The LARS ligament has an open weave that lies flat at the tunnel edge, and decreases the bending stress.

Conclusion

Experience has shown us that the synthetic ligament has severe limitations in withstanding the bending forces around bony tunnels. The reported failure rate in the literature of most of the synthetics has been unacceptably high.

What does the future hold for synthetics in the reconstruction of the PCL-deficient knee? It may well be that the synthetic of the future will be made of synthetic collagen. If it can be made strong enough, then the holy grail of ligament surgeons may well have been found. Until then the best option is still autogenous tissue.

Will the surgical technique improve? I believe that the two-bundle technique is the best available surgical option. It is not easy, and few surgeons have sufficient surgical cases to overcome the learning curve and become experienced. In several cases, I have used the split quadriceps tendon graft, placing the bone block in the tibia, and splitting the tendon portion to place it in two separate femoral tunnels. This same technique is possible with the Achilles tendon allograft.

In the future the graft material will be improved. It is possible that a synthetic collagen graft, or small intestine submucosa graft, will take the place of the polyester. Then an in-growth of tissue would ensure long-term viability of the graft.

II. The LARS Technique in Posterior Cruciate and Posterolateral Deficiency

J. P. Laboureau

Development and Design of the LARS® Ligament

The question was: Should the bad results of the first artificial ligaments lead us to definitely give up on the concept of their use? This has been the attitude of many. But the advantages they could bring to ligament surgery were sometimes so impressive that the reasons for the failures of the early days deserve to be carefully studied before any decision was made. Beside obvious technical surgical mistakes or bad indications, the ligament itself probably needed to be revised, taking into account the specific mechanical stresses and the biological environment.

Mechanical Stresses

A ligament does not work like a rope or a shoe-lace. A ligament is submitted to combined traction, flexion, and torsion. Flexion and torsion may vary with direction and placement of the tunnels. They can be diminished by surgical procedures, but cannot be eliminated. The synthetic ligaments must be designed for these types of stresses. The previous ligaments were all braided, woven, or knitted and somehow had fibers crossing each other.

Mechanical comparative studies of fatigue behavior were conducted on knee simulation devices combining traction, flexion, and torsion by P. Hunault (Sercovam 33611 Cestas, France) to see whether the ligament structure was important or not. It was demonstrated that the stresses are concentrated at the intraarticular exit of the tunnels where there is a junction between the mobile intraarticular and the fixed portions. The main factors for fatigue deterioration are the friction between the crossing longitudinal and transversal fibers, and the relative stiffness of this junction. It was concluded that this type of structure must be avoided. The intraarticular portion of a ligament must be very flexible to allow a well-balanced distribution of flexion and torsion stresses along the whole intraarticular length and not only at the junction of mobile and interosseous parts. In this portion transverse structures, which are useless, must be eliminated. Longitudinal fibers must be parallel, totally free, and not connected to each other. Moreover, the direction of torsion is different for ligaments of left and right knees: Looking from the notch, the torsion from extension to flexion of the PCL is counterclockwise for the right knee and clockwise for the left knee. (It is the opposite way for the ACL). Therefore, to avoid unnecessary stresses, the ligament, which is usually placed in flexion, must be pretwisted accordingly.

Studies were carried out on different types of knitted and "free fibers" ligaments made of the same material and with the same protocol. It was clearly demonstrated that on knitted ligaments ruptures occurred as soon as 5 million cycles, while a few partial ruptures were starting only at 10 million cycles on free fibers. Another advantage of the "free fibers" design is a gain of volume with less synthetic material for equivalent resistance; the design also offers a completely open structure for fibroblastic invasion and provides a flat structure behind the tibia, which is much better than a cylindrical body for flexion-tension stresses.

Biological Environment

Synthetic ligaments are designed to live in a fibroblastic and collagenic environment. Different types of fibers were first selected for their mechanical properties (resistance and hysteresis). Then these fibers were tested using human fibroblast in vitro cultures. On some materials, the fibroblasts degenerated and finally died. On some others the fibroblast culture was growing and the cells were invading the space between fibers. On each type of fibers the mass of cells per same volume unit of material was measured. The fiber that was finally selected was the one that had the best mechanical properties for the best fibroblast ingrowth. It is a specific type of polyester.

The raw fibers that come from different manufacturers are polluted by different substances. The lubricating substances were probably the main factor responsible for the acute synovitis encountered in the past. A cleaning process was developed to eliminate all these substances as well as to purify the fibers from any undesirable particles. Each ligament is submitted to several cycles of cleansing process until the micromeasures show no residue.

The LARS Posterior Cruciate Ligaments

The new generation LARS (Ligament Advanced Reinforcement System—LARS Company Arc-sur-Tille 21560 France) ligaments were designed to fulfill all of the above requirements and criteria. A ligament comprises two different parts: The intraarticular part is 40 mm long and made only of longitudinal parallel and totally independent fibers; the interosseous parts are made of the same fibers strongly united in these portions by a warp knitted process. Due to this strong connection in the bony tunnels, all the fibers will work together in the intraarticular part and will share the forces after all of them have been recruited (Fig. 14.5). The resistance of the ligament depends on the number of fibers: PC 60, PC 80, PC 100. Basic mechanical testing has been performed by the Institut Textile de France and Laboratoire National d'Essai in France, and Ecole Polytechnique de Montréal (Canada). Table 14.1 shows the mechanical behavior of the three ligaments under traction. After a maximum strain of about 9% the ligament comes back close to its initial length. This elongation will remain but will

Fig. 14.5. The different types of ligaments to be used to reinforce the torn PCL itself.

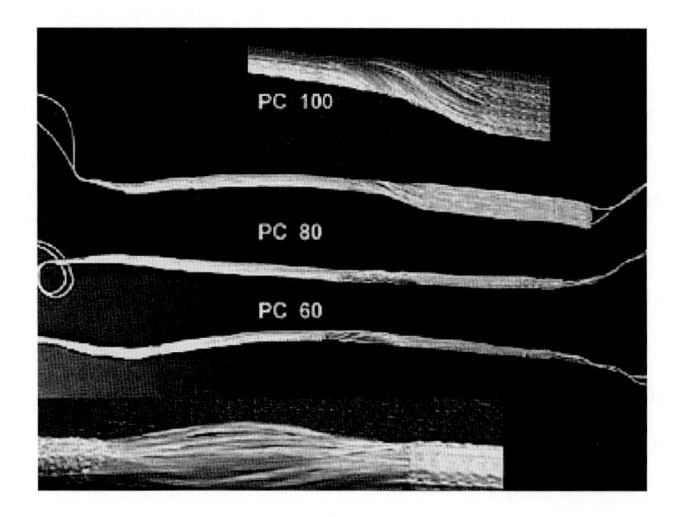

not increase in time for the same load. Residual resistance to traction after 10 million cycles combining traction, flexion, and torsion was 65% (average) of the initial resistance. Elongation depends on load but does not vary in time. For example, elongation on a 100-fiber ligament will reach 3% under a cyclic loading of 1,000 N during the first cycles, but this elongation will stabilize at 3% and will not increase by increasing the number of cycles.

This is a clinical significance of 3% of a 40-mm long PCL or 1.2 mm. One should not rely on time to expect any lengthening of a ligament placed with hypertension. A ligament should never be placed under tension, to allow full and tension-free complete motion. Surgically speaking, this means that with hypertensioning of the ligament, motion will be limited, or the ligament will be submitted to unnecessary permanent stresses and will end by rupturing. This explains the importance of the two main principles to be respected when using an artificial ligament: isometry and no tension. The PC 60 and PC 80 are cyclindrical and pass through a 6-mm bonny tunnel. The PC 100 is cyclindrical in its distal tibial part but flat in its proximal (femoral) part. This disposition gives to the intraarticular part a fan shape, reproducing the anatomy of the anterior bundle. The cylindrical part fits in a 7.5-mm tibial tunnel, and the flat part requires a series of 3.5-mm joining tunnels on a length of 2 cm. The PPLY 100 ligament is a special implant designed to allow with the same tibial (7.5 mm) tunnel the reconstruction of the PCL anterolateral bundle by one branch of the Y, and the posterolateral structures by the other branch of the Y (Fig. 14.6).

Three types of ligaments are designed to be used as a reinforcement with autogenous reconstructions to prevent their elongation during their restructuration phase. The ACTOR 10 is a cylindrical tube, comprising also free fibers in its intraarticular part. The autogenous transplant is placed inside the tube with the help of built-in leaders. The synthetic structure is very porous, especially in the intraarticular part, which permits biological

Table 14.1. The mechanical behavior of three ligaments under traction

LARS ligament	Resistance to rupture Average (Newtons)	Maximum strain	Residual strain after traction at 2,500 N
PC 60	2,500	8.5%	0.5%
PC 80	3,500	9.8%	0.5%
PC 100	4,500	9.8%	0.5%

Fig. 14.6. The PPLY 100 (Postero–Postero Lateral, shaped as a Y, 100 fibers) is designed to reconstruct both the anterolateral PCL bundle and posterolateral structures.

exchanges between transplant and environment. The transplant must be positioned and centered in the free fibers intraarticular part. The AC 30 RA and PTR 30 are small ligaments, with the usual LARS structure, which are designed to be placed inside the autogenous transplant with a canula or after opening the transplant as a "hot dog" (Fig. 14.7). The AC 30 RA is cylindrical in both bony parts. The PTR 30 is flat on one of its bony parts.

Histological Studies

The LARS was first studied in animals. In its present design, the ligament has been used in humans since 1992. A few revisions gave us the opportunity to do histological studies in humans. These studies show a perfect tolerance with no inflammatory reaction, no synovitis. The fibrous tissue coating the ligament is made of dense fibroblastic colonization with vessels. This colonization can be seen in between the fibers because of the very open structure, but also deeper inside the core of each fiber in between the microfilaments. Very few macrophages but also the synthesis of new collagen were noted. (Fig. 14.8). The examination using polarized light does not show any debris or particles coming from the synthetic fibers. Each fiber looks isolated from its neighbors by the fibroblastic invasion. Very

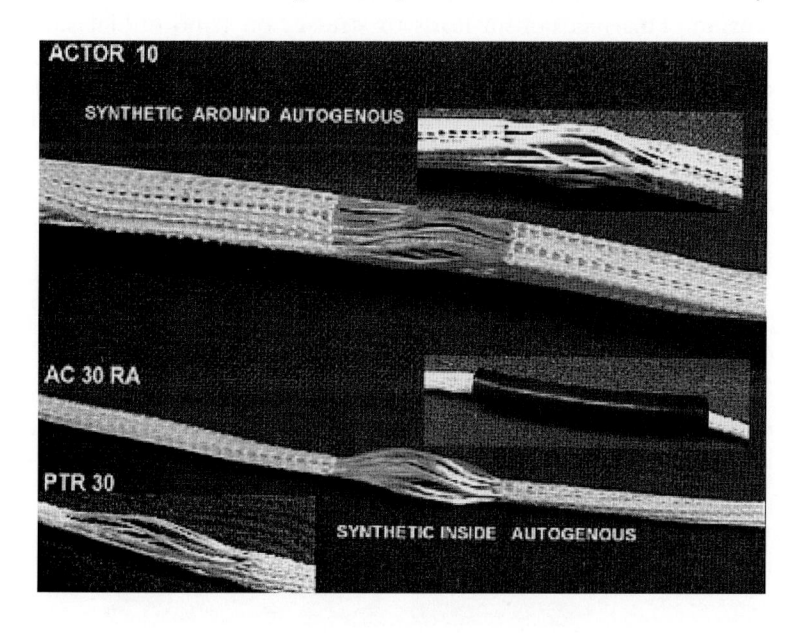

Fig. 14.7. The different types of scaffolds to be used with autogenous transplants when a good PCL stump is not available.

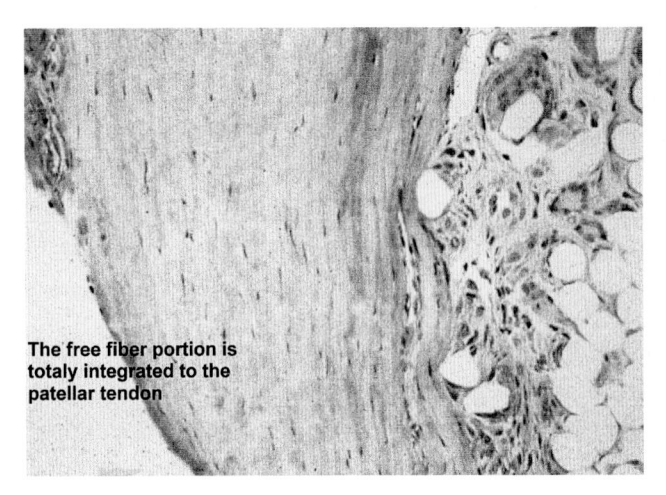

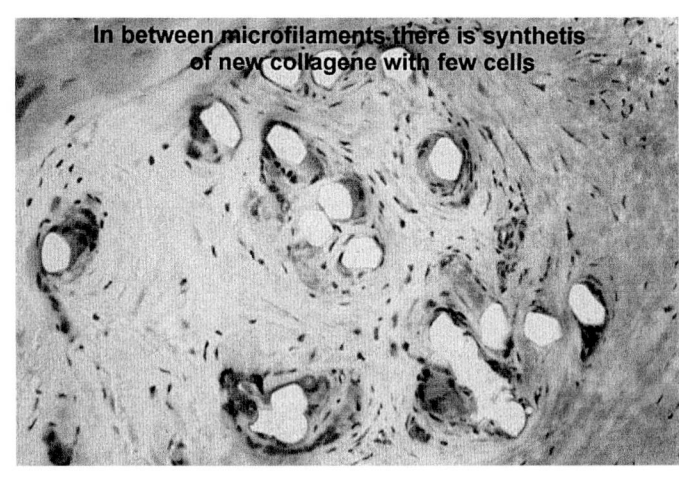

Fig. 14.8.

rarely some microparticles have been noticed in the intraarticular portion. But in that case they were stuck in the fibrous tissue and could not be released. This is probably why no chronic synovitis was clinically noted, even in the long term. As opposed to the "free fibers" portions, particles were noted with knitted structures, due to the friction between longitudinal and transversal fibers. These histological findings confirm what was noticed through the mechanical trials (Fig. 14.9).

Principles of the LARS Technique

Classic PCL reconstructions with autologous techniques are known to give inconstant results as long as a complete correction of the posterior tibial displacement is concerned. It is therefore logical that some surgeons think that a surgical reconstruction is not indicated, basing their attitude on the reputation of good functional tolerance of PCL deficiency. In fact, the data from literature as well as the experience of each of us are not so obviously in favor of this so-called tolerance. PCL deficiency disturbs the knee kinematics as severely as ACL deficiency. The permanent traction of the hamstrings produces the posterior tibial subluxation. This subluxation is reduced by the traction of the quadriceps in a first phase to active extension. This mechanism leads to stresses on femorotibial surfaces and patellofemoral cartilages as the patella must strongly press on the trochlea to bring the tibia forward. Clancy et al.[32] show cartilage damage two years after injury in 20% of the cases in the patellofemoral joint, and 70% in

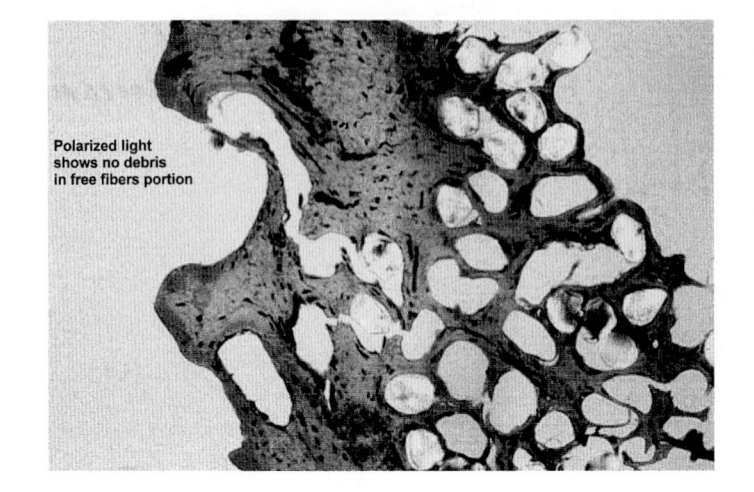

Fig. 14.9.

Table 14.2. Summary of results in four studies

	Pain	Effusion	Instability
Dandy[54]	70%	20%	45%
Dejour[33]	87%	20%	30%
Kennedy[55]	61%	67%	55%
Parolie[56]	52%	16%	20%

the femorotibial joint. The reputation of good tolerance is more than questionable when reviewing the literature, which shows pain, effusion, and instability in many cases (Table 14.2). The evolution to osteoarthritis is also confirmed by the same authors: 69% grade III and IV after 15 years for Dejour et al.[33] Keller et al.[34] report a study of 40 nonoperated patients 6 years after injury. This study shows a lower activity level in 65% and pain in 90%. At 10 years, there was 60% of grade II or III osteoarthritis, and 80% at 15 years. PCL injuries are far from harmless.

If we base the results of autografts on accurate radiolaximetry, their mechanical efficiency is poor. Good mechanical objective results, as published by Bianchi,[35] Bousquet et al.,[36] Kennedy and Galpin,[37] reach 50% with difficulty in the best case. These types of results would not be admitted for ACL reconstruction and would be called a failure.

The results of allografts are also disappointing (Covey et al.,[38] Harner et al.[39]), with the additional risk of viral disease transmission, which is not acceptable even though quite small. Due to their collagen transformation which weakens them, autogenous transplants require a postoperative period of rest. As opposed to what occurs in ACL, PCL reconstructions must be resistant from the early postoperative moments to resist hamstring action. Although one can increase the size of the transplant or combine several of them, the same inevitable decreasing of initial strength occurs and will end in elongation.

Scapinelli[40] has demonstrated the richness of PCL vascularization. We learned a long time ago that the PCL is able to heal, and this has been recently confirmed by MRI studies (by Shelbourne et al.[41] and Akisue et al.[42]) which show continuous but elongated PCL. The PCL heals, but with the tibia being posteriorly displaced, it becomes too long. This healing potential makes the PCL itself the best transplant, all the more so since it is the only ligament to have the mechanoreceptors for proprioception.

Therefore, the principle is to obtain a proper healing with no elongation, primary in acute injuries, secondary in chronic lesions with the tibia being recentered. This healing is always better in acute cases. In chronic cases the tibial attachments have to be stired up widely as the vascular pedicles are proximal. The use of an appropriate synthetic with specific ligaments allows us to recenter the knee, to get the healing process in the right position, and to maintain the ligament's length.

A reconstruction with one bundle is enough in acute injuries (less than 3 weeks). In chronic injuries our experience of two-bundles reconstruction started in 1978. Probably due to secondary global distensions in chronic cases, it is mandatory to have an anatomic and physiologic two-bundles reconstruction. This is the only way to stabilize the knee in all positions of flexion, extension, and rotation. The anterolateral bundle is slack in extension; the posteromedial is slack in flexion. The posterior displacement of the tibia is controlled by the anterior bundle when the knee is in flexion and the posterior bundle when the knee is in extension. Each bundle is progressively recruited during motion, and it is impossible to reproduce this mechanism with a single bundle.

Fig. 14.10. LARS PCL guide.

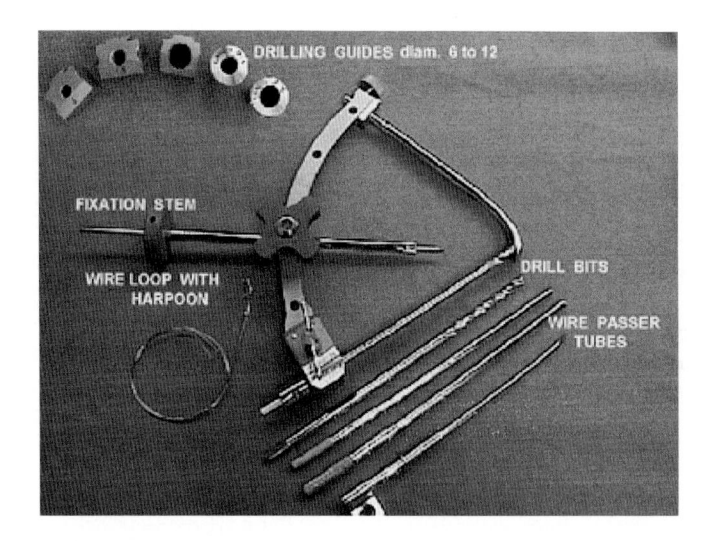

Surgical Technique

Instruments

The LARS instruments set comprises a tibial guide and different accessories to make the operation easy.

The *tibial guide* (Fig. 14.10) allows the placement and the drilling of the tibial tunnels while ensuring complete safety to the posterior nerves and vessels, without any posterior approach. The drilling guides allow the drill bits to be inserted through a small anterior incision and with an angle of about 45 degrees, until they reach the spatula placed behind the tibia. The contact between the drill bit and the spatula must be perceived. The drill bit must be pulled out and then passed again several times to remove all the bony debris of the posterior exit. The drill bit is then replaced by the wire passer tube. The wire passer tube is equipped with a metallic wire loop. The loop is itself equipped with a removable harpoon. The harpoon hooks itself into the hole of the spatula. When the spatula is pulled out of the knee, it automatically brings the wire loop. If there is some problem with hooking the harpoon due to the presence of soft tissue, a second solution is available. A simple wire loop, without harpoon, is pushed through another wire-passer which is designed to avoid the hole of the spatula. The stem of the spatula is hollow, and the wire loop will exit from the knee out of the base of the stem.

The *telescopic tubes* (Fig. 14.11) allow the passage of the drill bits through a microincision on the femoral side (especially when using an

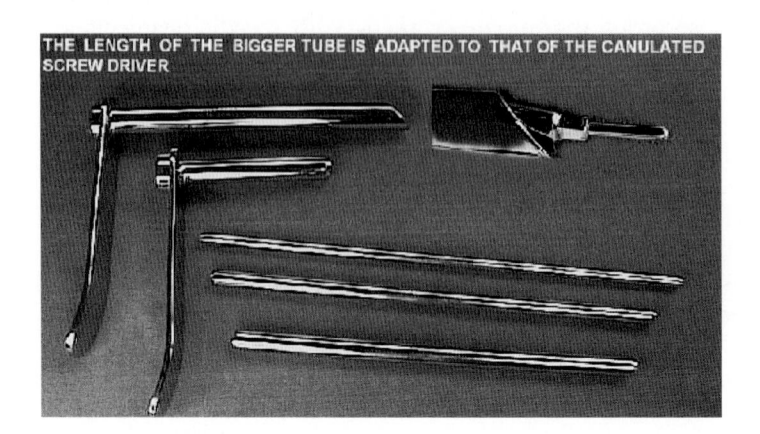

Fig. 14.11. Telescopic tubes.

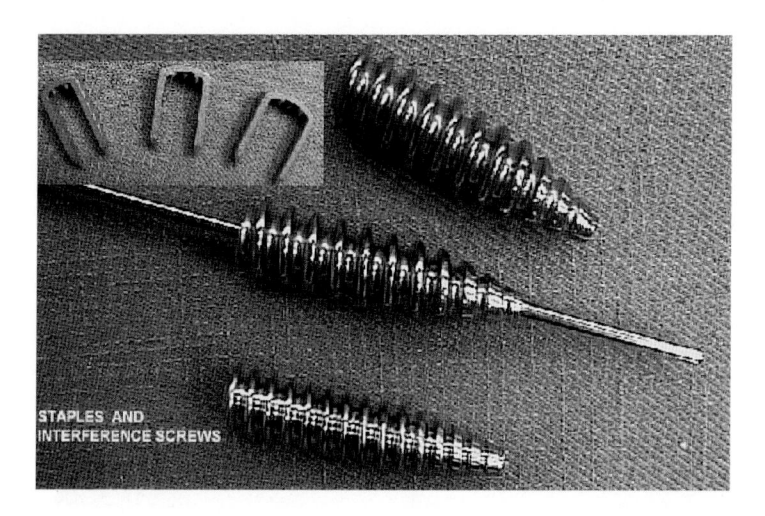

Fig. 14.12.

The fixation of the synthetic ligaments is made with *canulated interference screws* which are designed not to damage the ligament. The diameter of the screw must always be at least 1 mm greater than the diameter of the tunnel. (For example, a 6-mm tunnel requires at least a 7-mm screw.) After the ligament emerges from the tunnel, the fixation of the ligament is completed, particularly on the tibial side by a second transversal tunnel and another screw, or at least a staple (Fig. 14.12).

Surgical Procedures

For severe chronic or combined instabilities, or revisions, this technique should not be performed under arthroscopy control unless the surgeon has a great deal of experience. A clear, well-done, open surgery will always be better than an uncertain and long-lasting arthroscopic procedure. Both procedures are described.

Open Surgery

Step 1—Installation and approach. The patient is placed in the supine position. The thigh lies on a knee holder which is placed with the tourniquet as far proximal as possible. The knee must be able to be positioned in full extension and up to 100 to 110 degrees of flexion. The incision is medial parapatellar from the tibial plateau up to the quad tendon and passing laterally to the vastus medialis tendon. The patella must be laterally retracted. As opposed to what is done in other techniques, the stump of the PCL should never be removed. This is made possible by the use of the guide which does not require a direct view of the posterior tibia attachment. The first step consists in dissecting the remaining PCL from the ACL to which it is often adherent. Following the anterior aspect of the PCL, scissors cut all adherences and tibial attachments along the midline and 2 or 3 cm on each side. This step is most important in chronic injuries and is the most difficult step to complete under arthroscopy. This "reviving" will allow the reattachment of the PCL and adjacent structures when the tibia is reduced forward, with a shortening effect.

Step 2—Drilling of tibial tunnels. Tunnel for posteromedial bundle. The spatula of the guide is introduced through the notch, after removal of osteophytes, if any. The spatula passes on the medial side of the ACL and is pushed backward until its stem lies on the roof the notch, the knee being flexed at 90 degrees. The canulated fixation stem must be parallel to the tibial plateau (Fig. 14.13). The spatula is oriented medially to the midline, in such a way that the entrance point of the drill bit is at the middle part of the medial metaphyseal aspect of the tibia. When this proper orienta-

Fig. 14.13.

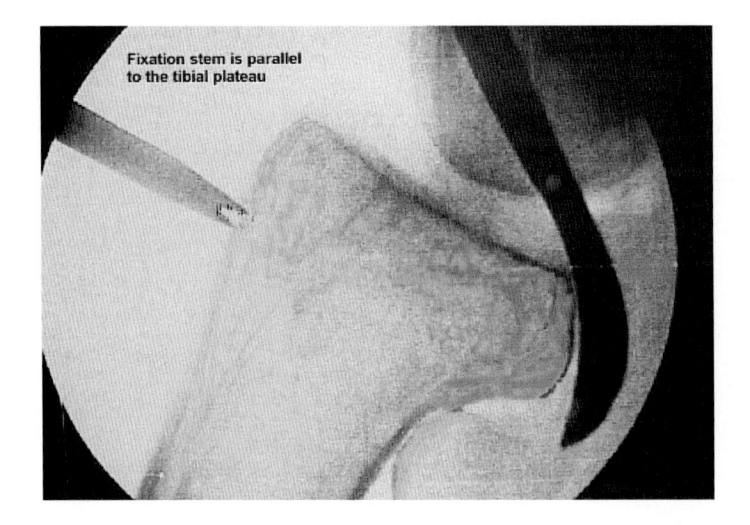

tion is obtained, the quide is affixed by means of the cannulated fixation stem and a 2.5-mm K-wire. A small incision is made vertically, centered by the position of the drill bit. The soft tissues are retracted from the cortex. There is a cutaneous bridge between the lower incision and the parapatellar approach. This bridge has to be respected to preserve the branches of the saphenous nerve (sensibility and proprioception). The 6-mm drill guide is affixed on the movable jaws of the guide. The sharp 6-mm drill bit is first used to attack obliquely the cortex, and is pushed until it bumps against the spatula. This first drill bit is replaced by the flat-ended drill bit, which is pushed and pulled back and forth to remove all the bony chips. The drill bit is then replaced by the wire passer tube and the harpoon. The harpoon is pushed through the center hole of the spatula. By pulling on the wire loop, one makes sure that the harpoon is correctly hooked. One can also use the second option: Use the specific wire passer tube which is curved at its extremity and push through a simple wire loop. The loop will appear at the other extremity of the spatula's cannulated stem. The wire loop has been brought outside the knee. The guide is dismounted.

Tunnel for anterolateral bundle (Fig. 14.14). A tube is placed in the first medial tunnel to protect the wire loop and to ensure that the second tunnel will not cross the first one. The guide and the spatula are put back in place. The tube of the first tunnel lies on the groove of the medial face of

Fig. 14.14. Anterolateral tunnel drilling.

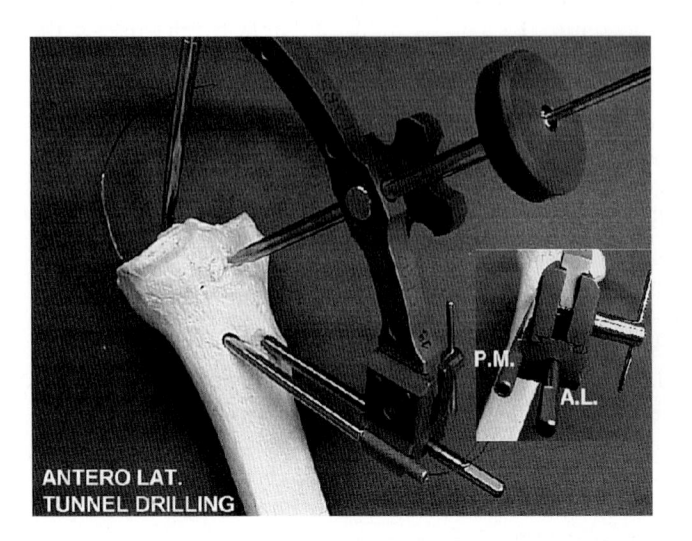

14. Cruciate Ligament Reconstruction with Synthetics

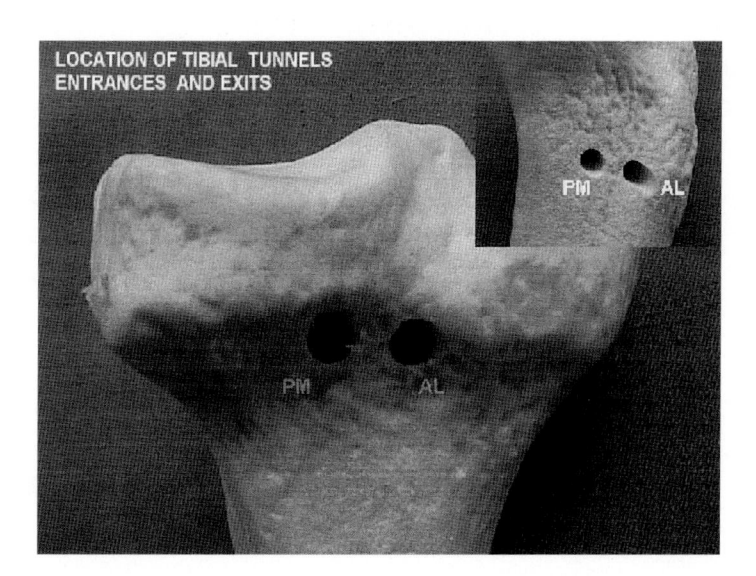

Fig. 14.15.

the drill guide. The spatula is oriented laterally from the midline and the guide affixed in that position (Fig. 14.15). Depending on the size of the bone the entrance of the second tunnel can be on the medial slope of the tibial tubercle on more distant on its lateral slope. If a ligament type PC 100 has been chosen for this anterior bundle, the 7.5-mm drill guide and corresponding drill bit and tube must be used. The same procedure as above will bring the wire loop outside the knee.

Step 3—Femoral tunnels. A good positioning of femoral tunnels is fundamental. The center of the anatomic PCL insertion corresponds to a point that can be geometrically defined for each knee (Fig. 14.16). This point is at 40% of a line drawn parallel to the Blumensat line and passing by the most prominent point of the posterior condyle on a x-ray where the two condyles are superimposed. This point corresponds also to the so-called isometric point as described by Ogata et al (Fig. 14.17).[43] This point can be determined on a preoperative x-ray and found intraoperatively by any graduated device or with the help of the image intensifier, which is highly recommended. A one-bundle reconstruction (acute cases) must be centered at this point. The tunnels of a two-bundles reconstruction must be located on each side of this point (Fig. 14.18): anterior and proximal for the anterolateral bundle, posterior and distal for the posteromedial bundle. The drilling is done inside out from the intraarticular aspect of the medial condyle. Directions of tunnels are most important to avoid any sharp angle of the implant and any risk of osteonecrosis and collapse of the condyle (Fig. 14.19). The anterior tunnel is directed anteriorly, proximally, and

Fig. 14.16. Geometrical definition of points E, C, and D.

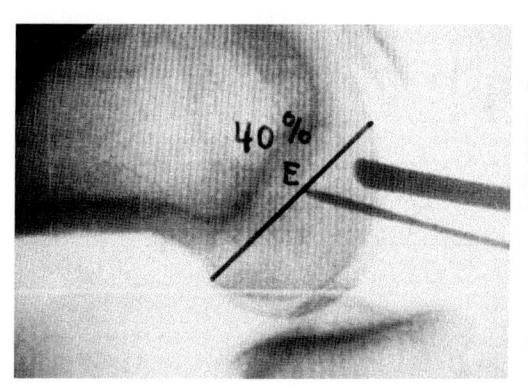

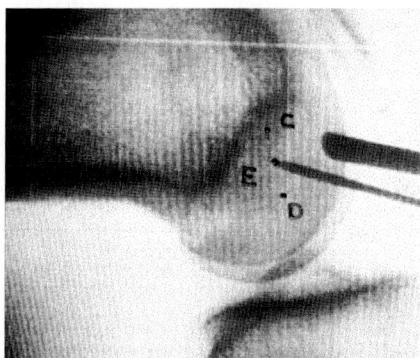

A. P. and lateral view of Ogata's points for femoral insertion

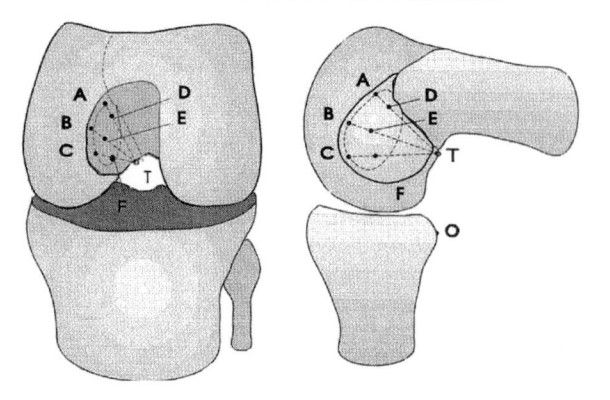

Fig. 14.17.

MEAN POSITIONING OF POINTS E, C, D

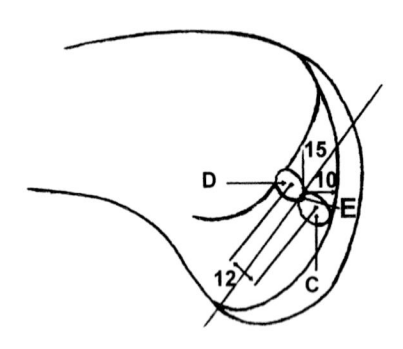

Fig. 14.18.

TUNNELS ORIENTATION

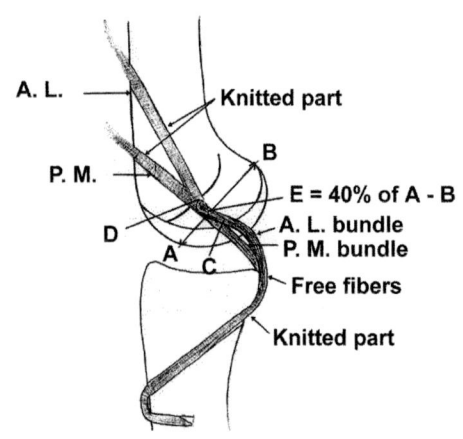

A. L.
P. M.
Knitted part
B
E = 40% of A - B
A. L. bundle
P. M. bundle
Free fibers
Knitted part

E - D - C = OGATA'S POINTS

Fig. 14.19.

medially to exit at the junction of the medial and anterior aspect of the femoral metaphysis. The posterior tunnel will exit more distally and medially just in the middle of the medial aspect. These exits are both extraarticular (Fig. 14.20). The less transverse and the closer to the axis of the femur the tunnels can be positioned, the less will be the risk of osteonecrosis, and the stresses given to the ligament in flexion and torsion will be minimized (Fig. 14.21).

Step 4—Passage of the ligament and femoral fixation. Wire passer tubes are placed in femoral tunnels. The wire loop and a blunt K-wire that will lead the interference screws are introduced outside-in. Each bundle is pulled through the corresponding tunnel inside-out. The junction of the free fibers and the knitted portion must be adjusted one mm inside the tunnel. Then using the tibial wire loops, each bundle is pulled through the corresponding tibial tunnel: anterior femoral bundle in the lateral tibial tunnel, and posterior femoral bundle in the medial tibial tunnel. The femoral fixation is immediately performed with the interference screws (8×30 or 7×30), which are placed outside-in with the help of the guiding K-wire and the telescopic tubes. The length of the bigger tube and the length of the cannulated screwdriver are adapted to allow the placement of the screw flush to the bone. It is even recommended to give another turn to the screw after removal of the bigger tube when the screwdriver has been stoped by the tube.

Step 5—Tensioning and tibial fixation. The anterolateral bundle must be tightened in flexion. This tensioning must be performed manually and must be limited to placing the ligament rectilinear and correcting the tibial posterior displacement. This is achieved when the ACL is back to its normal tensioning and obliquity. The ligament is then blocked is this position and one makes sure that complete extension can be obtained. Tibial fixation is performed with an interference screw (7×30 or 8×30). The screw must be placed at the upper face of the ligament and guided by a K-wire all the way up into the tunnel. The posteromedial bundle must be tightened in extension, and one makes sure that complete flexion can be obtained. Complete correction of the posterior sag and full motion must be again controlled and the tensioning of each bundle again readjusted if necessary. This step is fundamental, and one must take the necessary time until a perfect result is obtained. The stress on the ligaments being high, it is mandatory, especially on the tibial side, to complete the primary fixation with at least a staple or, better, with a second screw placed in a transversal lower tun-

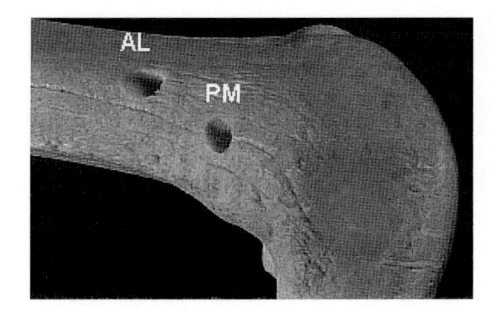

14. Cruciate Ligament Reconstruction with Synthetics

nel (Fig. 14.22). The parts of the ligaments which are preserved outside the main tunnels will eventually allow a revision if needed for retightening, new injury, or other indication. The operation is ended by a careful lavage, and closure is performed layer by layer on a vacuum drainage. No splint. A compression on machine can be successfully used to reduce the patient's discomfort and swelling.

Arthroscopic Procedure

This procedure is a mini-invasive "ligament synthesis". This procedure is recommended in mild isolated posterior laxities and mainly in acute injuries where a one-bundle "internal fixation" is enough. This internal fixation is a simple procedure that does not add any new damages to those of the injury itself.

Technical principles are the same as for open procedure. Particular points are:

- Do not use a pump in acute injuries.
- If there is an important liquid leak, which is rare, one must give up with the scope and do a mini-arthrotomy (watch for swelling of the calf).
- The portal for the spatula of the guide is medial parapatellar 1 or 2 cm above the lower extremity of the patella. The wire loop will exit from the same portal. The spatula must be placed behind the tibia on the middle line.
- Femoral tunnel: with a one-bundle reconstruction the femoral tunnel must be located at the "isometric" point. The use of the fluoroscope is highly recommended. This point is marked by a K-wire, which is introduced into the notch through the skin at the lateral edge of the patellar tendon, at the level of the joint-line. The K-wire is directed obliquely through the medial condyle to exit through the skin of the lower and anteromedial aspect of the thigh. A microincision allows the passage of the Telescopic tubes guided by the K-wire.

The cannulated drill bit is introduced through the bigger tube and will drill the tunnel from outside in. A wire loop is passed outside-in through the cannulated drill bit and will be caught by a small arthroscopic forceps. This forcep is introduced through the medial portal from which is already exiting the tibial wire loop. With the help of those two wire loops the ligament (PC 80) is passed first through the femur, then through the tibia.

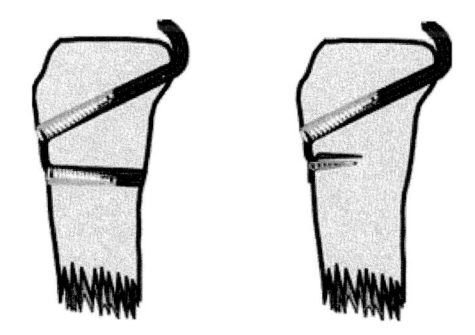

Fig. 14.20. Exit of femoral tunnels.

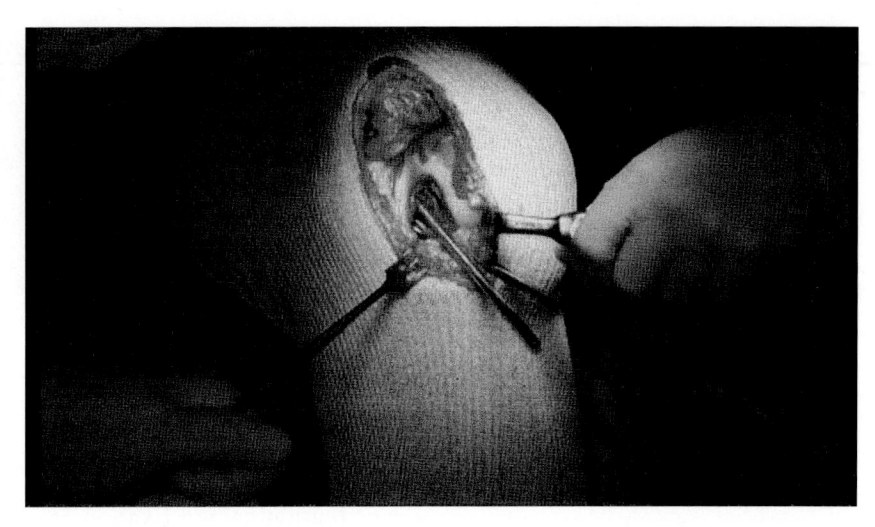

Fig. 14.21. Position of the drill bits for drilling anterior and posterior femoral tunnels.

Fig. 14.22. Securing tibial fixation with one staple or a second screw.

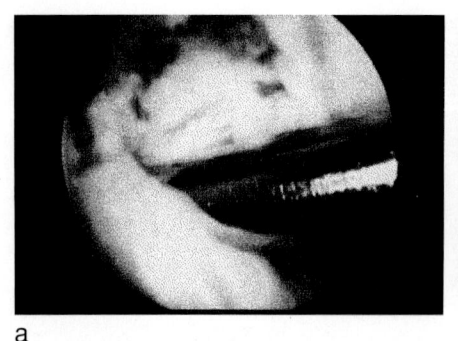

Fig. 14.23. (a) Spatula on the medial edge of the ACL. (b) Femoral and tibial wire loops. (c) PC 80 in place.

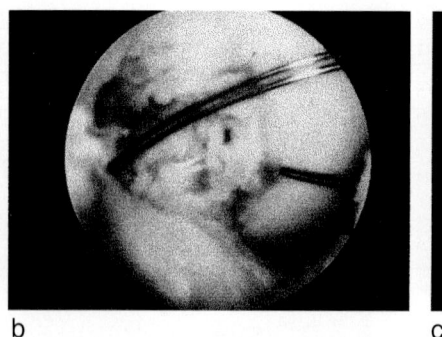

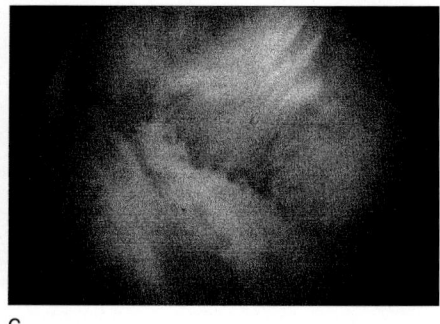

The scope controls the proper position of the free fibers in the joint (Fig. 14.23).

Tensioning and fixation are performed as in the open procedure. As the position described is supposed to be isometric, the single bundle can be tightened in flexion, which makes it easier to control the aspect of the ACL. But one must make sure that full extension is still easily obtained and adjust the tension accordingly.

Autogenous Option

For all the previously given reasons (vascularization, proprioception, real ligamentous structure) the PCL itself seems to us the best material. However, in some cases the PCL is totally absent (revisions) or poor. In these cases, in our opinion, the PCL must be reconstructed with an autogenous transplant reinforced with a synthetic ligament, as no fibroblastic ingrowth can be expected from nothing. At the beginning of our experience in PCL reconstruction with synthetic materials, we used to combine autogenous and synthetic material even if the PCL was looking elongated but strong and thick. We made two groups with or without autogenous transplant. After 2 years there was no difference between these two groups. So we actually think there is no need to harvest any structure except if there is no valuable PCL remaining structure.

But some still prefer, whatever the PCL looks like, the classic use of an autogenous transplant. When using autogenous material, we prefer the quadriceps tendon or gracilis and semitendinosus, which seem less damaging to the extensor apparatus than the patellar tendon. The extensor apparatus should be a main concern in PCL deficiency.

Specific LARS ligaments have been designed for autogenous reinforcement, to prevent postoperative elongation (ACTOR 10, AC 30 RA, PTR 30; see Fig. 14.7).

The tibial guide comprises drilling guides of 8- to 12-mm drill bits. For the passage of the wire loop, the wire passer tube must be centered by using the 6-mm drill guide. The autogenous transplant and the synthetic scaffold are fixed together in the femoral tunnel with an interference screw. On the tibial side, the autogenous transplant must be tightened, while the synthetic scaffold should not. The synthetic reinforcement must not avoid any work given to the transplant, but is there to limit the elongation. Therefore at the end of the operation one must get 1 to 2 mm of laxity with a firm endpoint. This is the condition to obtain a new ligamentous structure with collagenic and vascularization ingrowth.

Postoperative Care

The LARS techniques allow full and immediate mobilization and weight bearing.

14. Cruciate Ligament Reconstruction with Synthetics

Days 1–2:   Passive motion with CPM,
           Active motion as allowed by patient's pain,
           Isometric contractions of the quadriceps.

Days 2–3:   Weight bearing with crutches,
           Quad exercises,
           Mobilization,
           Removal of the drainage.

Physical therapy is then carried on every day for each muscular group. Active dynamic contractions of the quadriceps against resistance and in open chain must be prohibited. Isokinetic rehabilitation in closed chain follows isometric contractions. Rehabilitation of proprioception at 4 to 5 weeks. Resuming sports activities must be progressive as soon as proprioception is satisfactory. Full sports are generaly authorized after 3 to 4 months for all synthetic reconstructions and about 7 months when combined autogenous and synthetic have been used. The use of synthetic ligaments allows avoiding the use of a brace. Isolated PCL injuries represent about only 30% of the cases. More often they are combined with other ligament injuries, meniscal tears, and cartilage damage.

Posterolateral Laxities

Posterolateral laxities are quite frequent and must be diagnosed and treated simultaneously. In acute cases use a selective approach crossing the middle of the lateral collateral ligament incising the fascia lata and passing in front of the lateral gastrocnemius. One must suture what can be sutured and reinforce with a PC 60 or PC 80, reproducing the path of the popliteus tendon, to avoid the hyperexternal rotation of the tibial plateau and to resist the action of the biceps. Thus one can expect a proper healing in good position. The PCL guide features a specific device for positioning and drilling the tibial tunnel for posterolateral complex reinforcement (Fig. 14.24).

In chronic cases, if the patient is not involved in strenuous sports activities and the posterolateral laxity is moderate, a reconstruction using the PPLY 100 ligament is used. As described by J. Beacon,[44] one branch of the Y reconstructs the anterior bundle of the PCL, and the other branch reconstructs the path of the popliteus tendon (Fig. 14.25). If it is a major posterolateral laxity in a high-sports-level patient, a two-bundles recon-

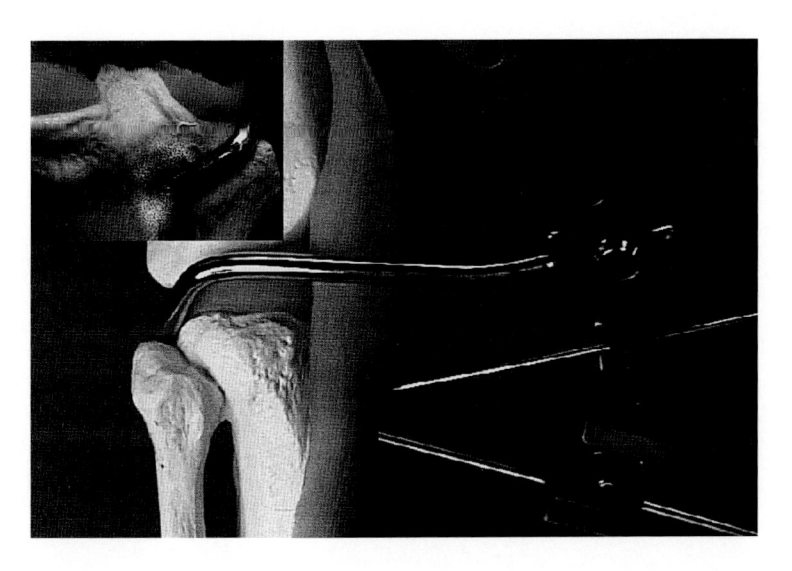

Fig. 14.24. LARS guide features a special arm for posterolateral reconstruction.

Fig. 14.25. Moderate combined PCL + PLI uses a PPLY 100 ligament for anterior PCL bundle and posterolateral reconstruction.

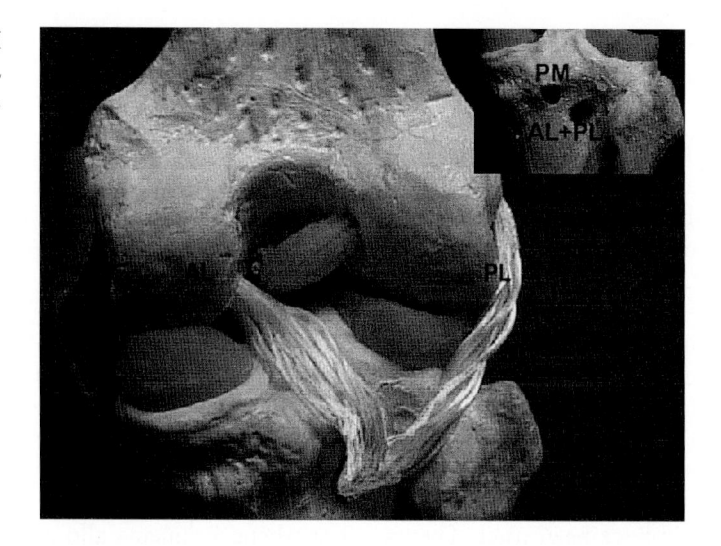

struction of the PCL and a separate reconstruction of the posterolateral structure using a PC 80 must be performed (Fig. 14.26). The tibial tunnel is 3 cm below the tibial plateau and 1 cm medial to fibular head. The posterolateral reconstruction crosses over the LCL and enters the lateral condyle just in front of and below its attachment.

Simultaneous Reconstruction of PCL and ACL

The PCL reconstruction must always be performed first. The tensioning of the PCL cannot be adjusted on the ACL. Therefore any hypercorrection of the posterior tibial displacement must be carefully avoided. Before fixation of the PCL tibial side, one must make sure that the posterior border of the condyle is not in front of the posterior edge of the tibial plateau when the knee is at 90 degrees of flexion. The use of a fluoroscope or intraoperative x-ray with the two condyles superimposed is very useful. A line is drawn down from the posterior edge of the condyle parallel to the posterior cortex of the tibia. If the tibial plateau is in front of this line, the tensioning of the PCL must be released. This line must be adjusted flush with the posterior border of the tibial plateau. Then the fixation is achieved. The ACL reconstruction can be performed only when the knee has been recentered, by PCL and/or posterolateral reconstruction.

Fig. 14.26. Position of the posterolateral reconstruction and the three tunnels behind the tibia.

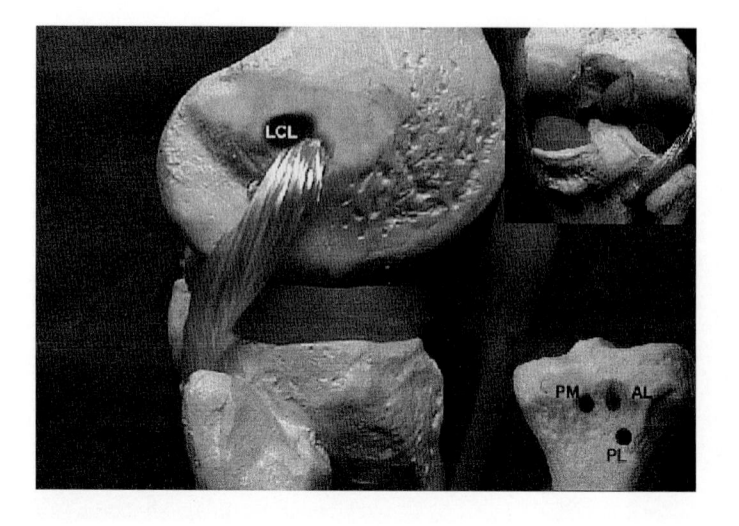

14. Cruciate Ligament Reconstruction with Synthetics

PCL and Knee Replacement

Some unimedial compartment osteoarthritis is related to PCL deficiency. In that case, it is much more conservative, instead of a total posterostabilized knee replacement, to perform a unicompartmental prosthesis combined with a PCL synthetic reconstruction (Fig. 14.27). Sometimes in old chronic cases there are severe damages of the patella femoral joint. But total knee replacement is not justified by other compartments. An isolated patellofemoral replacement can be performed in association with PCL reconstruction with good results.

Material

Since the new design of the LARS ligament, the author has operated on 119 patients, who are part of a prospective study. These patients have been examined after operation at 1 month, 3 months, 6 months, and then every year with clinical examination and dynamic stress x-rays using the Telos device at 100 and 250 N. The 119 patients were classified according to the classification system adopted by the International PCL Study Group and proposed by D. Cooper, related to combined injuries (Tab. 14.3).

This series comprises 28% of isolated PCL and 72% of combined injuries. Among the combined injuries, a posterolateral instability was found in 61 cases (51%), and 50 patients (42%) had a chondral or meniscal lesion. Seventy-two patients have a follow-up of more than 18 months (61 male, 11 female). The average follow-up at final review was 3.4 years (min 1.5 years, max 6 years); 20 patients had 5 years follow-up or more. Thirty-three were acute injuries (≤3 weeks), average age 26.5 years. Thirty-nine were chronic injuries, average age 31.6 years. Eleven of the 33 patients with acute injuries were competitors, 15 used to do recreational sports, the others being just active people. Only two of the 39 patients with chronic injuries were still competitors, and 10 were having some recreational sports activities. All of them were complaining of pain and effusions. Twenty-four patients (61%) were complaining of permanent impairment. Out of the 72 patients in this study (follow-up >1½ years, 21 were Level I (29%); 41 were Level II (57%) with 15 combined ACL; and 10 were Level III (14%). Among Level II and III, a PLI was found in 37 patients (51%). The preoperative posterior drawer was 6 to 10 mm in 14 patients (19%) and more than 10 mm in 58 (81%).

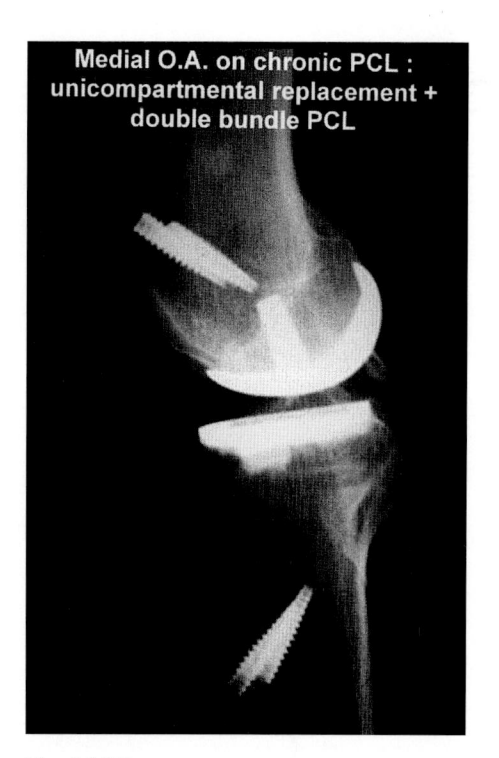

Fig. 14.27.

Table 14.3. Classification of patients in a prospective study of LARS ligament efficacy

Level I	Isolated PCL	n = 33
	Posterior drawn ≤ 10 mm	
Level II	PCL + posterolateral instability (PLI)	43
	PCL + ACL	22
	PCL + MCL	3
Level III	PCL + ACL + PLI	10
	PCL + ACL + MCL	2
	PCL + ACL + PLI + LCL	7
	PCL + PLI + MCL	1
Level IV	Knee dislocation. This level has been eliminated from this study due to other problems, neural or related to polytrauma.	3

Complications

Complications were rare.

- No acute synovitis.
- One chronic synovitis on a rupture of the anterolateral synthetic bundle. The synovitis disappeared after removal of the ligament and lavage.
- No infection.
- Four mobilizations under general anesthesia.
- Six deep veinous thrombosis.
- Two slight neurodystrophic syndrome, which were solved by medical treatment in 6 months.
- Five patients had to be revised for retightening of the ligament between 6 and 12 months postoperatively. Three were spontaneous slidings. Two happened after resuming sports activities. This sliding always occured in the tibial tunnel and was probably due to using an initial screw that was too small. It did not happen any more after the fixation was secured by a second transversal screw.
- Two intraarticular ruptures, one spontaneous, one traumatic. The spontaneous one was due to a technical mistake in the femoral placement that was too posterior. These two patients were revised with new ligaments successfully.
- No nerves or vessels injured intraoperatively.

Results

The results of the 72 patients with more than $1^{1}/_{2}$ years follow-up are reported in Table 14.4. Twenty-seven of the 33 patients with acute injuries (82%) had less than 3 mm posterior drawer (under 250 Newtons) and 94% less than 5 mm. Fifteen of the 39 patients with chronic injuries (38%) had the same 3-mm result and 25 (63%) less than 5 mm. Only three patients, all chronics, had 10 mm and had to be considered as total mechanical failures even though the preoperative posterior drawer had been reduced by about 50%. There was a slight increase of the posterior drawer between 1 month and 18 months postoperative (average 2 mm). But no significant change was noted between 18 months and the final review, which suggests that mechanical results became stable around 18 months postoperatively.

Mechanical Results Related to Level of Injury

Among the 21 Level I patients, 14 (66%) had an excellent mechanical result and were in group A (<3 mm), six were in group B (3–5 mm), one

Table 14.4. Mechanical results on posterior drawer at final review

PD in mm	Group A <3A		Group B 3–5		Group C 6–10		Group D >10	
Preop (72)	0		0		14	19%	58	81%
Postop acutes (33)	27	82%	4	12%	2	6%	0	
Postop chronics (39)	15	38%	10	25%	11	28%	3	8%

P.D., radio-laximetry; Telos, 250 N; $n = 72$

Table 14.5. IKDC global result

n = 72	Mechanical result			Functional result	
A	42	58.5%	} 78%	38	53%
B	14	19.5%		21	29%
C	13	18%		10	14%
D	3	4%		3	4%

in group C (6–10 mm). Among the 41 Level II patients, 24 were found in group A (58%), six in group B, nine in group C, two in group D. Among the 10 Level III patients, four were found in group A (40%), two were found in group B, three in group C, one in group D. Among the 37 patients with a combined PLI simultaneously treated, 23 were found in group A (62%).

Functional Results

Functional results were quoted according to the Arpege quotation on stability, range of motion, pain, and resistance to fatigue. Maximun score is 9. In acute cases the quotation was 8.9 for stability, 8.7 for range of motion, 8.7 for pain. In chronic cases the quotation was 8.8 for stability, 8.2 for range of motion, 7.8 for pain.

Using the IKDC evaluation system modified for PCL, global quotation on the 72 patients was 38 group A (53%), 21 group B (29%) 10 group C, 3 group D. See Table 14.5. The same IKDC quotation was applied separately on the 33 acutes and 39 chronics. See Tables 14.6 and 14.7. Pain and effusions were studied in detail on patients with chronic injuries. Preoperatively only three patients (7%) had never encountered any effusion; that number was 32 (82%) at final review. Twenty-one (54%) had occasional effusions after activities; that number was only 5 (13%) at final review. The main reason reported by all the patients for consulting was pain at different levels. At time of review eight patients (20%) had no pain at all, 27 (69%) had occasional pain after exercises or weather changes. Only two patients (5%) were still complaining of permanent pain, versus 61% preoperatively. At the time of review all 72 patients had recovered full extension. This recovery was obtained at an av-

Table 14.6. IKDC global acute PCL

n = 33	Mechanical result			Functional result		
A	27	82%	} 94%	22	67%	} 94%
B	4	12%		9	27%	
C	2	6%		1	37%	
D	0			1	13%	

Table 14.7. IKDC chronic PCL

n = 39	Mechanical result		Functional result	
A	15	38% ⎫	16	41%
B	10	25% ⎭ 63%	12	30%
C	11	28%	9	23%
D	3	8%	2	5%

erage of 6 weeks (min 1 week, max 4 months). 61 patients (85%) had recovered a flexion of 130 degrees or more; 10 had between 120 and 130 degrees; 1 had 110 degrees.

Results in correlation with activities are summarized in Table 14.8.

Among the 37 patients who presented a PLI, 32 had a simultaneous reconstruction of the posterolateral complex. Objective results were assessed by comparative stress x-ray in hyperexternal rotation and with the LARS rotational laxiometer. The use of this device, developed by J. Beacon,[44] has been evaluated by R.M. Bleday, and G. Fanelli et al.[45] and seems to be the most accurate for the moment. The instrument allows reproducible measurements of external rotation for different angles of flexion. These measurements must be always comparative. Among the 32 initially operated patients, 27 had an excellent mechanical result and 23 were in IKDC group A. All five of the patients who were not simultaneously operated on the PLI were partial or total failures and were found in group C or D. Four of them had to undergo a revision between 1 and 2 years after the initial procedure. Moreover, two misdiagnosed PLI were secondarily found in the patients of group C and D, and will have to be revised.

Fifteen patients of Level II injuries had a combined ACL + PCL without any other ligament injury association. The two ligaments were always simultaneously treated. Nine were found in group A and three were found in group B.

Table 14.8. Results in correlation with activities

	Competitors	Recreational	Active
Acutes n = 33			
Pre op	11	15	7
Review	9	17	6
Chronics n = 39			
Pre op	2	10	27
Review	1	26	12

Discussion

The results in acute cases look much better than in chronic cases. Surprisingly, the fact of being an acute or a chronic injury makes more difference than the level of injury. Particularly, the presence of ACL or posterolateral complex (PLC) injury did not practically affect the results if they were initially treated, especially in acute cases. Posterolateral structures and the PCL protect each other, as demonstrated by Markolf et al.[46] Neglecting posterolateral or other ligament reconstruction will lead to failure of PCL reconstruction sooner or later, as already noticed by Jeager,[47] Ferrari et al.,[48] and O'Brien et al.[49] Classic techniques are not simple and efficient enough to be easily combined with PCL reconstruction in the same surgical time (Witvoet et al.[50]).

Among the different combinations of injuries, a lesion of the lateral collateral ligament appeared to have a poor prognosis, but fortunately was quite rare. The worst situation was found in chronic injuries with a combined PCL + PLC + LCL in varus knees. In our opinion, it is now mandatory in these chronic combinations to perform an osteotomy at the same time as ligament reconstruction. It is suggested that multiple injuries should always be treated acutely and completely. It is in these cases that degenerative changes and secondary restraints appear quickly and end in a situation that is difficult to solve.

We compared the results of this series with a series of two-bundle reconstructions in 59 patients using $1/2$ T. and gracilis performed from 1977 to 1983. The procedure also used two tibial and two femoral tunnels. Good mechanical results were found in 29% of the acute cases and 26% of the chronic cases (versus 94% and 63% in the present study). Thus the difference between acute cases and chronic cases was not so important then as now. This is probably due to the fact that $1/2$ T. and gracilis two-bundles were not rigid enough to allow the proper healing of the PCL. Forty-six percent of the chronic patients reported having permanent pain and 78% had some effusion. In this old series as well as in the present one, functional results are parallel to results on the correction of the posterior tibial displacement. Correcting efficiently the posterior drawer must be the aim of PCL surgery.

Synthetic ligaments have the reputation of breaking sooner or later, even though they have been improved since the last series we published.[51,52] Nevertheless, we did not find in this series, or in a previous synthetic series operated from 1987 to 1991, any degradation of the result over time. The explanation was probably given by Loos et al. in 1981.[53] The PCL is extraarticular and well vascularized, and able to produce a strong fibrous healing. It is this potential that allows the good long-term result. It is therefore sufficient to correct the posterior laxity during a certain period of time. The study of the evolution of the posterior drawer by radio laximetry during the postoperative period suggests that this healing takes about $1\,1/2$ years to be definitely completed. Sufficient time is allowed by the life expectancy of the LARS ligament. This interpretation has been confirmed by histological findings on the few ligaments which were removed and by arthroscopic controls. The artificial ligaments were always totally surrounded by thick fibrous tissues and even invaded in the deep core by fibroblasts and new collagen (Fig. 14.8).

Conclusion

Considering the risk of severe degenerative changes in nonoperated PCL injuries, and considering the results that we are able to obtain now, it seems logical to operate on them. Our study suggests that PCL injuries in young and active patients should be operated acutely, where the results appear to be even better than for the ACL, for which surgery is widely accepted. The frequency and the importance of a combined PLI must lead us to systematically think about it. The concept of early "ligament synthesis" using synthetic ligaments which have proven their excellent tolerance and efficacy is now the procedure of first choice for many knee surgeons.

References

1. Blauth W, et al. Reconstruction of the cruciate ligament with special reference to synthetic substitute materials. Unfallchirurgie 1985;88(3):118–125.
2. Funk FJ Jr. Synthetic ligaments. Current status. Clin Orthop 1987;219:107–111.
3. Kdolsky R. Results of synthetic ligament augmentation of reconstruction of the anterior cruciate ligament. Wien Klin Wochenschr 1989;101(21):750–753.
4. Kennedy JC. Application of prosthetics to anterior cruciate ligament reconstruction and repair. Clin Orthop 1983;172:125–128.
5. Scharling M. Replacement of the anterior cruciate ligament with a polyethylene prosthetic ligament. Acta Orthop Scand 1981;52(5):575–578.
6. Bahuaud J, et al. Anterior ligament repair of the knee with Dacron using an arthroscopic approach. Acta Orthop Belg 1987;53(3):360–365.
7. Gillquist J. Important factors in the use of the Dacron ligament. Acta Orthop Belg 1987;53(3):353–355.
8. Dahhan P, et al. Use of carbon fibers to repair the anterior cruciate ligament of the knee. J Chir (Paris) 1981;118(4):275–280.
9. Goutallier D, et al. Use of carbon fibers in surgery of recent and old lesions of the antero-external cruciate ligament of the knee. Experimental studies in sheep and initial results in human practice. Acta Orthop Belg 1987;53(3):342–347.
10. Hejgaard N, et al. Alloplastic knee ligament replacement with carbon fiber. Ugeskr Laeger 1985;147(39):3060–3063.
11. Helbing G, et al. Cruciate ligament replacement—current status. Carbon fiber implants. Unfallchirurgie 1985;11(5):259–263.
12. Jenkins DH. The repair of cruciate ligaments with flexible carbon fibre. A longer term study of the induction of new ligaments and of the fate of the implanted carbon. J Bone Joint Surg 1978;60B(4):520–522.
13. Lemaire M. Reinforcement of tendons and ligaments with carbon fibers. Four years, 1300 cases. Clin Orthop 1985;196:169–174.
14. Mendes DG, et al. Carbon reconstruction in the anterior cruciate deficient knee. Orthopedics 1985;8(10):1244–1248.
15. Rusch RE, et al. The intra-articular use of bio-degradable polymer coated carbon filament as an adjunct to repair or reconstruction of cruciate ligaments. Acta Orthop Belg 1986;52(4):564–573.
16. Schweitzer G. Carbon fibre replacement of knee ligaments. S Afr Med J 1994;84(4):236.
17. Sim E, et al. Comparison of the value of magnetic resonance tomography and computerized tomography in the follow-up of augmentation-plasties with carbon fiber ligaments of the anterior cruciate ligament. Unfallchirurgie 1989;15(3);152–161.
18. Soudry M, et al. Carbon fiber reconstruction of knee ligaments. Harefuah 1985;108(2):76–79.
19. Strover AE, et al. The use of carbon fiber implants in anterior cruciate ligament surgery. Clin Orthop 1985;196:88–98.
20. Strum GM, et al. Clinical experience and early results of carbon fiber augmentation of anterior cruciate reconstruction of the knee. Clin Orthop 1985;196:124–138.
21. Wuschech H, et al. Technic of arthroscopic anterior cruciate ligamentplasty using the Lafil carbon ligment. Zentralbl Chir 1986;111(7):409–413.
22. Zarnett R, et al. The effect of continuous passive motion on knee ligament reconstruction with carbon fibre. An experimental investigation. J Bone Joint Surg 1991;73B(1):47–52.
23. Miles S. The Kennedy LAD ligament augmentation device. NATNEWS 1988;25(12):18–20.

24. Pagani PA, et al. Reconstruction of the anterior cruciate ligament with the Kennedy ligament augmentation device (L.A.D.) of reinforcement. Chir Organi Mov 1988;73(3):205–210.
25. Van Overschelde J, et al. Reconstruction of the anterior cruciate ligament using Kennedy's augmented plastic surgery. Acta Orthop Belg 1986;52(4):576–582.
26. Mockwitz J, et al. Ideal indications for the use of a textile ligament of high strength Trevira in alloplastic replacement of the anterior cruciate ligament in chronic isolated rupture. Unfallchirurgie 1988;14(5):276–282.
27. Fujikawa K. Reconstruction of the cruciate ligament of the knee using an artificial ligament. Kango Gijutsu. 1985;31(1):99–101.
28. Zacherl J. Replacement of the anterior cruciate ligament with an artificial Leeds-Keio ligament. Wien Klin Wochenschr 1993;105(5):147–151.
29. Apel R, et al. Alloplastic replacement of the anterior cruciate ligament by a polytetrafluoroethylene ligament. Animal experimental study. Unfallchirurgie 1986;12(6):291–294.
30. Somerville DW. Gore-Tex cruciate ligament replacement: recent experience in the Royal Navy. J R Nav Med Serv 1989;75(3):165–169.
31. Vogel U. The cruciate ligament prosthesis made of polytetrafluoroethylene: 2 years' results. Helv Chir Acta 1989;56(4):587–590.
32. Clancy WG, Shelbourne KD, Zoellner GB, Keene JS, Reider B, Rosenberg TD. Treatment of knee joint instability secondary to rupture of the posterior cruciate ligament. Report of a new procedure. J Bone Joint Surg 1983;65(A):310–322.
33. Dejour H, Walch G, Peyrot J, Eberhard P. Histoire naturelle de la rupture du ligament croisé postérieur. Rev Chir Orthop 1988;74:35–43.
34. Keller PM, Shelbourne KD, McCarroll JR, Rettig AC. Nonoperatively treated isolated posterior cruciate ligament injuries. Am J Sports Med 1993;21:132–136.
35. Bianchi M. Acute tears of the posterior cruciate ligament: clinical study and results of operative treatment in 27 cases. Am J Sports Med 1983;11:308–314.
36. Bousquet G, Girardin P, Cartier JL, De Jesse A, Eberhard P. Le traitement chirurgical de la rupture chronique du ligament croisé postérieur. A propos de 78 cas. Rev Chir Orthop 1988;74 (Suppl 1):188–190.
37. Kennedy JC, Galpin RD. The use of the medial head of the gastrocnemius muscle in the posterior cruciate-deficient knee. Indications, technique, results. Am J Sports Med 1982;10:63–73.
38. Covey DC, Sapega AA, Martin RC. Arthroscope-assisted allograft reconstruction of the posterior cruciate ligament. Technique and biomechanical considerations. J Orthop Tech 1993;(1–2):91–98.
39. Harner CD, Maday MG, Miller MD, et al. Posterior cruciate ligament recontruction using fresh-frozen allograft tissue. Indications, techniques, results and controversies. Scientific Exhibit, American Academy of Orthopaedic Surgeons 60th Annual Meeting, San Francisco, CA, February 1993.
40. Scapinelli R. Studies on the vasculature of the human knee joint. Acta Anat 1968;70(3):26–31.
41. Shelbourne KD, et al. MRI evaluation of PCL injuries. Assessment of healing potention. AOA March 1998, livre des résumés p. 201.
42. Akisue T, et al. Evaluation of the healing of injured PCL analysis of instability and MRI. AOA March 1998, livre des résumés p. 201.
43. Ogata K, McCarthy JA. Measurements of length and tension patterns during reconstruction of the posterior cruciate ligament. Am J Sports Med 1992;20:351–355.
44. Beacon J. Rotational stability of the PCL and measurement. The First International Symposium on the PCL, Brocket Hall, Hertfordshire, England, 25–26 November 1993.
45. Bleday RM, Fanelli GC, et al. Instrumented measurement of the posterolateral corner. Arthroscopy: 1998;14(5):489–494.
46. Markolf KL, Wascher DC, Finerman GAM. Direct in vitro measurement of forces in the cruciate ligaments. J Bone Joint Surg 1993;75(A)3:387–394.
47. Jaeger JH. Les laxités chroniques postéro-externes du genou. Les Journées de traumatologie de Colmar. Maîtrise Orthopédique June 1992;15:1–13.
48. Ferrari DA, Ferrari JD, Coumas J. Posterolateral instability of the knee. J Bone Joint Surg 1994;76(B)2:187–192.

49. O'Brien SL, Warren R, Pavlov H, Panariello R, Wickiewicz TL. Reconstruction of the chronically insufficient anterior cruciate ligament with the central third of the patellar ligament. J Bone Joint Surg 1991;73(A):278–286.
50. Witvoet J, Christel P, Pasquier G. Résultats du traitement chirurgical des laxités chroniques postéro-externes du genou. A propos de 40 cas. Rev Chir Orthop 1989;75(Suppl 1):144–145.
51. Laboureau JP et al. Ligamentoplastie à deux faisceaux du LCP par voie antérieure pure. Cahiers d'enseignement de la SOFCOT, 41, Ligaments artificiels: 92–98.
52. Laboureau JP et al. A two bundles artifical plasty of the posterior cruciate ligament. Surgical technique and results of an experience of eight years. 7th International Symposium Advances in Cruciate Ligament Reconstruction of the Knee: Autogenous vs. Prosthetic. 1990, Indian Wells/Palm Desert, CA.
53. Loos WC, Fox JM, Blazina ME, Del Pizzo W, Friedman MJ. Acute posterior cruciate ligament injuries. Am J Sports Med 1981;9:86–92.
54. Dandy DJ, Pusey RJ. The long term results of unrepaired tears of the posterior cruciate ligament. J Bone Joint Surg 1982;64:92–94.
55. Kennedy JC, Roth JH, Walker DM. Posterior cruciate ligament injuries. Orthop Dig 1979;7:19–31.
56. Parolie JM, Bergfeld JA. Long-term results of nonoperative treatment of isolated posterior cruciate ligament injuries in the athlete. Am J Sports Med 1986;14:35–38.

Chapter Fifteen

Arthroscopically Assisted Combined Anterior Cruciate Ligament/Posterior Cruciate Ligament Reconstruction

Gregory C. Fanelli and Daniel D. Feldmann

The dislocated knee is a severe injury resulting from violent trauma. It results in disruption of at least three of the four major ligaments of the knee and leads to significant functional instability. Vascular and nerve damage, as well as associated fractures, may contribute to the challenge of caring for this injury. Historical treatment was primarily limited to immobilization. However, with the advent of better surgical instrumentation and technique, the management of combined anterior and posterior cruciate (ACL/PCL) ligament tears associated with medial or lateral collateral ligament (MCL/LCL) disruption has become primarily surgical.

This chapter presents the basic knee anatomy, mechanisms and classifications of injury, evaluation, treatment, postoperative rehabilitation, and our experience with treating the ACL/PCL multiple-ligament–injured knee.

Anatomy

Stability of the knee is due to several anatomic structures. The articulation of the femorotibial joint is maintained in part by the bony anatomy of the femoral condyles and the tibial plateau. The menisci serve to increase the contact area between femur and tibia and thus increase stability of the joint. The four major ligaments (ACL, PCL, MCL, LCL) and the posterior medial and posterior lateral corners are the most significant ligamentous stabilizers of the knee. In addition to these static anatomic structures, dynamic anatomic structures, such as the musculature that crosses the knee joint, also play a role in stabilization. In any knee injury, examination must include evaluation of all these anatomic structures.

When evaluating a dislocated knee, it is imperative to evaluate the structural integrity of any remaining ligamentous structure; consequently, the functions of these structures must be well understood. The ACL primarily prevents anterior translation of the tibia relative to the femur, and accounts for about 86% of the total resistance to anterior tibial translation.[1] It is also involved in limiting internal and external rotation of the tibia relative to the femur when the knee is in extension.[2] The ACL will also limit varus and valgus stress in the face of either an LCL or MCL injury.

The PCL may be considered the primary static stabilizer of the knee given its location near the center of rotation of the knee and its relative strength.[3] The PCL has been shown to provide 95% of the total restraint to posterior tibial displacement forces acting on the tibia.[1] The PCL works in concert with structures of the posterior lateral corner, and injury to both structures is required to significantly increase posterior translation.[4]

The MCL and LCL act alone to resist valgus and varus stresses, respectively, at 30 degrees of knee flexion. Together, they act in a secondary fashion to limit anterior and posterior translation, and rotation of the tibia on the femur. The anatomy of the posterior lateral corner of the knee is complex; its major structures are (1) the LCL, (2) the arcurate complex, (3) the popliteal tendon, and (4) the popliteal-fibular ligament.[5] The posterolateral corner primarily resists posterior lateral rotation of the tibia relative to the femur, but also contributes to resisting posterior tibial transation. The posteromedial corner of the knee consists primarily of the posterior oblique portion of the MCL and associated joint capsule. These structures provide resistance to valgus stress and posterior medial tibial translation. Evaluation of traumatic knee dislocation must include these anatomic structures; typically, three areas or more are injured in knee dislocation. Failure to recognize and treat capsular and ligamentous injury, besides the obvious ACL/PCL injury, will result in less than optimal results.[6–8]

Neurovascular structures are also at risk of injury. The popliteal fossa is defined by the tendons of the pes anserinus and semimembranosus medially and the biceps tendon laterally. The space is closed distally by the medial and lateral heads of the gastrocnemius and proximally by the hamstrings. Within this space the popliteal artery and vein and the tibial and peroneal branches of the sciatic nerve are located. The popliteal artery may be most at risk to injury in knee dislocations. The popliteal artery is tethered proximally at the adductor hiatus as it exits from Hunter's canal, and distally as it passes under the soleus arch, making it vulnerable to injury in these areas. This artery is considered to be an "end artery" of the lower limb; if it is injured, the surrounding geniculate arteries are not sufficient to maintain collateral blood flow to the lower extremity. The popliteal vein is in close association with the artery, but seems to be less at risk during injury than the popliteal artery. From a surgical standpoint, the popliteal vessels are located directly posterior to the posterior horns of the medial and lateral meniscus, and dissection in this area may put these structures at risk if they are not adequately protected. The sciatic nerve divides into its peroneal and tibial divisions within the popliteal space. These nerves are less likely to be injured with knee dislocation, probably because they are not tethered as the popliteal artery is. The peroneal nerve does seem to be at higher risk, as its course around the fibular head functionally decreases its potential excursion, and violent varus injuries may result in traction injury to this nerve. Its location must be identified during dissections to reconstruct the posterolateral corner.

Classification

Classification of knee dislocation is primarily based on the direction the tibia dislocates relative to the femur.[9,10] This results in five different categories: anterior, posterior, lateral, medial, and rotatory. The anterior-medial and lateral and posterior-medial and lateral dislocations are classified as "rotatory" dislocation. Other factors to be considered include whether (1) the injury is open or closed, (2) the injury is due to high-energy or low-energy trauma, (3) the knee is completely dislocated or subluxed, and (4) there is neurovascular involvement. Furthermore, one should be acutely conscious of the fact that a complete dislocation may spontaneously reduce, and any triple-ligament knee injury constitutes a frank dislocation.[7,11,12]

Reports vary, but anterior and/or posterior dislocation appear to be the most common direction of dislocation. Frassica and coworkers[13] found a 70% incidence of posterior, 25% anterior, and 5% rotatory dislocations in their series. Green and Allen[14] reported a 31% anterior, 25% posterior, and 3% rotatory dislocation in their series. Rotatory dislocations occur less frequently; however, the posterolateral dislocation seems to be the most common combination. This particular pattern may be irreducible secondary to the medial femoral condyle becoming "buttonholed" through the anteromedial joint capsule. In addition, the MCL invaginates into the joint space, blocking reduction. This buttonholing results in a skin furrow along the medial joint line, as the subcutaneous tissue attachments to the joint capsule drag the skin into the joint.[15] Attempts at reduction in this scenario make the skin furrow more pronounced.

The actual incidence of different directional dislocation is not as important as correctly diagnosing the direction of injury and how it relates to potential neurovascular injury. Hyperextension injuries (or posterior dislocations), because of the tethered popliteal artery and vein, may have the highest incidence of associated vascular injury; however, any dislocation, if initial displacement is severe enough, will result in injury to the popliteal artery. The common peroneal nerve is less at risk because it has a greater excursion than the popliteal vessels, but it is still susceptible when a varus force is applied to the knee. Posterolateral dislocation is associated with a high incidence of injury to the common peroneal nerve.[16,17]

Open knee dislocations are not uncommon. Reported incidence is between 19% and 35% of all dislocations.[17,18] An open knee dislocation, in general, carries a worse prognosis secondary to severe injury to the soft tissue envelope. Furthermore, an open injury may require an open ligament reconstruction, or staged reconstruction, as arthroscopically assisted techniques cannot be performed in the acute setting with these open injuries.

Distinguishing between low- and high-energy injuries is important. Low-energy or low-velocity injuries, usually associated with sports injuries, have a decreased incidence of associated vascular injury. High-energy or high-velocity injuries, secondary to motor vehicle accidents (MVAs) or falls from a height, tend to have an increased incidence of vascular compromise. With decreased pulses in an injured limb and the history of a high-energy injury, one should obtain vascular studies urgently.

Mechanism of Injury

The mechanism of injury for the two most common knee dislocation patterns, anterior and posterior, are reasonably well described. Kennedy[16] was able to reproduce anterior dislocation by a hyperextension force acting on the knee. At 30 degrees of hyperextension, Kennedy found that the posterior capsule failed. When extended further, to about 50 degrees, the ACL, PCL and popliteal artery fail. There is some question whether the ACL or the PCL fails first with hyperextension[16,19]; however, in our clinical experience, both the ACL and PCL fail with dislocation.[7] Other series demonstrated both ACL and PCL tears with complete knee dislocation.[13,20,21]

A posterior-directed force applied to the proximal tibia when the knee is flexed to 90 degrees is thought to produce a posterior dislocation, the so-called dashboard injury.[20] Medial and lateral dislocations result from varus/valgus stresses applied to the knee. A combination of varus/valgus stress with hyperextension/blow to proximal tibia will likely produce one of the rotatory dislocations.

Associated Injuries

Several anatomic structures are at risk in the dislocated knee. The four major ligaments of the knee as well as the posterior medial and lateral corners can be compromised. Vascular and nerve injuries are common. There may also be associated bony lesions, avulsion fractures of the ACL or PCL, frank tibial plateau or distal femur condylar fractures, or ipsilateral tibial or femoral shaft fractures.

There is evidence in the literature that a frank dislocation may not result in complete rupture of three of the four major ligaments[16,17,22]; however, this seems to be the exception rather than the rule. Several authors in their series have found that a frank dislocation of the knee invariably results in rupture of at least three of the four major ligaments. Sisto and Warren[21] found that all knees in their series had three or more ligaments compromised. In Frassica et al.'s[13] series, all 13 patients treated operatively were found to have ACL, PCL, and MCL disruptions. In Fanelli et al.'s[7] series, 19 of 20 were found to have a third component (posterior lateral corner or MCL) in addition to complete ACL and PCL disruption. With a frank dislocation of the knee, careful ligament examination is necessary to fully diagnose the extent of the injury.

The incidence of vascular compromise in knee dislocations has been estimated to be about 32%.[14] When limited to anterior or posterior dislocation, the incidence may be as high as 50%.[23] Recent studies confirm the significant incidence of arterial injury, reaffirming the need for careful vascular evaluation.[13,21,24,25] The popliteal artery is an "end-artery" to the leg, with minimal collateral circulation through the genicular arteries. Furthermore, the popliteal vein is responsible for the majority of venous outflow from the knee. If either structure is compromised to the point of prolonged obstruction, ischemia and eventual amputation are often the result.[26,27]

Two mechanisms have been described for injury to the popliteal artery. One is a stretching mechanism, seen with hyperextension, until the vessel ruptures. This may occur secondary to the tethered nature of the artery at the adductor hiatus and the entrance through the gastroc-soleus complex. This type of injury should be suspected with an anterior dislocation. Posterior dislocations may cause direct contusion of the vessel by the posterior plateau, resulting in intimal damage. Under no circumstance should compromised vascular status be attributed to arterial spasm; in this circumstance, there is often intimal damage and impending thrombosis formation. Cone[28] points out that initial examination may be normal; however, thrombus formation can occur hours to days later,[28–31] and other series have demonstrated delayed thrombus formation.[13,21] Furthermore, bicruciate ligament ruptures presenting as a reduced dislocated knee may have as high an incidence of arterial injury as a frank dislocation.[12]

Popliteal vein injury occurs much less frequently, or at least historically had not been reported. Despite this, venous occlusion must also be recognized and appropriately treated. Historically, whether to repair venous injury was controversial. Ligating the popliteal vein, a common practice in the Vietnam conflict, led to severe edema, phlebitis, and chronic venous stasis changes. Venous repair was thought to lead to thrombophlebitis and pulmonary embolism. Currently, if obstruction to outflow is recognized, surgical repair of the popliteal vein is warranted.[32]

Injury to either the peroneal nerve or the tibial nerve has been documented,[16,17,21–25,33] with an incidence of about 20% to 30%. The nervous structures about the knee are not as tightly anchored as the popliteal vessels; this probably accounts for the lower incidence of injury compared to neighboring vascular structures. The mechanism of injury is usually one of stretch. The peroneal nerve seems to be more frequently involved than the tibial nerve, probably due to its anatomic location. With any varus loading of the knee, the peroneal nerve is placed under tension. In Shields et al.'s[17] series, posterior dislocation caused the majority of the nerve injuries.

Given the fact that knee dislocation is usually secondary to violent trauma, associated fractures are common; the incidence may be as high as 60%.[22] Tibial plateau fractures and avulsion fractures from the proximal tibia or distal femur are common.[13,18,21,22,33,34] Recognition of these injuries is also important, as additional bony involvement has implications for definitive treatment. Associated distal femur fractures and proximal tibial fractures treated with intramedullary nailing make bone tunnel placement for ACL and PCL reconstruction difficult. With violent trauma, any conceivable fracture or avulsion may occur with a dislocated knee; however, there is a suggestion that medial and lateral dislocations are associated with some increased frequency of minor bony lesions.[35]

Fracture-dislocations represent a separate entity in the spectrum of pure knee dislocation to tibial plateau fractures. Pure knee dislocation requires only a soft tissue reconstruction to gain stability; tibial plateau fractures require purely bony stabilization. Fracture-dislocations of the knee often involve both bony and ligamentous repair or reconstruction, adding an element of complexity to their treatment.[10,36] Long-term outcome of fracture-dislocation injuries to the knee joint falls somewhere between tibial plateau fractures and pure dislocations, with tibial plateau fractures doing the best and dislocations the worst.[36]

Initial Evaluation and Management

General Considerations and Physical Examination

Obvious deformity may be present on initial examination. However, in a polytrauma patient who is intubated and sedated, the injury may escape initial evaluation (Fig. 15.1). Abrasions or contusions about the knee, gross crepitus, or laxity may allude to injury in an otherwise normal-appearing knee. This importance of immediate recognition of knee dislocation or fracture dislocation lies not with the treatment of instability, but with the recognition of potential vascular injury and possible vascular compromise.[12] Neurovascular status must be assessed on both lower extremities. Neurologic examination may be difficult in the polytrauma patient, and is not as important initially as serial neurologic examination. Vascular examination is more pressing, as ischemia for longer than 8 hours usually results in amputation.[14] In the reduced knee, a white, cool limb that is obvious on physical examination and denotes arterial damage requires immediate arteriogram. However, normal pulses, Doppler signals, and capillary refill do not rule out an arterial injury.[28] Thrombosis may occur hours to days later, necessitating serial examination. If there is any question of perfusion of the limb, arteriogram is warranted.

The presence of a "dimple sign" on the anteromedial surface should be recognized. This indicates a posterolateral dislocation and is associated with a high incidence of irreduceability and potential skin necrosis.[37] In this circumstance, open reduction is warranted.

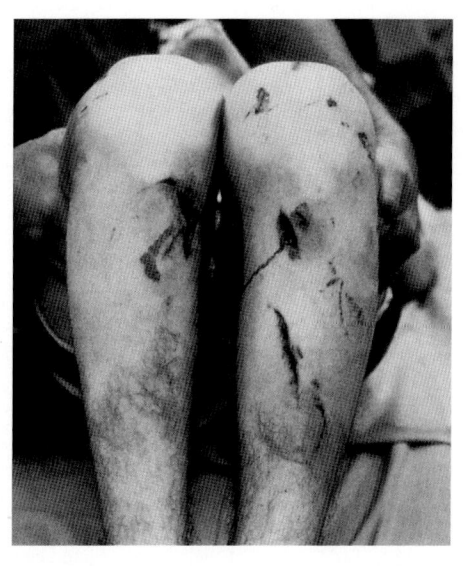

Fig. 15.1. Multiple trauma patient with left knee ACL/PCL/MCL tears, and right knee ACL/posterolateral corner tears. (From Fanelli et al.,[8] with permission.)

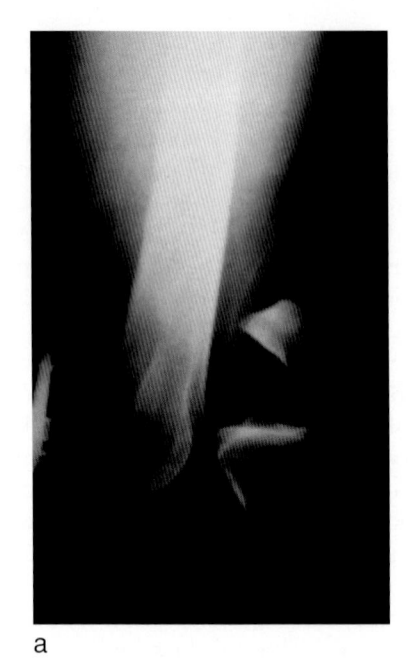

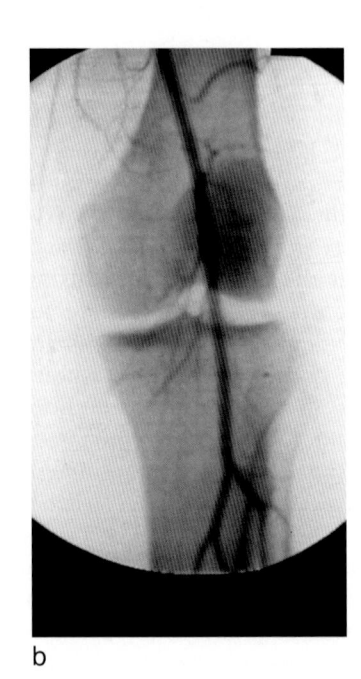

Fig. 15.2. (a) Lateral radiograph of anterior tibiofemoral knee dislocation. (b) Arteriogram after closed reduction of dislocated knee documenting vascular status of the extremity.

Imaging Studies

Prior to any manipulation, anteroposterior (AP) and lateral radiographs of the affected extremity should be completed. This is important to confirm the direction of dislocation and any associated fractures, and aids in planning the reduction maneuver. In the presence of cyanosis, pallor, weak capillary refill, and decreased peripheral temperature following reduction, arteriography must be considered (Fig. 15.2). Venography may be required if the clinical picture indicates adequate limb perfusion but obstruction of outflow.

After the acute management of the dislocated knee, magnetic resonance imaging (MRI) may be obtained subacutely to confirm, and aid in planning the reconstruction of, compromised ligamentous structures.

Reduction

An unreduced dislocated knee constitutes an orthopedic emergency, and reduction should be undertaken as soon as possible, preferably in the emergency department. Prior to manipulation, adequate AP and lateral x-ray evaluation should be performed. This allows for determination of the direction of the dislocation and any associated fractures, and assists in planning the reduction maneuver. In the isolated knee dislocation, intravenous morphine or conscious sedation is usually required. Slow, gradual longitudinal traction is applied to the leg from the ankle, and the proximal tibia is manipulated in the appropriate direction to effect a reduction. Once reduced, x-ray evaluation to confirm tibiofemoral congruency must be done, as well as repeated neurovascular examination. The limb should be placed in either a long leg splint or extension knee immobilizer. It is imperative to perform x-ray evaluation after placement in the splint or brace, as posterior subluxation of the tibia on femur is common. A "bump" consisting of a towel or pad behind the gastroc-soleus complex to aid in reduction maintenance may be required.

As mentioned, the dimple sign indicates a posterolateral dislocation, and closed reduction should probably not be attempted. The medial femoral condyle buttonholes the medial joint capsule, causing interposition of soft tissue in the joint, warranting open reduction.[10,15]

Vascular Injuries

A full spectrum of vascular injuries may be encountered. The clinical picture varies from an uncomplicated, bicruciate ligament injury with possible intimal damage with a normal physical examination to a polytrauma patient, with a closed head injury, intraabdominal bleeding, and dislocated knee with vascular compromise. Life-threatening injuries must be addressed first. Keeping this in mind, the orthopedic surgeon needs to be wary of the total time frame of ischemia to the limb. If there is any suspicion of arterial damage, a vascular consult should be obtained immediately. Reduction should always be performed to see if this restores flow to the limb. If the total ischemic time approaches 6 hours,[14] there is an urgency to restore flow to the lower extremity. An intraoperative angiogram during vascular exploration and shunting may be required at the expense of high-quality preoperative angiogram.[10] Mechanism of injury should also be noted. A high-energy injury (MVA, fall from height) may be more suspicious for vascular injury, and one may elect to obtain arteriograms despite a normal vascular examination.[12]

In the situation of an isolated dislocated knee with suspected arterial injury (asymmetric pulses, Doppler, or ankle-brachial index), arteriography should still be performed as the simple presence of pulses does not rule out vascular damage.[28] Any suspicion warrants a vascular surgery consult. If the limb is well perfused, and all indices are normal, one may elect to forgo a formal arteriogram, if there are frequent neurovascular checks to the lower extremity. Despite the historical preference to obtain an arteriogram in the presence of a knee dislocation as a screening tool, it has been demonstrated that arteriography following significant blunt trauma to the lower extremity with normal vascular examination has a low yield rate for detecting surgical vascular lesions.[12,38–40]

Popliteal vein injury is also possible, although rare. If the clinical picture warrants, a venogram may be useful.

Absolute Surgical Indications

As previously mentioned, a state of irreducibility and vascular injury warrants immediate surgical intervention. One should consider four-compartment fasciotomy of the limb when ischemic time is greater than 2.5 hours. Inability to maintain reduction also mandates early ligamentous reconstruction to stabilize the knee so as to avoid potential recurrent vascular compromise. Open dislocations and open fracture-dislocations warrant immediate surgical debridement to decontaminate the wound. An external fixator may be a reasonable option in the case of an open dislocation with a large soft tissue defect or an open fracture-dislocation. In this circumstance, access to soft tissue would be maintained for surgical debridement.

Definitive Surgical Management

Historical Management

Knee dislocations were initially managed conservatively with a cylinder cast for several months.[41,42] Early reports by Kennedy[16] and Meyers and Harvey[18] reported reasonable outcomes for nonoperatively treated knee dislocations. However, there was a suggestion that surgically stabilized dislocated knees would fare better in the long term. A report by Almekinders and Logan[24] compared surgically stabilized knees with conservative treatment and concluded that conservative treatment was comparable to surgical treatment. Despite similar outcomes, the conservatively treated knees were grossly unstable compared to surgically stabilized knees. Their study was retrospective from 1963 to 1988 and the typical surgical treatment during this period was in most cases open direct repair of the ligaments. Sisto and Warren[21] found similar results comparing four conservatively treated knees to 16 direct suture repair of torn ligaments. Frassica et al.[13] also evaluated early (within 5 days of injury) direct repair (with or without augmentation) of torn ligamentous structures in 13 of 17 patients. They concluded better results where obtained with early versus later direct repair of torn ligaments. This study supports surgical management of the dislocated knee, and introduces the concept of benefit from a ligamentously stable knee.

Within the last decade, the technique of arthroscopic assisted ACL/PCL reconstruction has become popular. Several advancements have made these techniques possible: (1) better procurement, sterilization, and storage of allograft tissue; (2) improved arthroscopic surgical instrumentation; (3) better graft fixation methods; (4) improved surgical technique; and (5) improved understanding of the ligamentous anatomy and biomechanics of the knee. Few reports of combined ACL/PCL reconstruction are available in the literature, but surgical reconstruction appears to afford at least the same results as, if not better than, direct repair of the ligaments. Shapiro and Freedman[25] reconstructed seven ACL/PCL injuries with primarily allograft Achilles tendon or bone–patellar tendon–bone. They found that three patients had excellent results, three had good results, and one had fair results. Furthermore, average KT-1000 was +3.3 mm side-to-side difference, with very little varus/valgus instability or significant posterior drawer. All seven of their patients were able to return to school or the workplace.

Fanelli et al.[6] reported on 20 ACL/PCL arthroscopically assisted ligament reconstructions. In their study group, there was one ACL/PCL tear, 10 ACL/PCL/posterior lateral corner tears, seven ACL/PCL/MCL tears, and two ACL/PCL/MCL/posterior lateral corner tears. Achilles tendon allografts and bone–patellar tendon–bone autografts were used in PCL reconstructions, and auto- and allograft bone–patellar tendon–bone was used in ACL reconstruction. An additional component, not previously mentioned with any consistency in the literature, was the addressing of the associated MCL or posterior lateral corner injury. It is imperative to address these injuries as well, or the results of ACL/PCL reconstruction alone will be less than optimal.

Postoperatively, significant improvement was found utilizing the Lysholm, Tegner, and Hospital for Special Surgery knee ligament rating scales, and the KT-1000 arthrometer. Overall postoperatively, 75% of patients had a normal Lachman test, 85% no longer displayed a pivot shift, 45% restored a normal posterior drawer test, and 55% displayed grade I posterior laxity. All 20 knees were deemed functionally stable and all patients returned to desired levels of activity. These authors concluded that results of reconstruction are reproducible, and that appropriate reconstruction will produce a stable knee.

Noyes and Barber-Westin[43] evaluated surgically reconstructed ACL/PCL tears (all had additional MCL or LCL/PCL reconstruction) at an average of 4.8 years. Seven patients had acute knee dislocations and four had chronically unstable knees secondary to knee dislocations. At follow-up, five of the seven acute knee injury patients had returned to their preinjury level of activity. Three of the four chronic knee injury patients were asymptomatic with activities of daily living. Arthrometric measurements at 20 degrees showed less than 3-mm side-to-side difference with anterior to posterior translation in 10 of the 11 knees; at 70 degrees, there were nine knees that had less than 3-mm side-to-side difference in anterior-posterior translation. These authors concluded that simultaneous bicruciate ligament reconstruction is warranted to restore function to the knee.

Fanelli Sports Injury Clinic Experience

Our practice is at a tertiary care regional trauma center. There is a 38% incidence of PCL tears in acute knee injuries, with 45% of these PCL-injured knees being combined ACL/PCL tears.[44,45] Careful assessment, evaluation, and treatment of vascular injuries are essential in these acute multiple ligament injured knees. There is an 11% incidence of vascular injury associated with these acute multiple-ligament–injured knees at our center.[39]

Our preferred approach to combined ACL/PCL injuries is an arthroscopic ACL/PCL reconstruction using the transtibial technique, with collateral/capsular ligament surgery as indicated. Not all cases are amenable to the arthroscopic approach, and the operating surgeon must assess each case individually. Surgical timing is dependent on vascular status, reduction stability, skin condition, systemic injuries, open versus closed knee injury, meniscus and articular surface injuries, other orthopedic injuries, and the collateral/capsular ligaments involved.

Surgical Timing

Most ACL/PCL/MCL injuries can be treated with brace treatment of the MCL followed by arthroscopic combined ACL/PCL reconstruction in 4 to 6 weeks after healing of the MCL. Certain cases may require repair or reconstruction of the medial structures and must be assessed on an individual basis.

Combined ACL/PCL/posterolateral injuries should be addressed as early as safely possible. ACL/PCL/posterolateral repair-reconstruction performed between 2 and 3 weeks postinjury allows sealing of capsular tissues to permit an arthroscopic approach, and still permits primary repair of injured posterolateral structures.

Open multiple ligament knee injuries/dislocations may require staged procedures. The collateral/capsular structures are repaired after thorough irrigation and debridement, and the combined ACL/PCL reconstruction is performed at a later date after wound healing has occurred. Care must be taken in all cases of delayed reconstruction that the tibiofemoral joint is reduced.

Fig. 15.3. Prepared Achilles tendon allograft. The graft is tubed for easy passage, and to fill the bone tunnels; no. 5 braided permanent sutures are woven through the ends of the graft for traction and/or fixation.

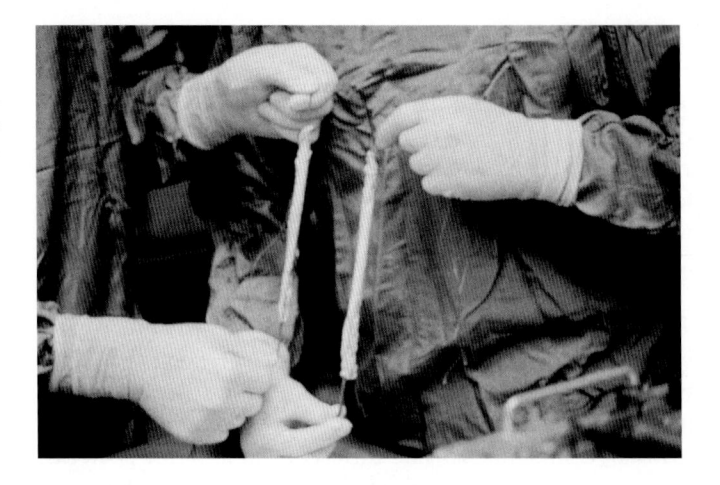

The surgical timing guidelines outlined above should be considered in the context of the individual patient. Many patients with multiple ligament injuries of the knee are severely injured multiple-trauma patients with multisystem injuries. Modifiers to the ideal timing protocols outlined above include the vascular status of the involved extremity, reduction stability, skin condition, open or closed injury, and other orthopedic and systemic injuries. These additional considerations may cause the knee ligament surgery to be performed earlier or later than desired. We have previously reported excellent results with delayed reconstruction in the multiple-ligament–injured knee.[6,7]

Graft Selection

The ideal graft material should be strong, provide secure fixation, be easy to pass, be readily available, and have low donor-site morbidity. The available options in the United States are autograft and allograft sources. Our preferred graft for the PCL is the Achilles tendon allograft because of its large cross-sectional area and strength, absence of donor-site morbidity, and easy passage with secure fixation (Fig. 15.3). We prefer Achilles tendon allograft or bone–patellar tendon–bone allograft for the ACL reconstruction. The preferred graft material for the posterolateral corner is a split biceps tendon transfer, or free autograft (semitendinosus) or allograft (Achilles tendon) tissue when the biceps tendon is not available.[46] Cases requiring MCL and posteromedial corner surgery may have primary repair, reconstruction, or a combination of both. Our preferred method for MCL and posteromedial reconstructions is a posteromedial capsular advancement with autograft or allograft supplementation as needed.

Surgical Approach

Our preferred surgical approach is a single-stage arthroscopic combined ACL/PCL reconstruction using the transtibial technique with collateral/capsular ligament surgery as indicated. The posterolateral corner is repaired, and then augmented with a split-biceps tendon transfer, biceps tendon transfer, semitendinous free graft, or allograft tissue. Acute medial injuries not amenable to brace treatment undergo primary repair, and posteromedial capsular shift or allograft reconstruction as indicated. The surgeon must be prepared to convert to a dry arthroscopic procedure or open procedure if fluid extravasation becomes a problem (Fig. 15.4).

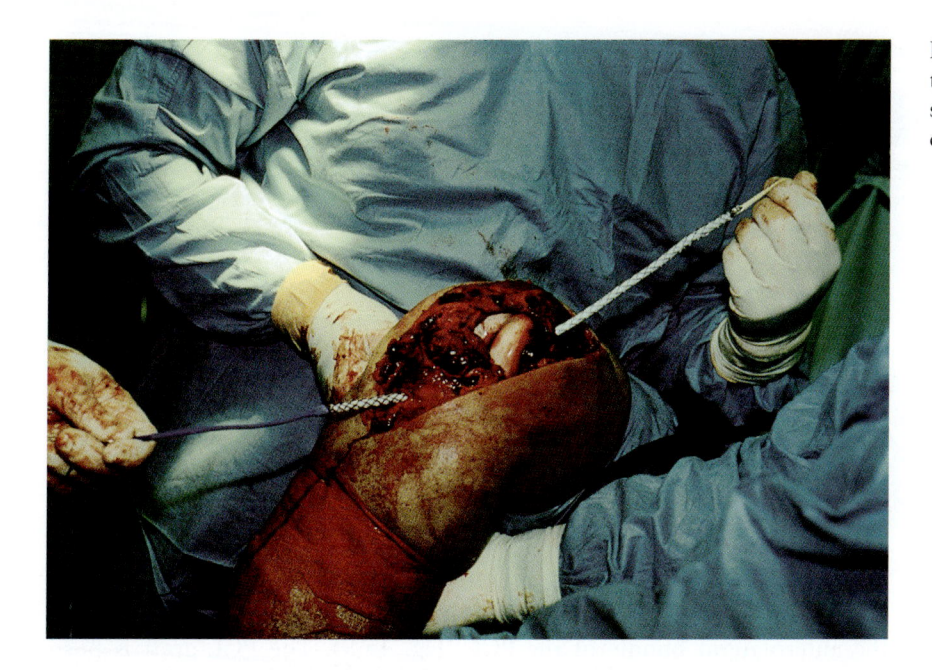

Fig. 15.4. Open ACL/PCL/MCL reconstruction in dislocated knee. Severe capsular destruction prevented an arthroscopic procedure from being performed.

Surgical Technique

The principles of reconstruction in the multiply injured knee are to identify and treat all pathology, ensure accurate tunnel placement, use anatomic graft insertion sites, use strong graft material, ensure secure graft fixation, and plan a postoperative rehabilitation program.

The patient is positioned supine on the operating room table. The surgical leg hangs over the side of the operating table, and the healthy leg is supported by the fully extended operating table. A lateral post is used for control of the surgical leg. We do not use a leg holder. The surgery is done under tourniquet control unless prior arterial or venous repair contraindicates the use of a tourniquet. Fluid inflow is by gravity. We do not use an arthroscopic fluid pump.

Allograft tissue is prepared prior to bringing the patient into the operating room. Arthroscopic instruments are placed with the inflow in the superior lateral portal, arthroscope in the inferior lateral patellar portal, and instruments in the inferior medial patellar portal. An accessory extracapsular extraarticular posteromedial safety incision is used to protect the neurovascular structures and to confirm the accuracy of tibial tunnel placement (Fig. 15.5).

The notchplasty is performed first and consists of ACL and PCL stump debridement, bone removal, and contouring of the medial wall of the lateral femoral condyle and the intercondylar roof. This allows visualization of the over-the-top position, and prevents ACL graft impingement throughout the full range of motion. Specially curved PCL instruments (Arthrotek, Warsaw, IN) are used to elevate the capsule from the posterior aspect of the tibia (Fig. 15.6).

The PCL tibial and femoral tunnels are created with the help of the Fanelli PCL/ACL drill guide (Arthrotek) (Fig. 15.7). The transtibial PCL tunnel goes from the anteromedial aspect of the proximal tibia 1 cm below the tibial tubercle to exit in the inferior lateral aspect of the PCL anatomic insertion site (Fig. 15.8). The PCL femoral tunnel originates externally between the medial femoral epicondyle and the medial femoral

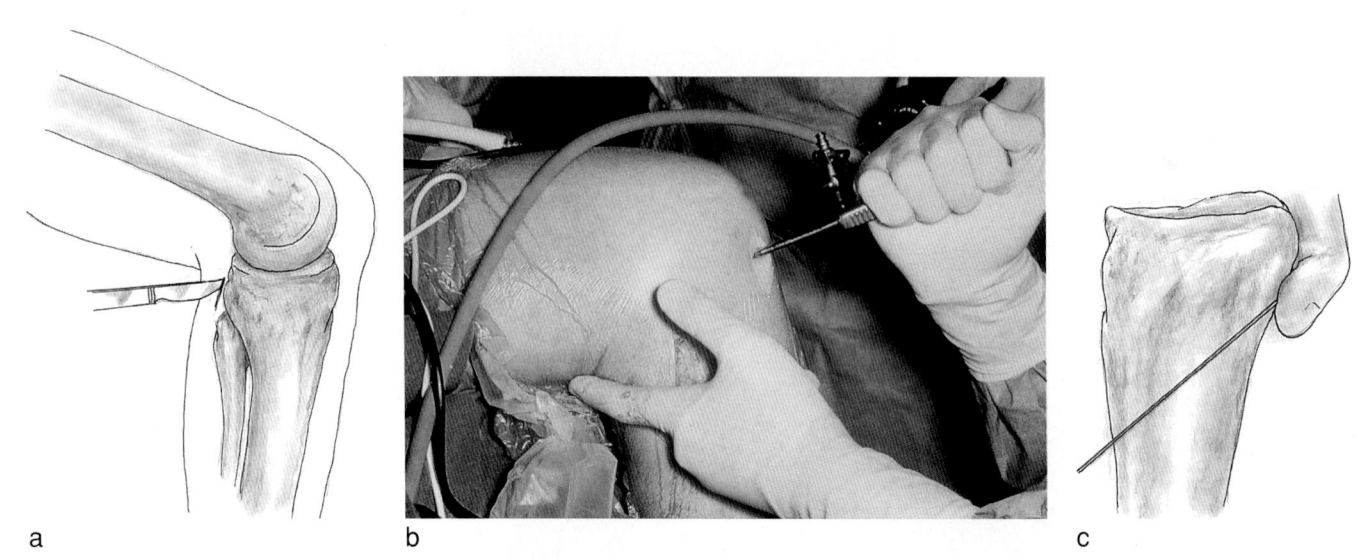

Fig. 15.5. A 1- to 2-cm extracapsular extraarticular posteromedial safety incision allows the surgeon's finger to protect the neurovascular structures, confirm the position of the PCL instruments and drill guide on the posterior aspect of the proximal tibia, and ensure the accuracy of PCL tibial tunnel placement both in the medial-lateral and proximal-distal planes. (a) Drawing of extracapsular-extraarticular posteromedial safety incision. (b) Photograph of surgeon's finger monitoring instrument position through extracapsular posteromedial safety incision. (c) Drawing of surgeon's finger palpating guide wire through posteromedial safety incision during tibial tunnel drilling.

condylar articular surface to emerge through the center of the stump of the anterolateral bundle of the PCL (Fig. 15.9). The PCL graft is positioned and anchored on the femoral or tibial side, and left free on the opposite side.

The ACL tunnels are created using the single-incision technique. The tibial tunnel begins externally at a point 1 cm proximal to the tibial tubercle on the anteromedial surface of the proximal tibia to emerge through the center of the stump of the ACL tibial footprint. The femoral tunnel is positioned next to the over-the-top position on the medial wall of the lateral femoral condyle near the ACL anatomic insertion site. The tunnel is created to leave a 1- to 2-mm posterior cortical wall so interference fixation can be used. The ACL graft is positioned and anchored on the femoral side, with the tibial side left free. Attention is then turned to the posterior lateral corner.

Our preferred technique for posterolateral reconstruction is the split-biceps tendon transfer to the lateral femoral epicondyle (Fig. 15.10). The requirements for this procedure include an intact proximal tibiofibular joint, intact posterolateral capsular attachments to the common biceps tendon, and an intact biceps femoris tendon insertion into the fibular head. This technique creates a new popliteofibular ligament and LCL, tightens the posterolateral capsule, and provides a post of strong autogenous tissue to reinforce the posterolateral corner.

Fig. 15.6. Fanelli curved PCL instruments enable the surgeon to elevate the capsule from the posterior proximal tibia (Arthrotek, Warsaw, IN).

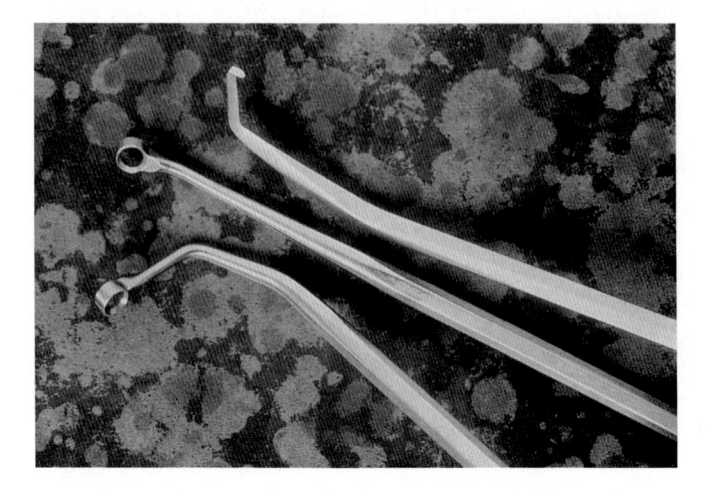

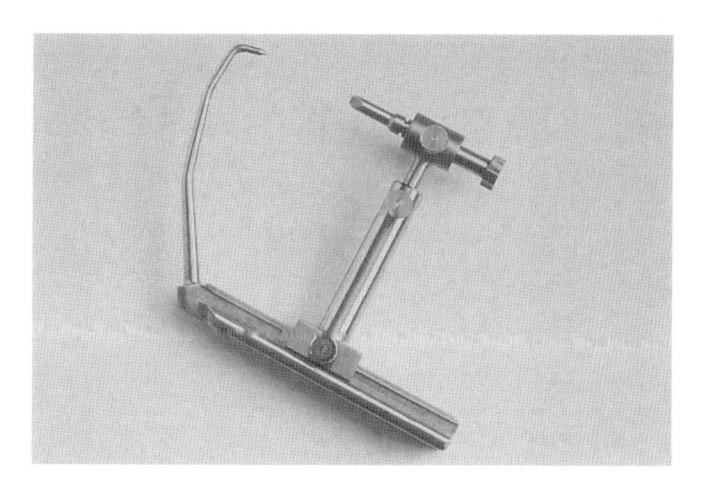

15. Arthroscopically Assisted Combined ACL/PCL Reconstruction

Fig. 15.7. The Fanelli PCL/ACL drill guide enables the surgeon to drill the PCL and ACL tibial and femoral tunnels.

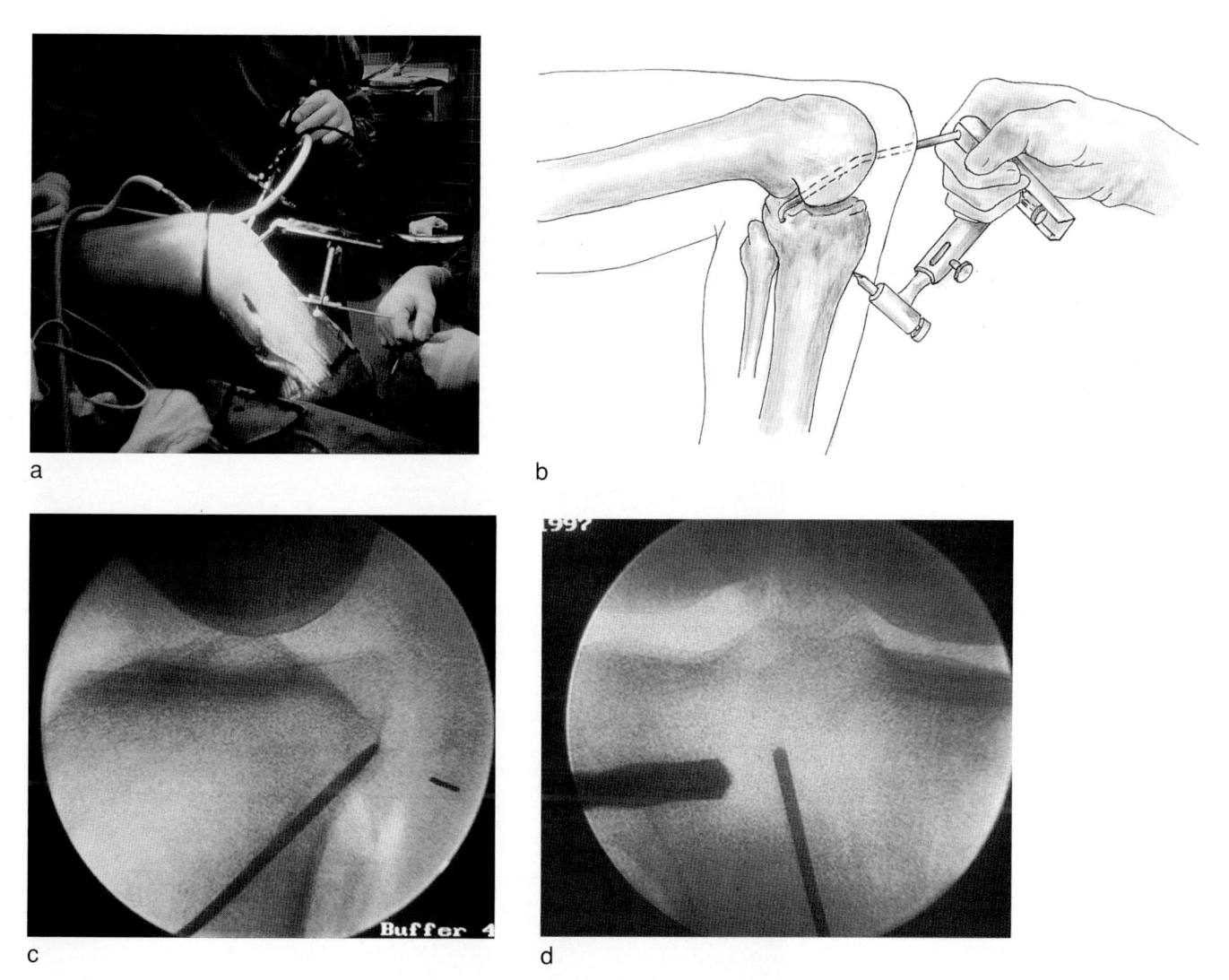

Fig. 15.8. The PCL tibial tunnel should exit at the inferior and lateral aspect of the anatomic insertion site of the PCL. (a) Fanelli PCL/ACL drill guide positioned to place guide wire for the PCL tibial tunnel. Note posteromedial safety incision. (b) Drawing of PCL/ACL drill guide positioned so that the guide wire exits inferior and lateral in the PCL anatomic insertion site. Lateral (c) and anterior-posterior (d) radiographs demonstrating correct position of guide wire prior to PCL tibial tunnel drilling. (From Fanelli et al.,[8] with permission.)

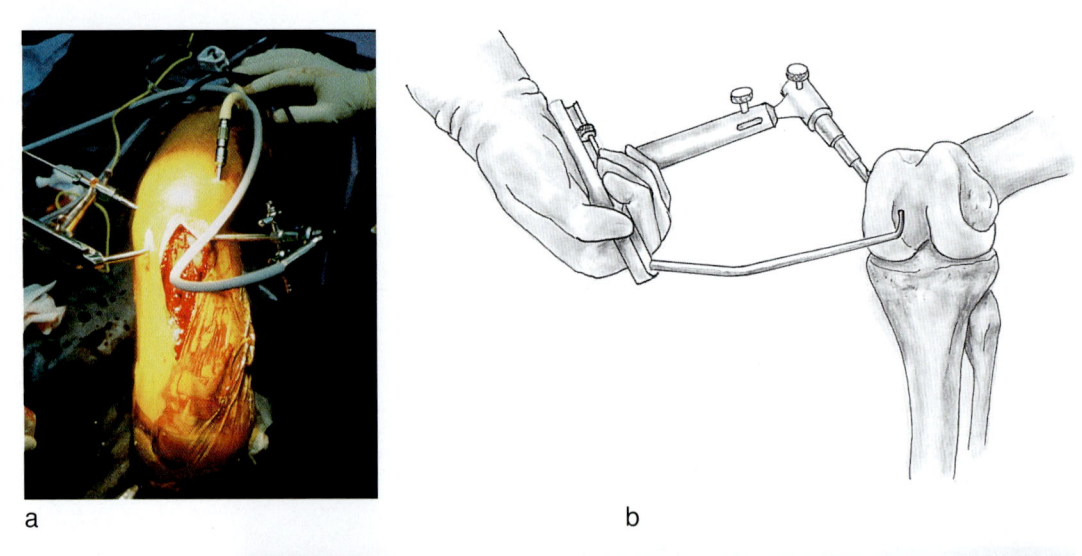

Fig. 15.9. The PCL femoral tunnel enters the joint through the center of the anterolateral fiber region or bundle of the PCL anatomic insertion site. The single-bundle PCL reconstruction technique reproduces the anterolateral bundle of the PCL. Femoral drill hole is in the center of the anterolateral bundle of the PCL. (a) Fanelli PCL/ACL drill guide positioned to place guide wire for the PCL femoral tunnel. (b) Drawing of PCL/ACL drill guide positioned to pass guide wire for femoral tunnel creation. (c) Guide wire exiting through center of the anterior lateral bundle of the PCL. (d) PCL femoral tunnel emerges through the center of the stump of the anterior lateral bundle of the PCL. (From Fanelli et al.,[8] with permission.)

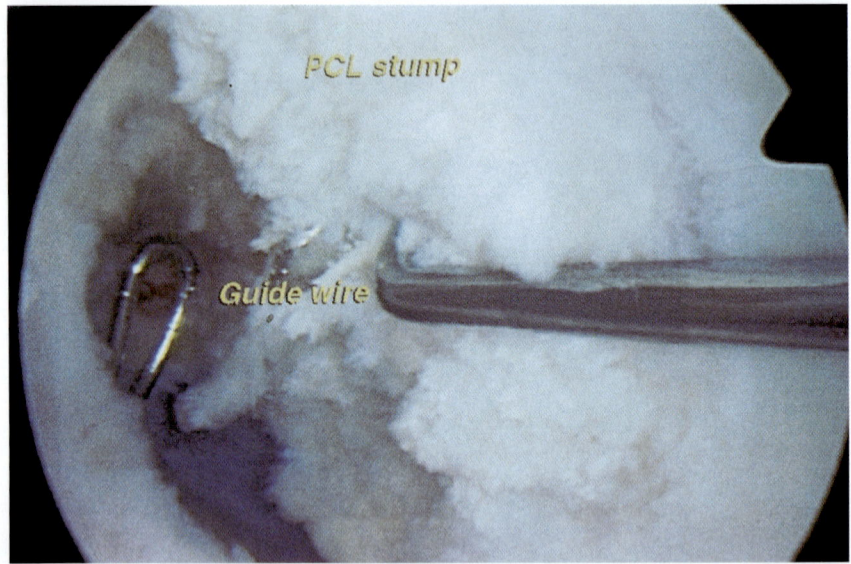

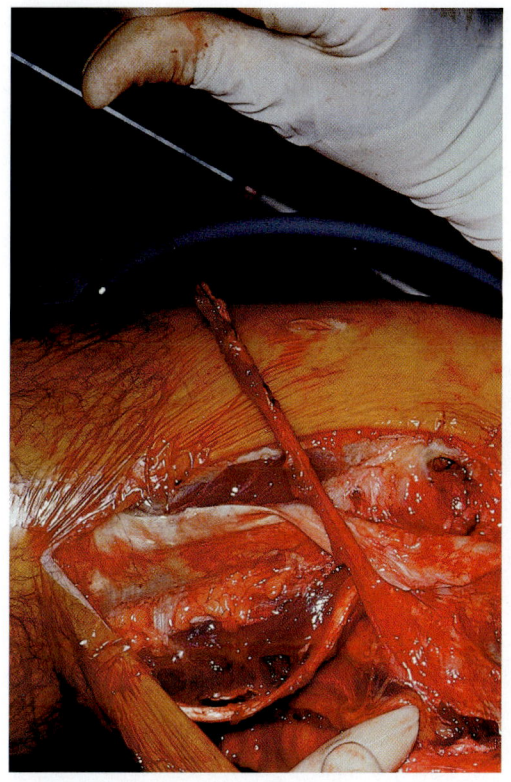

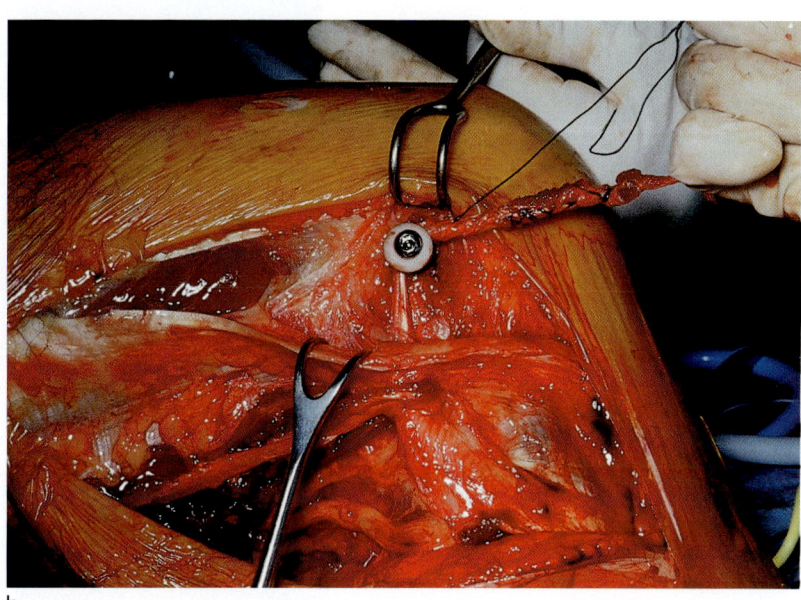

15. Arthroscopically Assisted Combined ACL/PCL Reconstruction

Fig. 15.10. (a and b) The posterolateral reconstruction is performed using a split biceps tendon transfer combined with a posterolateral capsular shift. This technique re-creates the function of the popliteofibular ligament and lateral collateral ligament, and eliminates the posterolateral capsular redundancy. (a) The anterior two-thirds of the long head and common biceps femoris tendon are isolated from the short head muscle. The tendon is detached proximally, and left attached distally to its anatomic insertion site. (b) The peroneal nerve is protected. The split biceps tendon is passed medial to the iliotibial band, and secured to the lateral femoral epicondyle approximately 1 cm anterior to the fibular collateral ligament femoral insertion using a cancellous screw and spiked ligament washer.

A lateral hockey-stick incision is made. The peroneal nerve is dissected free and protected throughout the procedure. The long head and common biceps femoris tendon is isolated, and the anterior two-thirds is separated from the short head muscle. The tendon is detached proximal and left attached distally to its anatomic insertion site on the fibular head. The strip of biceps tendon should be 12 to 14 cm long. The iliotibial band is incised in line with its fibers, and the fibular collateral ligament and popliteus tendons are exposed. A drill hole is made 1 cm anterior to the fibular collateral ligament femoral insertion. A longitudinal incision is made in the lateral capsule just posterior to the fibular collateral ligament. The split biceps tendon is passed medial to the iliotibial band, and secured to the lateral femoral epicondylar region with a screw and spiked ligament washer at the above-mentioned point. The residual tail of the transferred split biceps tendon is passed medial to the iliotibial band, and secured to the fibular head. The posterolateral capsule that had been previously incised is then shifted and sewn into the strut of transferred biceps tendon to eliminate posterolateral capsular redundancy.

Posteromedial and medial reconstructions are performed through a medial hockey-stick incision (Fig. 15.11). Care is taken to maintain adequate skin bridges between incisions. The superficial MCL is exposed, and a longitudinal incision is made just posterior to its posterior border. Care is taken not to damage the medial meniscus during the capsular incision. The interval between the posteromedial capsule and medial meniscus is developed. The posteromedial capsule is shifted anterosuperiorly. The medial meniscus is repaired to the new capsular position, and the shifted capsule is sewn into the MCL. When superficial MCL reconstruction is indicated, it is performed with allograft tissue or semitendinosus autograft. This graft material is attached at the anatomic insertion sites of the superficial MCL

Fig. 15.11. (a) Posteromedial reconstruction performed through a medial hockey-stick incision using (b) posteromedial capsular shift procedure.

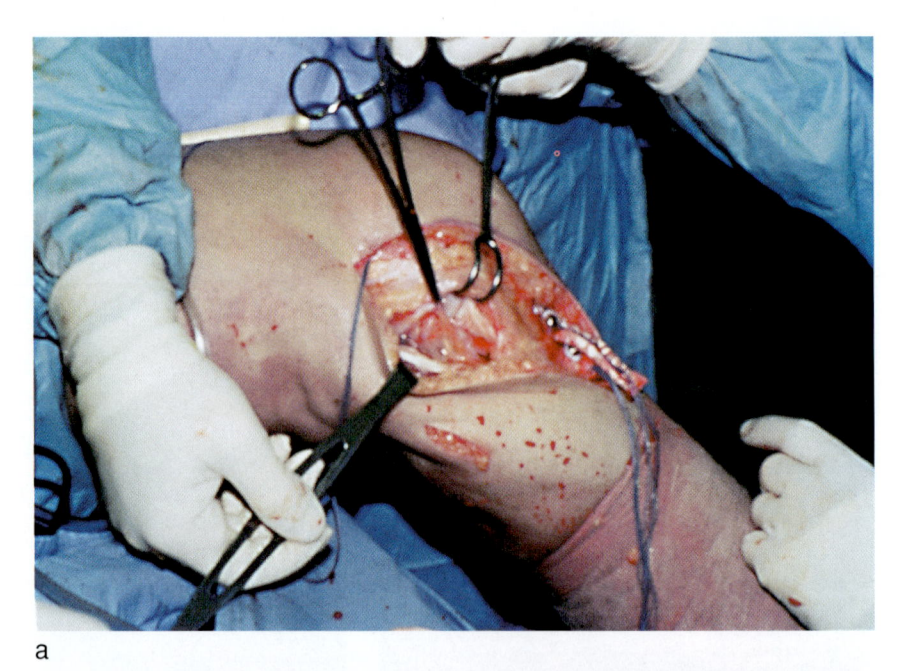

on the femur and tibia. The posteromedial capsular advancement is performed, and sewn into the newly reconstructed MCL. The order of reconstruction is summarized in Table 15.1.

Graft Tensioning and Fixation

The PCL is reconstructed first followed by the ACL followed by the posterolateral complex and/or posterior medial corner. Tension is placed on the PCL graft distally, and the knee is cycled through a full range of motion 25 times to allow pretensioning and settling of the graft. The knee is placed in 70 to 90 degrees of flexion, a firm anterior drawer force is applied to the proximal tibia to restore the normal tibial step-off, and fixa-

15. Arthroscopically Assisted Combined ACL/PCL Reconstruction

Table 15.1. Order of ACL/PCL/PLC/MCL reconstruction

1. PCL tibial tunnel
2. PCL femoral tunnel
3. PCL graft passage and femoral or tibial side fixation
4. ACL tibial tunnel
5. ACL femoral tunnel
6. ACL graft passage and femoral fixation
7. Posterolateral complex reconstruction/repair
8. Medial side reconstruction/repair

tion is achieved on the tibial or femoral side of the PCL graft with a screw and spiked ligament washer and bioabsorbable interference screw. The knee is then placed in 30 degrees of flexion, the tibia is internally rotated, slight valgus force is applied to the knee, and final tensioning and fixation of the posterolateral corner is achieved. The knee is returned to 70 to 90 degrees of flexion, a posterior drawer force is applied to the proximal tibia with tension on the ACL graft, and final fixation is achieved of the ACL graft with a bioabsorbable interference screw and spiked ligament washer backup fixation. Reconstruction and tensioning of the MCL and posteromedial corner are performed after the ACP, PCL, and PLC reconstructions, and are done in 30 degrees of knee flexion. Graft tensioning and final fixation are summarized in Table 15.2.

Technical Hints

The posteromedial safety incision protects the neurovascular structures, confirms accurate tibial tunnel placement, and allows the surgical procedure to be done at an accelerated pace. The single-incision ACL reconstruction technique prevents lateral cortex crowding, and eliminates multiple through-and-through drill holes in the distal femur, reducing the stress riser effect. It is important to be aware of the two tibial tunnel directions, and to have a 1-cm bone bridge between the PCL and ACL tibial tunnels (Fig. 15.12). This will reduce the possibility of fracture. We have found it useful to use primary and backup fixation. Primary fixation is with resorbable interference screws, and backup fixation is performed with a screw and spiked ligament washer. Secure fixation is critical to the success of this surgical procedure (Fig. 15.13).

Postoperative Rehabilitation

The knee is kept in full extension and a non–weight-bearing status is maintained for 6 weeks. Progressive range of motion occurs after postoperative

Table 15.2. Order of tensioning and final graft fixation

1. PCL	Seventy to 90 degrees of knee flexion restoring normal tibial step-off
2. Posterolateral complex	Thirty degrees of knee flexion, internal rotation, and anterior directed force applied to the proximal tibia (must reduce posterolateral spin)
3. ACL	Seventy to 90 degrees of knee flexion maintaining normal tibial step-off and neutral position; tension and fixation performed at same knee flexion angle as PCL
4. MCL/medial side	Thirty degrees of knee flexion after PCL, ACL, and lateral side have been tensioned and fixation performed

Fig. 15.12. A model showing the tibial tunnel positions for combined ACL/PCL reconstructions. It is essential to have an adequate bone bridge between the two tunnels. (From Fanelli et al.,[8] with permission.)

week 6. The brace is unlocked at the end of 6 weeks and the crutches are discontinued after progression to full weight bearing has been achieved. Progressive closed kinetic chain strength training and continued motion exercises are performed. The brace is discontinued after week 10. Return to sports and heavy labor occurs after the ninth postoperative month when sufficient strength and range of motion has returned. It should be noted that a loss of 10 to 15 degrees of terminal flexion can be expected in these complex knee ligament reconstructions. This does not cause a functional problem for these patients, and is not a cause for alarm. Here is a summary of the postoperative rehabilitation program:

0–6 weeks:	Long leg brace locked in full extension, non–weight bearing
7–10 weeks:	Begin range of motion; progress to full weight bearing with crutches
11–24 weeks	Progressive range of motion; progressive strength training
25–36 weeks	Advanced strength training
37 weeks	Return to sports and heavy labor if strength and range of motion are appropriate

Complications

Potential complications in treatment of the multiple-ligament–injured knee include failure to recognize and treat vascular injuries (both arterial and venous), iatrogenic neurovascular injury at the time of reconstruction, iatrogenic tibial plateau fractures at the time of reconstruction, failure to recognize and treat all components of the instability, postoperative medial femoral condyle osteonecrosis, knee motion loss, and postoperative anterior knee pain.

We have performed 110 PCL reconstructions using the transtibial tunnel technique described in this chapter; 107 of these were arthroscopically assisted reconstructions and 57 were combined ACL/PCL reconstructions with posterior lateral or posterior medial reconstructions. Our complications include postoperative adhesions requiring arthroscopic lysis and manipulation in three cases, and removal of painful hardware in five cases.

15. Arthroscopically Assisted Combined ACL/PCL Reconstruction

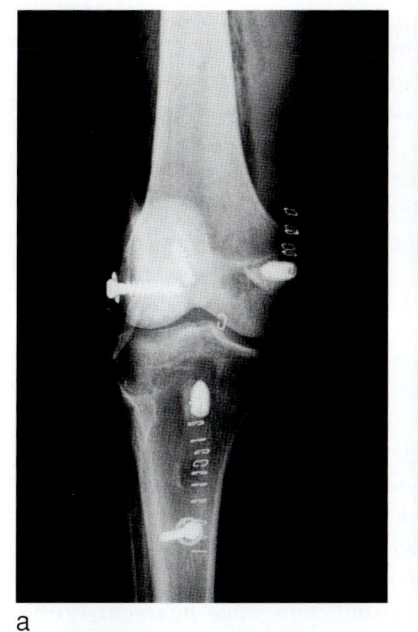

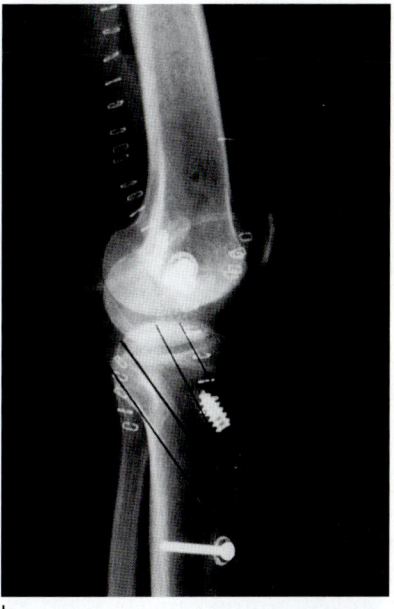

Fig. 15.13. Anteroposterior (AP) (a) and lateral (b) radiographs after combined ACL/PCL/posterolateral complex reconstruction. Note position of tibial tunnel on radiograph. Tibial tunnel guide wire exits at the apex of the tibial ridge posteriorly, which places the graft at the anatomic tibial insertion site after the tibial tunnel is drilled. (c) Arthroscopic view following combined ACL/PCL reconstruction using allograft bone–patellar tendon–bone for the ACL, and allograft Achilles tendon for the PCL.

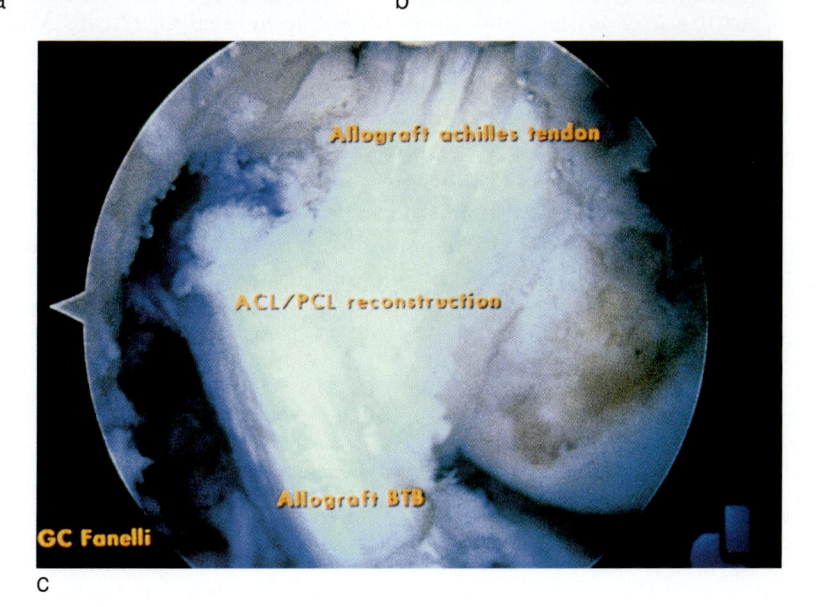

Results

We have previously published the results of our arthroscopically assisted combined ACL/PCL and PCL/posterolateral complex reconstructions using the reconstructive technique described in this chapter.[6–8] Our results indicated that we were able to restore functional stability in all knees, and all patients in these series were able to return to their preinjury level of activity. We were able to predictably restore normal tibial step-off and posterior drawer in 9 of 20 of the reconstructed knees, and to achieve grade I posterior drawer with 5-mm decreased tibial step-off in 11 of 20 reconstructed knees. All knees had grade III posterior drawer tests preoperatively, and negative step-off. Fifteen of 20 knees had normal postoperative Lachman tests improved from grade III preoperative Lachman tests, and 20 of 20 knees had elimination of the preoperative pivot shift. Posterolateral instability was corrected in all cases. We have recently presented our results of 3- to 8-year follow-up of multiple ligament reconstructions analyzed with stress radiography demonstrating the longevity of these reconstructions[47–49]; 45.2% of knees demonstrated 0 to 3 mm side-to-side difference, 32.3% of knees demonstrated 4 to 5 mm side-to-side difference, and 19.4% of knees demonstrated 6 to 7 mm side-to-side difference with posterior proximal tibial force using lateral stress radiography. All knees had grade III preoperative posterior laxity. All preoperative pivot shifting phenomena were eliminated. Posterolateral instability was consistently eliminated.

Conclusion

Combined ACL/PCL injuries are complex injuries requiring a systematic approach to evaluation and treatment. Gentle reduction and documentation and treatment of vascular injuries are primary concerns in the acute dislocated/multiple ligament injured knee. Arthroscopically assisted combined ACL/PCL reconstruction is a reproducible procedure. Knee stability is improved postoperatively when evaluated with knee ligament rating scales, arthrometer testing, and stress radiographic analysis. Acute MCL tears when combined with ACL/PCL tears may in certain cases be treated with bracing. Posterolateral corner injuries combined with ACL/PCL tears are best treated with primary repair as indicated combined with reconstruction using a post of strong autograft (split biceps tendon, biceps tendon, semitendinosus), or allograft (Achilles tendon, bone–patellar tendon–bone) tissue. Surgical timing depends on the ligaments injured, the vascular status of the extremity, reduction stability, and the overall health of the patient. We prefer the use of allograft tissue for reconstruction in these cases because of the strength of these large grafts, and the absence of donor-site morbidity.

References

1. Butler DL, Noyes FR, Grood ES. Ligamentous restraints to anterior-posterior drawer in the human knee. A biomechanical study. J Bone Joint Surg 1980;62A:259–270.
2. Wilson SA, Vigorita VJ, Scott WN. Anatomy. In: Scott WN, ed. The Knee. St. Louis: Mosby, 1994.
3. VanDommelen BA, Fowler PJ. Anatomy of the posterior cruciate ligament. A review. Am J Sports Med 1989;17:24–29.
4. Gollehon DL, Torzilli PA, Warren RF. The role of the posterior lateral corner and cruciate ligaments in the stability of the human knee. A biomechanical study. J Bone Joint Surg 1987;69A:233–242.
5. Seebacher JR, Inglis AE, Marshall JL, et al. The structure of the posterolateral aspect of the knee. J Bone Joint Surg 1982;64A:536–541.
6. Fanelli GC, Gianotti BF, Edson CJ. Arthroscopically assisted combined anterior and posterior cruciate ligament reconstruction. Arthroscopy 1996;12(1):5–14.
7. Fanelli GC, Gianotti BF, Edson CJ. Arthroscopically assisted combined posterior cruciate ligament/posterior lateral complex reconstruction. Arthroscopy 1996;12(5):521–530.
8. Fanelli GC, Gianotti BF, Edson CJ. The posterior cruciate ligament arthroscopic evaluation and treatment. Arthroscopy 1994;10(6):673–688.
9. Ghalambor N, Vangsness CT. Traumatic dislocation of the knee: a review of the literature. Bull Hosp Jt Dis 1995;54(1):19–24.
10. Good L, Johnson RJ. The dislocated knee. JAAOS 1995;3(5):284–292.

11. Shelbourne KD, Porter DA, Clingman JA, et al. Low-velocity knee dislocation. Orthop Rev 1991;20:995–1004.
12. Wascher DC, Dvirnak PC, Decoster TA. Knee dislocation: initial assessment and implications for treatment. J Orthop Trauma 1997;11(7):525–529.
13. Frassica FJ, Sim FH, Staeheli JW, et al. Dislocation of the knee. Clin Orthop Rel Res 1991;263:200–205.
14. Green A, Allen BL. Vascular injuries associated with dislocation of the knee. J Bone Joint Surg 1977;59A:236–239.
15. Wand JS. A physical sign denoting irreducibility of a dislocated knee. J Bone Joint Surg 1989;71B:862.
16. Kennedy JC. Complete dislocation of the knee joint. J Bone Joint Surg 1963;45A:889–904.
17. Shields L, Mital M, Cave EF. Complete dislocation of the knee: experience at the Massachusetts General Hospital. J Trauma 1969;9:192–215.
18. Meyers MH, Harvey JP Jr. Traumatic dislocation of the knee joint: a study of eighteen cases. J Bone Joint Surg 1971;53A:16–29.
19. Girgis FG, Marshall JL, Monajem A. The cruciate ligaments of the knee joint. Clin Orthop Rel Res 1975;106:216–231.
20. Roman PD, Hopson CN, Zenni EJ Jr. Traumatic dislocation of the knee: a report of 30 cases and literature review. Orthop Rev 1987;16:917–924.
21. Sisto DJ, Warren RF. Complete knee dislocation: a follow-up study of operative treatment. Clin Orthop Rel Res 1985;198:94–101.
22. Meyers MH, Moore TM, Harvey JP Jr. Traumatic dislocation of the knee joint. J Bone Joint Surg 1975;57A:430–433.
23. Welling RE, Kakkasseril J, Cranley JJ. Complete dislocations of the knee with popliteal vascular injury. J Trauma 1981;21:450–453.
24. Almekinders LC, Logan TC. Results following treatment of traumatic dislocation of the knee joint. Clin Orthop Rel Res 1991;284:203–207.
25. Shapiro MS, Freedman EL. Allograft reconstruction of the anterior and posterior cruciate ligaments after traumatic knee dislocation. Am J Sports Med 1995;23(5):580–587.
26. Ashworth EM, Dalsing MC, Glover JL, et al. Lower extremity vascular trauma: a comprehensive, aggressive approach. J Trauma 1988;28:329.
27. Rich NM, Hobson RW, Wright CB. Repair of lower extremity venous trauma: a more aggressive approach required. J Trauma 1974;14:639.
28. Cone JC. Vascular injury associated with fracture-dislocations of the lower extremity. Clin Orthop Rel Res 1989;243:30–35.
29. Grimley RP, Ashton F, Slaney G, et al. Popliteal arterial injuries associated with civial knee trauma. Injury 1981;13:1–6.
30. O'Donnell TF Jr, Brewster DC, Darling RC, et al. Arterial injuries associated with fractures and/or dislocations of the knee. J Trauma 1977;17:775–784.
31. Savage R. Popliteal artery injury associated with knee dislocation: improved outlook? Am Surg 1980;46:627–632.
32. Rich NM, Hobson RW, Collins GJ, et al. The effect of acute popliteal venous interruption. Ann Surg 1976;183:365–368.
33. Wright DG, Covey DC, Born CT, et al. Open dislocation of the knee. J Orthop Trauma 1995;9(2):135–140.
34. Malizos KN, Xenakis T, Mavrodontidis AN, et al. Knee dislocations and their management. Acta Orthop Scand 1997;68(suppl 275):80–83.
35. McCoy GF, Hannon DG, Barr RJ, et al. Vascular injury associated with low-velocity dislocations of the knee. J Bone Joint Surg 1987;69B:285–287.
36. Moore TM. Fracture-dislocation of the knee. Clin Orthop Rel Res 1981;156:450–453.
37. Hill JA, Rana NA. Complications of posterolateral dislocation of the knee: case report and literature review. Clin Orthop Rel Res 1981;154:212–215.
38. Applebaum R, Yellin AE, Weaver FA, et al. Role of routine arteriography in blunt lower-extremity trauma. Am J Surg 1990;160:221–225.
39. Fanelli GC. Paper presented at the American Academy of Orthopaedic Surgeons 66th annual meeting, Anaheim, CA, February 1999.
40. Trieman GS, Yellin AE, Weaver FA, et al. Evaluation of the patient with a knee dislocation: the case for selective arteriography. Arch Surg 1992;127(9):1056–1063.
41. Myles JW. Seven cases of traumatic dislocation of the knee. Proc R Soc Med 1967;60:279.

42. Taylor AR, Arden GP, Rainey MA. Traumatic dislocations of the knee: a report of forty three cases with special reference to conservative treatment. J Bone Joint Surg 1972;54B:94.
43. Noyes FR, Barber-Westin SD. Reconstruction of the anterior and posterior cruciate ligaments after knee dislocation. Am J Sports Med 1997;25(6):769.
44. Fanelli GC. PCL injuries in trauma patients. Arthroscopy 1993;9:291–294.
45. Fanelli GC, Edson CJ. PCL injuries in trauma patients, part II. Arthroscopy 1995;11:526–529.
46. Fanelli GC, Feldmann DD. The use of allograft tissue in knee ligament reconstruction. In: Parisien JS, ed. Current Techniques in Arthroscopy, 3rd ed. New York: Thieme, 1998.
47. Fanelli GC. Paper presented at the American Academy Of Orthopaedic Surgeons 66th annual meeting, Anaheim, CA, February 1999.
48. Fanelli GC, Maish DR, Edson CJ. Stress radiographic analysis of arthroscopically assisted PCL reconstruction: 3 to 8 year follow-up. Paper presented at the Arthroscopy Association of North America Annual Meeting, Vancouver, Canada, April 1999.
49. Fanelli GC, Maish DR, Edson CJ. Stress radiographic analysis of arthroscopically assisted PCL reconstruction: 3 to 8 year follow-up. Paper presented at the ISAKOS Congress, Washington, DC, May 29–June 2, 1999.

Chapter Sixteen

Surgical Treatment of Posterolateral Instability

Roger V. Larson and Michael H. Metcalf

Injury to the posterior cruciate ligament (PCL) is thought to account for 3% to 20% of all knee ligament injuries.[1] The true incidence of PCL injuries remains unknown because many isolated PCL injuries may go undetected.[2,3] It is generally accepted that isolated ruptures of the PCL do not generally cause functional instability and are best managed by nonoperative treatment.[4-7] When functional instability is present, the situation is usually not an isolated PCL injury, but a combined ligamentous injury frequently involving the posterolateral corner.[8-11] Posterolateral instability of the knee in combination with PCL insufficiency is frequently the cause of functional instability, and management of this instability requires addressing not only the PCL injury, but the associated posterolateral corner injury.

It is also generally accepted that the results of PCL reconstruction are not as good as the results of anterior cruciate ligament (ACL) reconstruction. Several reasons have been postulated why PCL reconstruction techniques do not generally restore normal laxity and motion patterns to the postoperative knee. The postulated reasons include the anisometry of the normal PCL and the inability to reconstruct the PCL with a single graft,[12-14] the fairly constant daily forces on the normal PCL, and graft tissue that may not be as strong as necessary in this position. Another important reason why PCL reconstruction may not yet be as effective as ACL reconstruction is the failure to recognize and address associated injuries that may coexist, particularly to the posterolateral corner. Since isolated PCL ruptures do not generally cause functional instability and are generally treated nonoperatively, when functional instability is present there is usually associated laxity, often of the posterolateral corner. Since posterolateral corner instability cannot be treated by PCL reconstruction alone, it is a very common scenario that when PCL insufficiency is treated operatively a posterolateral reconstruction is performed simultaneously. It has also been shown by Cooley et al.[15] that when posterolateral laxity is present associated with PCL insufficiency, normal motion patterns are not restored by PCL reconstruction alone but can be restored by combined PCL and posterolateral corner reconstruction. Their study also showed that strain on an intraarticular PCL graft can be reduced and the environment for survival enhanced if an appropriate extraarticular backup procedure is added.

Biomechanics

The PCL is the primary restraint to posterior tibial displacement at all angles of flexion. The posterolateral corner and the posterior capsule of the knee joint are important secondary restraints to posterior tibial displacement, particularly with the knee near full extension. The PCL is not an isometric ligament.[12,16,17] The area within the femoral footprint that strains less than 10% through a full range of motion is so small that a finite-sized graft cannot be placed and be supportive through a full range of motion.[12] The PCL instead functions as at least two distinct ligaments, an anterolateral (AL) band that is supportive in flexion, but becomes somewhat loose near full knee extension, and a smaller anteromedial (AM) band that functions in a reciprocal manner, providing support to posterior tibial displacement near full extension. The anteromedial band becomes lax and nonsupportive at greater flexion angles when the anterolateral band is functioning. One reason why the posteromedial band of the PCL may be smaller than the anterolateral band is that near terminal extension the posterior capsule and posterolateral corner of the knee become taut and supportive, decreasing the contribution needed from the PCL.

It is generally accepted that when performing a single-graft PCL reconstruction, the anterolateral band should be reconstructed.[18,19] This is appropriately tensioned with the knee near 70 to 90 degrees of flexion, thus providing support when the knee is in that position. A graft in this position, however, will tend to become lax as the knee is fully extended. But if the posterior capsule and posterolateral corner are competent, they can provide support near terminal extension, thus substituting for the posteromedial band of the PCL. It is therefore usually appropriate when performing an anterolateral band ACL reconstruction to also provide a posterolateral reconstruction to support the knee near full extension. This is of particular importance if there has been any injury to the posterolateral corner, which is often the case when operative intervention is necessary.

Anatomy

The posterolateral corner of the knee is an area of complicated and variable anatomy. The anatomic structures of importance include the lateral collateral ligament, arcuate ligament, the popliteofibular ligament, the posterolateral capsule, the popliteus tendon, the short lateral ligament, and the fabellofibular ligament. From the standpoint of reconstruction of the posterolateral corner the most consistent and important structures to become familar with are the popliteus tendon, the popliteofibular ligament, and the lateral collateral ligament.[20–23] These structures are demonstrated in Fig. 16.1. The popliteofibular ligament is that portion of the popliteus tendon that attaches to the posterior aspect of the fibular head rather than to the popliteus muscle belly. This portion of the popliteus tendon is static in nature and can comprise greater than 50% of the bulk of the popliteus tendon. The popliteofibular ligament is well positioned to resist posterolateral tibial rotations as well as varus rotations.

Isometry of the Fibular Collateral and Popliteofibular Ligaments

A study by Sidles et al.[12] has evaluated the isometry of the lateral side of the knee, particularly the fibular head relative to the lateral femur. The study demonstrated that the entire fibular head is isometric to the lateral femoral epicondyle through a functional range of knee motion. There was

16. Surgical Treatment of Posterolateral Instability

Fig. 16.1. The relationship between the lateral collateral ligament, the popliteus tendon, and popliteofibular ligament. The popliteofibular ligament is the static portion of the popliteal tendon that attaches to the posterior aspect of the fibular head.

Fig. 16.2. The fibular head is isometric to the lateral femoral epicondyle. The posterior aspect of the fibular head is slightly more isometric to the anterior aspect of the epicondyle (dark areas), and the anterior aspect of the fibular head is slightly more isometric to the posterior aspect of the epicondyle (dotted area).

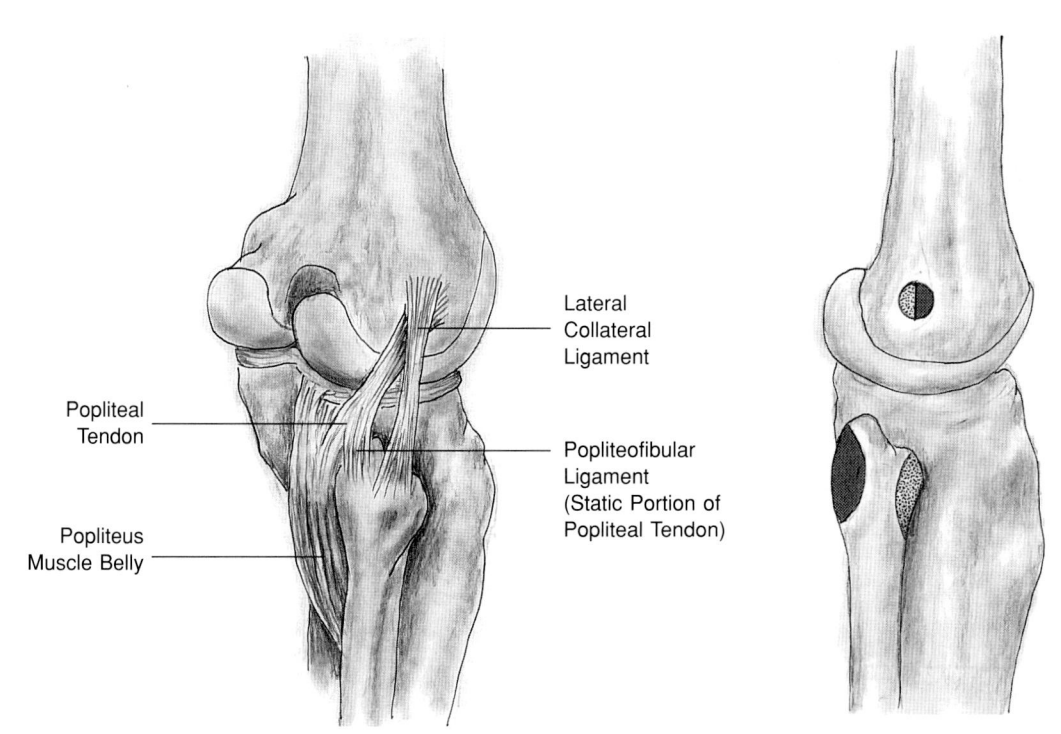

a slight decrease in strain from the posterior aspect of the fibular head to the anterior epicondyle and from the anterior aspect of the fibular head to the posterior epicondyle (Fig. 16.2). These data demonstrate that a graft taken from any position on the fibular head to the lateral femoral epicondyle will remain supportive through a functional range of knee motion. It can also be noted that the posterior aspect of the fibular head relative to the anterior femoral epicondyle represents the popliteofibular ligament or static portion of the popliteal tendon. This static portion of the popliteal tendon remains supportive through a functional range of motion.[22] The portion of the popliteus tendon that attaches to the popliteus muscle belly is not isometric, but tension can be altered by the muscle belly itself. A lateral collateral ligament (LCL) graft attached at the fibular head and the lateral femoral epicondyle will remain supportive through a functional range of motion.

Mechanism of Injury to the Posterolateral Corner

The posterolateral corner provides a restraint to varus rotations, posterolateral tibial rotations, and to posterior tibial translations near full terminal extension. These structures are therefore injured with excessive varus stresses, posterolateral (external) tibial rotations, and hyperextension.

Symptoms of Posterolateral Instability

The symptoms of posterolateral instability can be quite variable and depend on the extent of the laxity, tibiofemoral mechanical alignment, the presence or absence of varus laxity, the presence or absence of hyperextension and the patient's activity level. In a low-demand patient with a valgus mechanical alignment and minimal hyperextension, the symptoms may

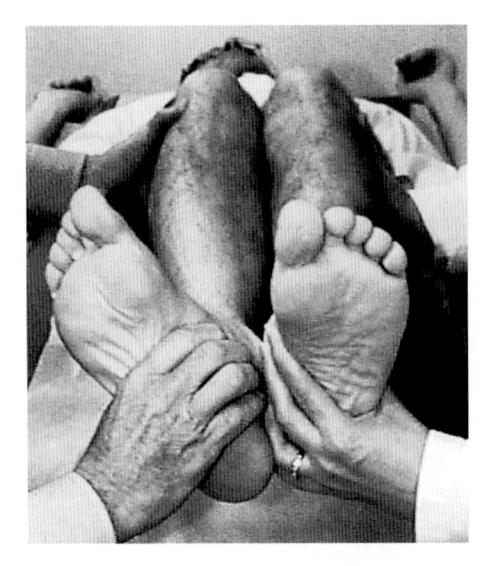

Fig. 16.3. The dial test. An equal amount of external rotation torque is applied to both ankles and the amount of external rotation measured. The test is done at 30 and 90 degrees of knee flexion. It can be done with the patient supine as demonstrated or in the prone position.

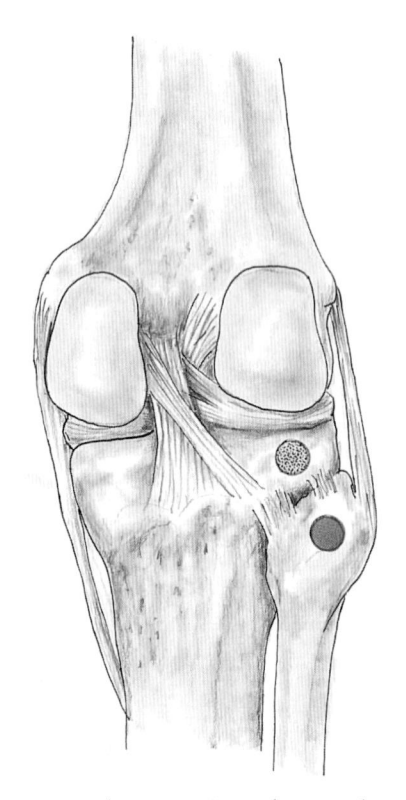

Fig. 16.4. The exit points of posterolateral reconstruction through the posterolateral tibia (dotted) and fibular head (dark area) are demonstrated. The fibular head exit point is further from the axis of rotation and further posterior than is that from the posterolateral tibia.

be minimal. On the other hand, a patient with hyperextension, associated varus laxity, and a varus alignment may present with a varus thrust that is markedly disabling with everyday ambulation. All factors, therefore, need to be recorded and considered before an appropriate treatment recommendation can be made.

Physical Examination of Posterolateral Instability

Posterolateral instability often has a component of varus laxity, particularly if the injury to the posterolateral corner was by a varus mechanism. Varus laxity, however, is not essential to the diagnosis of posterolateral instability. The most important physical finding with posterolateral instability is increased external tibial rotation present at both 30 and 90 degrees of knee flexion. This can be demonstrated by the "dial test," where the femur is stabilized and the tibia, ankle, and foot are externally rotated and compared to the normal side at both 30 and 90 degrees of flexion (Fig. 16.3). If more than 5 degrees of increased external rotation can be demonstrated, an injury to the posterolateral corner can be presumed. This can be present with or without significant injury to the PCL. Usually the increase in external rotation is significantly more than 5 degrees with posterolateral corner injuries.

Optimizing Graft Placement for Posterolateral Reconstruction

Reconstructive procedures for posterolateral instability involve placing tissue from the posterior aspect of the fibular head to the lateral femoral epicondyle or from the posterolateral corner of the tibia to the lateral femoral epicondyle. The attachment points of these two options are demonstrated in Fig. 16.4. Tissue placed in these locations will resist external tibial rotations. There are several reasons why tissue connecting the lateral femoral epicondyle to the posterior fibular head is favorable to tissue connecting the lateral femoral epicondyle to the posterolateral tibia. Tissue passed from the posterior aspect of the fibular head to the lateral femoral epicondyle is isometric. Tissue from the posterolateral corner of the tibia to the lateral femoral epicondyle is not isometric. If a graft from the posterolateral tibia to the epicondyle is tensioned near full extension it may provide appropriate support near full extension, but the tissue will become lax as the knee goes into greater flexion angles. In the normal situation tissue in the position of a "popliteus bypass" would be tensioned by an intact popliteus muscle belly.

Tissue from the lateral femoral epicondyle to the posterior fibular head has additional biomechanical advantages over tissue from the epicondyle to the posterolateral corner of the tibia. As can be seen in Fig. 16.5, if the axis of rotation of the tibia relative to the femur is considered to be near the PCL tibial attachment site, then the lever arm of a graft attached to the posterior fibular head is greater by approximately 50% than that of a graft attached at the posterolateral tibia. The graft attached to the posterior fibular head is proportionately more effective in resisting posterolateral rotations. Wroble et al.,[24] in a ligament cutting study, have demonstrated that significant posterolateral instability could not be produced until the LCL/popliteofibular complex had been sectioned, further validating the im-

16. Surgical Treatment of Posterolateral Instability

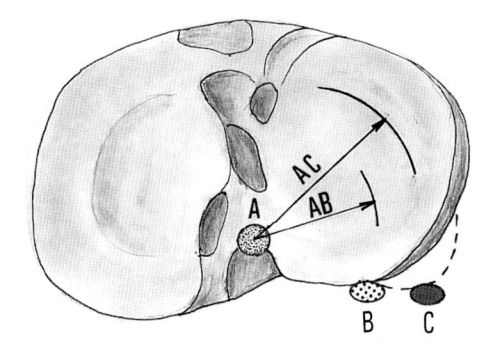

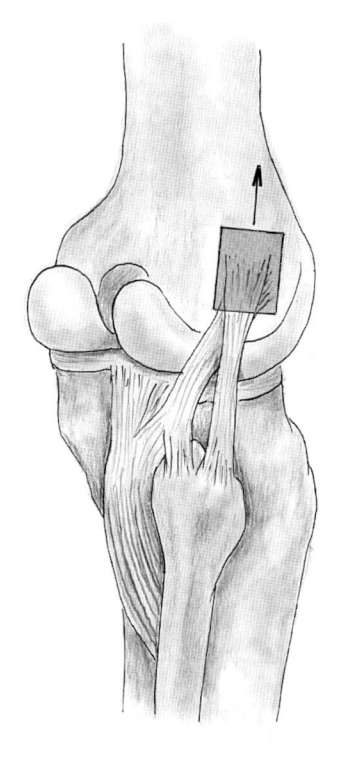

Fig. 16.5. The lever arm for resisting posterolateral rotations of a graft exiting the posterolateral tibia (AB) and the posterior fibular head (AC). The lever arm to the posterior fibular head is greater by approximately 50% than the lever arm of a graft exiting the posterolateral tibia.

Fig. 16.6. If the lateral collateral, popliteus, popliteofibular complex is intact but stretched, it can be tightened by proximal advancement or recession of their combined insertions.

portance of these structures. It is therefore appropriate that a posterolateral reconstruction should include tissue from the posterior aspect of the fibular head to the lateral femoral epicondyle. In cases of severe laxity, it may also be appropriate to include tissue to the posterolateral corner of the tibia, recognizing that appropriately tensioned tissue in this position will add additional support near terminal extension, but may become somewhat lax as the knee flexes.

Options for Treating Posterolateral Instability

Advancement or Recession of the Lateral Femoral Epicondyle

It has been advocated that an advancement or recession of the lateral femoral epicondyle will tighten the lateral and posterolateral structures, thus treating varus and posterolateral laxity. This in fact can be an appropriate procedure when structures that attach at the epicondyle, including the popliteus tendon, LCL, and popliteofibular ligament, are present and have only suffered stretch injuries. If these structures, however, are not in continuity or are deficient, then this procedure is not likely to be effective alone in addressing the problem of posterolateral instability.

The advancement of the epicondyle in a proximal direction may have an effect on the isometry of the ligaments attached to the bone block. Advancing the bone block proximally, however, does not change the isometry as significantly as would anterior or posterior displacement of the bone block. It is usually more appropriate, however, to recess the epicondyle at the anatomic position to avoid deviation from the normal anatomic attachment site. The advancement of a bone block including the epicondyle is demonstrated in Fig. 16.6.

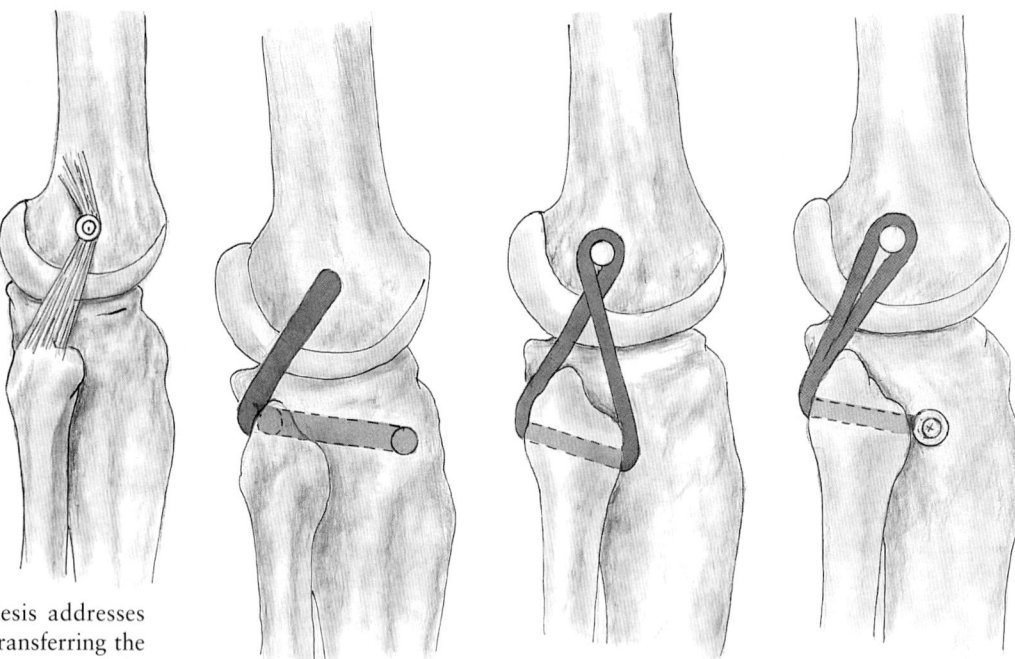

Fig. 16.7. The biceps tenodesis addresses posterolateral instability by transferring the biceps tendon to the anterior aspect of the lateral epicondyle. This creates a restraint from the posterior fibular head to the anterior epicondyle, thus restoring the popliteofibular ligament. It also serves to reef the posterolateral capsule by leaving posterior attachments to the biceps tendon intact.

Fig. 16.8. The popliteus bypass procedure is schematically represented. Tissue is taken through a tibial drill hole to the posterolateral corner of the tibia, and then to the anterior femoral epicondyle.

Fig. 16.9. A free semitendinosus graft can be utilized to reconstruct the lateral collateral and popliteofibular ligaments. The graft is placed anterior to posterior through a fibular head drill hole, and then taken in a figure-of-8 course to the lateral femoral epicondyle. The figure-of-8 course optimizes isometry in each limb of the graft. The graft exiting the posterior fibular head and attaching at the anterior epicondyle represents a reconstruction of the popliteofibular ligament.

Fig. 16.10. When posterolateral instability is present without varus instability, the free semitendinosus graft procedure can be modified to pass both limbs of the graft from the posterior fibular head to the anterior aspect of the epicondyle, thus placing more tissue in the position of the popliteofibular ligament. The distal end of the graft can be fixed either to the fibular head or passed through the fibular head from posterior to anterior and secured to the anterolateral tibia as shown.

Reconstructions for Posterolateral Instability

The most common reconstructions for posterolateral instability include a biceps tendon tenodesis, a popliteus bypass procedure, a free graft reconstruction to the fibular head, and combinations of these procedures. The biceps tendon tenodesis popularized by Clancy[25] is demonstrated in Fig. 16.7. In this procedure the biceps femoris tendon or a portion thereof is left attached to the fibular head. The tendon is freed up and advanced anteriorly to the anterior aspect of the lateral femoral epicondyle, where it is fixed with a soft tissue screw and washer. The posterior capsular attachments of the biceps are left intact to provide a reefing effect. This repair is isometric and has been shown to prevent abnormal posterolateral rotation. It has the potential complications, however, of breaking loose from the femoral attachment site and of overconstraining external rotation. There is also concern with sacrificing the only lateral knee flexor originating proximally. Variations of the biceps tenodesis utilizing only a portion of the tendon have been described.

The popliteus bypass procedure popularized by Mueller[26] utilizes free graft material through a transtibial tunnel exiting at the posterolateral corner (Fig. 16.8). This tissue is then taken from the posterolateral corner of the tibia to the anterior aspect of the lateral femoral epicondyle. This tissue provides a restraint to posterolateral tibial rotation, but is anisometric. Tissue in this position when appropriately tensioned will be supportive near full extension, but may be less supportive as the knee passes into greater flexion angles. It also does not have as favorable a biomechanical advantage to resisting posterolateral rotation as does tissue attaching to the fibular head. The popliteus bypass procedure also requires considerably more dissection than does tissue placed to or through the fibular head.

A semitendinosus free graft as described by Larson et al.[27] can also be utilized through a fibular head drill hole (Fig. 16.9). A semitendinosus graft can be passed through the fibular head and routed to the lateral femoral

16. Surgical Treatment of Posterolateral Instability

epicondyle for fixation. If the fixation is around a screw and washer, a figure-of-eight course of the graft will optimize the isometry in each limb. The tissue passing from the posterior aspect of the fibular head to the anterior aspect of the epicondyle provides an isometric reconstruction of the popliteofibular ligament portion. Both the anterior and posterior bands are isometric and provide a restraint to varus rotations. When no varus laxity is present, the procedure can be modified to reconstruct only the popliteofibular ligament. In this case both strands of the graft can be passed from the posterior aspect of the fibular head to the lateral femoral epicondyle (Fig. 16.10). This provides more graft tissue in the position resisting posterolateral rotations.

In cases of gross posterolateral and varus laxity, combinations of the above procedures may be undertaken, including tightening of existing tissues by reefing or epicondylar recession, graft placement from the fibular head to the lateral femoral epicondyle, and graft placement from the posterolateral tibia to the lateral femoral epicondyle (Fig. 16.11).

Technique of Lateral Collateral and Popliteofibular Ligament Reconstructions Utilizing a Semitendinosus Free Graft

The use of an autogenous semitendinosus free graft to reconstruct the LCL and popliteofibular ligament has been very effectively used to augment primary repairs of these structures. The procedure can also be utilized alone for treating varus and posterolateral laxity of moderate degree in both acute and chronic situations. It is most often done in combination with an ACL or PCL reconstruction where the injury has involved the lateral and posterolateral structures.

The procedure involves the harvest of a semitendinosus tendon, the removal of muscle tissue from the graft, and the use of a Bunnell stitch of large nonabsorbable suture in each end of the graft. A curvilinear incision is then made on the lateral aspect of the knee extending from the lateral femoral epicondyle to the posterior aspect of the fibular head. The fibular head is exposed and a tunnel is created in the area of maximum diameter from anterior to posterior. The tunnel is created by passing a guide pin and then a cannulated drill, usually 4.5 to 6 mm in diameter. At this location the drilling is aimed directly toward the peroneal nerve and due caution needs to be taken to isolate and protect the nerve during this portion of the procedure. The free semitendinosus graft is then passed through the fibular head drill hole.

An incision is then made in line with the fibers of the iliotibial band directly overlying the lateral femoral epicondyle. The epicondyle can usually be easily palpated. The posterior band of the semitendinosus graft is then passed beneath the biceps tendon and iliotibial band to the epicondyle. The band exiting the anterior aspect of the fibular head is passed beneath the iliotibial band also to the epicondyle. The epicondyle is exposed by using a bovie through the periosteum followed by a perosteal elevator. The grafts are then fixed to the epicondyle by placing a low-profile titanium screw and washer combination. The grafts are passed in a figure-of-eight course around the femoral screw (Fig. 16.9). In cases where minimal varus laxity is present, both bands of the graft can be passed from the posterior fibular head to the epicondyle in the position of the popliteal fibular ligament (Fig. 16.10). The washer at the epicondyle needs to be very low profile to avoid contacting and irritating the overlying iliotibial band. The position of the fibular head drill hole and epicondylar fixation is demonstrated radiographically in Fig. 16.12.

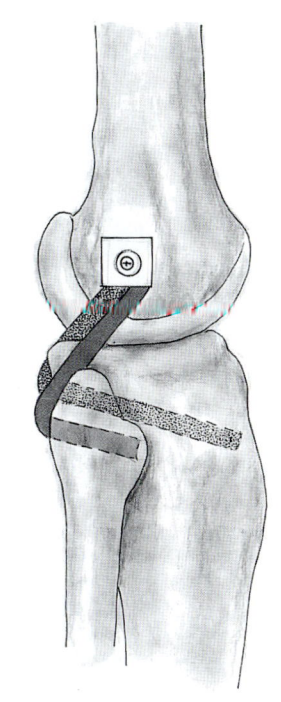

Fig. 16.11. In cases of chronic instability, it is often necessary to perform both a popliteofibular ligament reconstruction and a popliteus bypass procedure. Here a graft with a bone block is secured to the epicondyle and tissue is attached to both the posterior aspect of the fibular head and the posterolateral tibia.

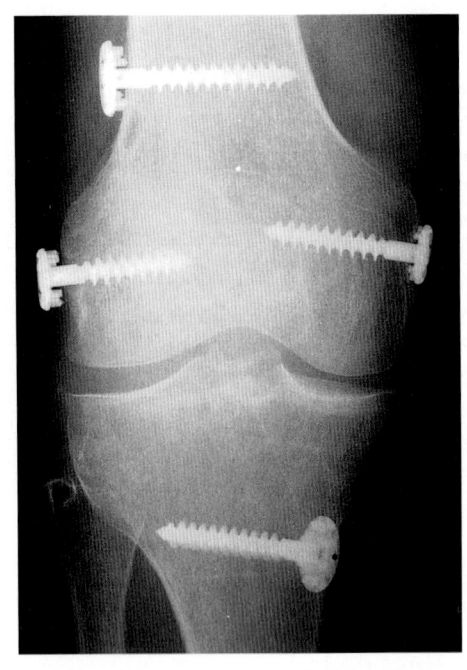

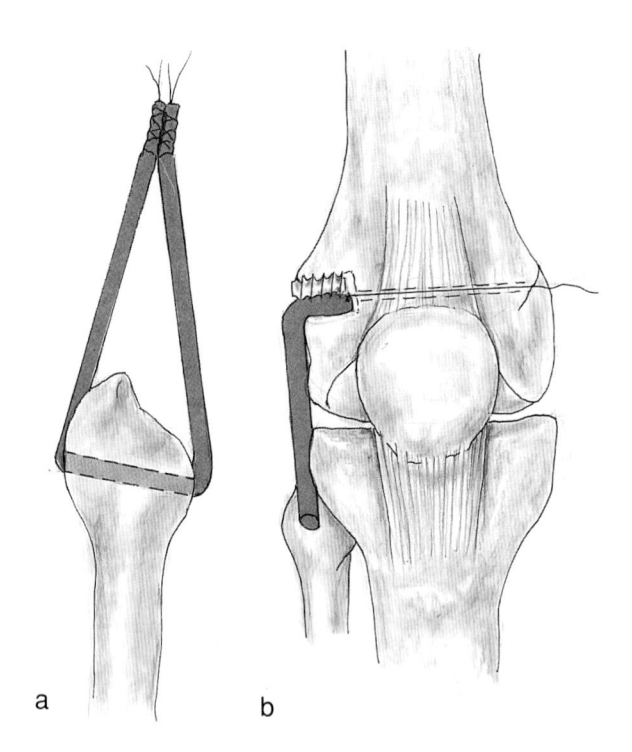

Fig. 16.12. Radiograph demonstrating fixation of a free semitendinosus lateral collateral and popliteofibular ligament reconstruction. The graft is passed from anterior to posterior through the fibular head drill hole, and then fixed to the lateral femoral epicondyle with a screw and washer. This patient also underwent an anterior cruciate ligament (ACL) reconstruction and primary attachment of the medial collateral ligament (MCL) at the medial epicondyle.

Fig. 16.13. Alternate fixation for a semitendinosus free graft lateral collateral ligament (LCL)/popliteal fibular ligament reconstruction is demonstrated above from a lateral (a) and anterior (b) vantage. The looped end of the graft is passed through the fibular head drill hole and the free ends of the grafts are fixed at the epicondyle by pulling them tightly into a socket, and then fixing them with an interference-fit screw. This modification creates less bulk beneath the iliotibial band and makes hardware removal unnecessary.

Alternate Femoral Fixation

To avoid the problem of iliotibial band irritation by screw and washer fixation at the epicondyle, an optional method can be utilized for femoral fixation. After passing the graft through the fibular head, the free ends are brought to the epicondyle as previously described. The grafts are shortened to allow approximately 20 mm of penetration into a socket at the epicondyle. A socket is created at the epicondyle approximately 25 mm in depth and of the appropriate diameter to tightly fit the graft. The ends of the grafts are tied together and a passing suture is attached to the graft. A suture passing guide pin is utilized to pull the passing suture through to the medial side of the knee. The graft ends are then pulled tightly into the femoral socket to the desired tension. The grafts are fixed to the epicondyle with an interference-fit screw (Fig. 16.13). The passing suture is then removed. This optional fixation provides excellent initial strength and avoids the need for subsequent hardware removal.

The described semitendinosus free graft procedure provides a restraining force to varus and posterolateral rotations. It is done with minimal surgical exposure. By attaching to the fibular head as opposed to the posterolateral tibia, better isometry of the graft tissue is obtained. A graft in this position also has a mechanical advantage over grafts attached to the posterolateral tibia. In cases where significant chronic instability exists, it is sometimes advisable to place grafts both to the fibular head and to the posterolateral corner of the tibia. In chronic cases where varus alignment is present, a correction of varus alignment should precede any attempts of soft tissue reconstruction.

Tibial Osteotomy in Treatment of Posterolateral Instability

It is essential when treating chronic varus or posterolateral instability to have a valgus mechanical axis at the knee. If a patient has a varus align-

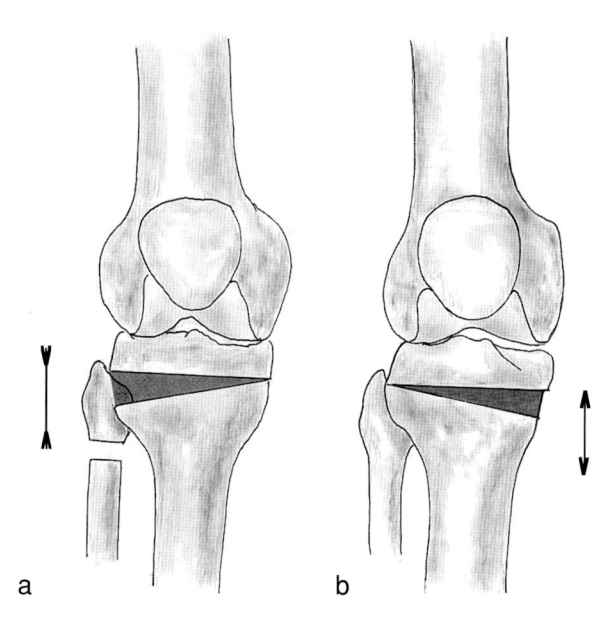

Fig. 16.14. In cases of chronic posterolateral instability with varus alignment, an osteotomy should precede soft tissue reconstruction. The osteotomy can performed by either a closing wedge technique (a) or an opening wedge technique (b).

ment and varus thrust at the knee, soft tissue reconstructions laterally are highly unlikely to be able to correct the problem. It is most appropriate in this situation to stage the treatment by first performing an osteotomy to create a valgus mechanical axis. This usually requires creating an anatomic axis of approximately 7 degrees and not the overcorrection often needed for the treatment of medial compartment osteoarthritis. Valgus producing osteotomy can be accomplished by either a closing or opening wedge (Fig. 16.14). The advantage of performing an opening wedge osteotomy is that it is not necessary to perform a fibular osteotomy, which has the potential to alter ligament relationships on the lateral side of the knee. The importance of staging the reconstruction is that frequently a valgus mechanical axis will solve the problem of a varus thrust and eliminate functional instability. It is not infrequent that a secondary soft tissue reconstruction is unnecessary once the appropriate mechanical axis has been created.

Conclusion

When PCL injuries result in functional instability, the reason is frequently the presence of associated posterolateral instability. Posterolateral instability is caused by an injury to the posterolateral anatomic structures and is not created by an isolated injury to the PCL. Since the posterolateral instability is not caused by the PCL injury, it stands to reason it cannot be corrected by reconstruction by the PCL alone. The problem of posterolateral instability, however, can be corrected by appropriate extraarticular techniques. Since most patients with isolated PCL injuries function quite well with nonoperative management, most patients who require surgical intervention have combined injuries often involving the posterolateral corner. Therefore, most patients undergoing PCL reconstructive surgery require an extraarticular procedure to return knee motion patterns closer to normal. The chances of success of an intraarticular PCL reconstruction can also be enhanced by addressing associated injuries to the posterolateral corner. An appropriate extraarticular procedure will reduce the strain on an intraarticular graft and reduce chances of failure.

References

1. Miyasaka KC, Daniel DM, Stone ML. The incidence of knee ligament injuries in the general population. Am J Knee Surg 1991;4:3–8.
2. Hughston JC, Bowden JA, Andrews JR, Norwood LA. Acute tears of the posterior cruciate ligament. Results of operative treatment. J Bone Joint Surg 1980;62A:438–450.
3. Dandy DJ, Pusey RJ. The long-term results of unrepaired tears of the posterior cruciate ligament. J Bone Joint Surg 1982;64B:92–94.
4. Fowler PJ, Messieh SS. Isolated posterior cruciate ligament injuries in athletes. Am J Sports Med 1987;15:553–557.
5. Parolie JM, Bergfeld JA. Long-term results of nonoperative treatment of isolated posterior cruciate ligament injuries in the athlete. Am J Sports Med 1986;14:35–38.
6. Satku K, Chew CN, Seow H. Posterior cruciate ligament injuries. Acta Orthop Scand 1984;55:26–29.
7. Torg JS, Barton TM, Pavlov H, Stine R. Natural history of the posterior cruciate ligament-deficient knee. Clin Orthop 1989;246:208–216.
8. Baker CL, Norwood LA, Hughston JC. Acute combined posterior cruciate and posterolateral instability of the knee. Am J Sports Med 1984;12:204–208.
9. DeLee JC, Riley MB, Rockwood CA Jr. Acute posterolateral rotatory instability of the knee. Am J Sports Med 1983;11:199–207.
10. Fleming REJ, Blatz DJ, McCarroll JR. Posterior problems in the knee: posterior cruciate insufficiency and posterolateral rotatory insufficiency. Am J Sports Med 1981;9:107–113.
11. Hughston JC, Jacobson KE. Chronic posterolateral rotatory instability of the knee. J Bone Joint Surg 1985;67A:351–359.
12. Sidles JA, Larson RV, Garbini JL, Downey DL, Matsen FAI. Ligament length relationships in the moving knee. J Orthop Res 1988;6:593–610.
13. Ogata K, McCarthy JA. Measurement of length and tension patterns during reconstruction of the posterior cruciate ligament. Am J Sports Med 1992;20:351–355.
14. Galloway MT, Grood ES, Mehalik JN, Levy M, Saddler M, Noyes FR. Posterior cruciate ligament reconstruction: an in vitro study of femoral and tibial graft placement. Am J Sports Med 1996;24:415–426.
15. Cooley VJ, Larson RV, Harrington RM. Effect of lateral ligament reconstruction on intra-articular posterior cruciate ligament graft forces and knee motion. University of Washington Orthopaedic Research Report. 1996:37–41.
16. Grood ES, Hefzy MS, Lindenfeld TN. Factors affecting the region of most isometric femoral attachments. Am J Sports Med 1989;17:197–207.
17. Covey DC, Sapega AA, Sherman GM. Testing for isometry during reconstruction of the posterior cruciate ligament. Anatomic and biomechanical considerations. Am J Sports Med 1996;24:740–746.
18. Harner CD, Xerogeanes JW, Livesay GA, et al. The human posterior cruciate ligament complex: an interdisciplinary study. Am J Sports Med 1995;23:736–745.
19. Race A, Amis AA. PCL reconstruction. In vitro biomechanical comparison of "isometric" versus single and double-bundled "anatomic" grafts. J Bone Joint Surg 1998;80B:173–179.
20. Maynard MJ, Deng X, Wickiewicz TL, Warren RF. The popliteofibular ligament. Rediscovery of a key element in posterolateral stability. Am J Sports Med 1996;24:311–316.
21. Veltri DM, Warren RF. Anatomy, biomechanics, and physical findings in posterolateral knee instability. Clin Sports Med 1994;13:599–614.
22. Veltri DM, Deng XH, Torzilli PA, Maynard MJ, Warren RF. The role of the popliteofibular ligament in stability of the human knee. A biomechanical study. Am J Sports Med 1996;24:19–27.
23. Seebacher JR, Inglis AE, Marshall JL, Warren RF. The structure of the posterolateral aspect of the knee. J Bone Joint Surg 1982;64A:536–541.
24. Wroble RR, Grood ES, Cummings JS, Henderson JM, Noyes FR. The role of the lateral extraarticular restraints in the anterior cruciate ligament–deficient knee. Am J Sports Med 1993;21:257–262.
25. Clancy WGJ. Repair and reconstruction of the posterior cruciate ligament. In:

Chapman MW, ed. Operative Orthopedics, Vol 3. Philadelphia: JP Lippincott, 1988:1651–1655.
26. Mueller W. Die Rotationsinstabilitat am kniegelenk. Hefte Unfallhk 1990;125:51–68.
27. Larson RV, Sidles JA, Beals TC. Isometry of the lateral collateral and popliteofibular ligaments and a technique for reconstruction. University of Washington Orthopaedic Research Report. 1996:42–44.

Chapter Seventeen

Surgical Treatment of Medial Ligament Injuries Associated with Posterior Cruciate Ligament Tears

Fred Flandry and Jack C. Hughston

The structures composing the posteromedial corner of the knee play important roles in medial and lateral stability and protection of the medial meniscus and the anterior cruciate ligament (ACL), and, by their stabilizing functions, protection of the joint against mechanical arthritic degeneration over time.[1-11] Accordingly, we have emphasized and continue to reinforce the importance of repair or reconstruction of severely disrupted and functionally unstable medial ligaments.

This chapter addresses surgery for severe acute and chronic injuries to the ligaments in the posteromedial corner of the knee in the context of a posterior cruciate ligament (PCL) injury. As such, it is primarily a technical monograph on one specific aspect of an overall procedure. Other aspects, and the treatment of the PCL injury itself, are not discussed here.

Operative Versus Conservative Management

We believe that surgery is indicated for severe instability patterns evident on initial examination and further validated by ligament examination under anesthesia. Some confusion surrounds the need for surgical repair or reconstruction of medial ligament injuries. We feel that this dilemma has arisen largely from confusing terminology, in particular the distinction between a grade III ligament tear and a grade III instability. If a grade III tear of the tibial collateral ligament is present, but the instability is grade II or less, we agree with other authors[7,12-15] that nonoperative management is appropriate. When grade III instability is present, we[5,6,16,17] and others[7] have shown, based on long-term follow-up data, that surgery is essential to a satisfactory long-term result.

Clinical Presentation

Most patients with an acute injury have difficulty ambulating. No effusion is present because the capsule is so compromised by the injury that all fluid extravasates from the joint. By 24 hours after injury, this extravasated fluid results in periarticular edema and induration. In a straight medial instability, the abduction stress test is markedly positive (often without a perceptible end point) and relatively painless. Although not as markedly abnormal in extension, the test remains significantly positive. In acute injuries, the posterior drawer at 90 degrees in neutral is negative, despite a torn PCL. This finding has led some to the erroneous conclusion that the PCL is intact,[5] when in fact it is torn, because an intact arcuate ligament prevents posterior translation. In the chronic situation, the arcuate becomes stretched out and eventually allows the posterior drawer to occur as well.

Surgical Indications

We advocate surgery for acute straight medial instability in all cases where the instability is determined to be grade III. Every effort should be made to perform the procedure within the first week (although we will perform the procedure as long as 2 weeks after injury). The only exceptions to operative treatment are those patients in whom surgery is contraindicated for other reasons. In those patients, the limb should be immobilized with the tibiofemoral joint in a reduced position until the surgery can be performed. If not splinted, the joint may scar in a subluxated position, making reconstruction much more difficult. Chronic cases of a symptomatic grade III instability also constitute an indication for surgical reconstruction.

Surgical Procedures Described in the Literature

The positions advocated in the early 1980s against surgery for medial ligament injuries,[13] the studies suggesting that medial ligaments were of trivial importance to knee stability,[18] and the emergence and emphasis on minimally invasive cruciate-only reconstructive procedures[19] have resulted in few contemporary papers being written on this topic. The procedures described in classic studies are rarely performed today and are primarily of historic interest. Further, even fewer studies specifically address injuries to the medial ligaments in the presence of a PCL tear. Most medial ligament surgical technique papers, therefore, describe cases of anteromedial rotatory instability. Arguably, the technical aspects of the procedures for posteromedial repair or reconstruction would be unchanged regardless of the status of either cruciate ligament. Reviewing some of the classic techniques does give insight into the evolution of the authors' current preferred approach to this injury. We assume that there is a core knowledge of the basic, functional, and surgical anatomy of the capsular structures in the posteromedial corner of the knee, as outlined in Chapter 3. The appreciation of those anatomic principles is essential in forming a rational treatment strategy.

Nicholas[20] described a "five-in-one" procedure performed and studied in 52 patients. The procedure consisted of a

- total meniscectomy
- posterior and proximal advancement of the tibial collateral ligament and medial mid-third capsular ligament on the femur
- mobilization and distal and forward advancement of the posterior oblique ligament on the advanced mid-third medial ligaments
- forward and distal advancement of the posterior vastus medialis into the posteromedial capsule as a dynamic arm
- distal advancement of the pes anserinus tendons on their insertion.

In his series, 26 patients returned to sports brace free, 15 returned to sports wearing a derotational brace, and nine procedures failed.

O'Donoghue[21] reported on 60 patients in whom he advanced the tibial collateral ligament and capsular tissues as a single cuff distally and anteriorly on the tibia. Significant improvement was documented in 45 patients. There were only two failures (disability unchanged from preoperative status) in 60 cases.

Hughston and Barrett[4] reported long-term follow-up on 132 knees of average 7.85 years, treated by a methodology that remains the basis of our current surgical techniques almost two decades later. Emphasis was placed on restoring the anatomic and functional integrity of the meniscus-posterior oblique-semimembranosus unit. In that series, 73% had a satisfactory objective outcome, yet 89% were satisfied both subjectively and functionally. Of the injured athletes, 94% were able to return to their preinjury level of participation.

Authors' Preferred Technique

We divide capsular sprains into acute and chronic injuries for the purposes of the surgical technique employed. This division is based on the ability to effectively retension torn ligaments by various suture techniques. That ability is influenced by the temporal phase of healing. Our concept of the basic science of ligament healing coupled with direct surgical observation and clinical outcomes is as follows.

As healing occurs, inflammation-mediated resorption, fibroblast-mediated repair, and remodeling of scar tissue formed according to stresses experienced by the healing tissues take place simultaneously. The relative balance of these activities, however, differs with time.

In the first to second week, infiltration by inflammatory cells that would result in sufficient resorption to affect the mechanical properties of the injured ligamentous tissue has not occurred. Slightly beyond this period, resorptive activity begins to peak as reparative activity increases. At this time, the injured ligaments are swollen, indurated, and friable. Sutures placed in ligaments to repair, reattach, or retension will either pull out as they are tensioned or tied or will pull through in the early postoperative period.

In the period from 4 to 6 weeks, reparative activity has adequately restored the mechanical properties such that ligaments will again hold sutures and can be retensioned. As resorption and repair processes begin to wane, remodeling activity peaks. Necessary to this remodeling phase is adequate and appropriate stress to the tissues.

As important as, if not more so than, the stresses of motion and routine activities, is the reinstitution of normal dynamic stresses. Ligaments cannot experience normal stresses if they are not attached. Ligaments cannot experience normal stresses if they are not in continuity and do not have normal tension. Ligaments cannot experience normal stresses if they are not receiving normal dynamic input from the semimembranosus. Should these last three conditions occur, normal mechanoreceptor feedback from these tissues will not be provided to the central nervous system.

If surgery can be undertaken within 2 weeks of the injury, an acute repair is performed. If surgery is delayed much beyond this time, we will probably wait until 6 weeks from the injury, and then, if indicated by ongoing functional disability, perform a chronic reconstruction. Regardless of whether an acute repair or chronic reconstruction is performed, the principles of the initial approach are similar.

Surgical Approach

The patient is positioned supine with the knee flexed and the hip abducted and externally rotated (Fig. 17.1a). Initial exposure of injury pathology is performed under tourniquet control. The tourniquet is then released before

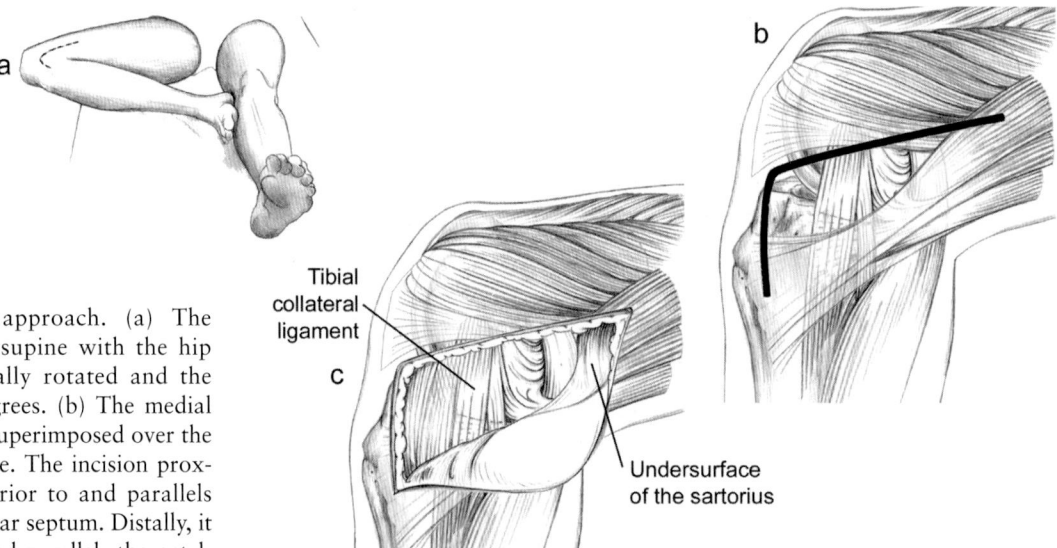

Fig. 17.1. Surgical approach. (a) The patient is positioned supine with the hip abducted and externally rotated and the knee flexed to 90 degrees. (b) The medial hockey-stick incision superimposed over the underlying musculature. The incision proximally is slightly anterior to and parallels the medial intermuscular septum. Distally, it is slightly medial to and parallels the patellar tendon extending onto the medial face of the proximal tibia. (c) Reflection of the sartorius. If the sartorius is not traumatically disrupted, though it frequently is, the sartorial fascia distally may be incised in line with the skin incision and reflected (with the sartorius muscle) along with the posterior flap. Regardless, an extensile exposure of the medial capsular ligaments, posteromedial corner, and posterior capsule (as far as the PCL and fovea centralis) is done. (Copyright © 2000 by the Hughston Sports Medicine Foundation, Inc.)

the repair or reconstruction is completed, minimizing tourniquet time, which in our hands is commonly between 10 and 20 minutes.

Two incisions are used: a medial hockey-stick incision (Fig. 17.1a,b,c) and a posteromedial arthrotomy incision. The posteromedial incision exactly parallels the posteromedial arthrotomy incision (Fig. 17.2e). It is smaller and, as such, is more appropriate for surgeons who are more experienced in exploring capsular injuries in this region or for cases in which a minimum extent of pathology is anticipated (unlikely with PCL injuries).

The medial hockey-stick incision begins on the mid-medial thigh at a point proximal to the plane of the posterior border of the leg (knee flexed to 90 degrees) and anterior to the medial epicondyle/medial intermuscular septum, all of which are visible or palpable landmarks (Fig. 17.1b). The incision extends distally, gently curving to turn distal just proximal to the medial border of the patellar tendon. It then continues distally over the medial face of the proximal tibia to a point slightly distal to the plane of the posterior border of the thigh. The incision is taken down to, but not through, the sartorial fascia, and a flap is created through this areolar and relatively avascular plane. Care should be exercised to identify and protect the infrapatellar branch of the saphenous nerve. If this nerve branch is not protected, painful neuromas can result. In cases where the sartorius is undamaged, the sartorial fascia and distal sartorius can be reflected with the posterior flap (Fig. 17.1c).

The tibia must be slightly internally rotated before the medial ligaments are repaired. This position can easily be maintained by resting the lateral forefoot on a sandbag (Fig. 17.2a) and will reduce anteromedial rotation of the medial tibial plateau by as much as 1 cm. Failure to achieve this 1-cm correction of rotational subluxation will lead to insufficient tensioning of the medial ligaments and a lax and inadequate repair. When the knee is flexed to 90 degrees, a narrow areolar interval extends from the "saddle" (medial epicondyle/adductor tubercle), which is easily palpable and indicates the location of the posteromedial arthrotomy incision (Fig. 17.2c). The posteromedial arthrotomy is a window to the mid-third capsular ligament, posterior oblique ligament (POL), and medial meniscus. More importantly, it becomes the interval in which the mid-third capsular ligament and the POL (and indirectly the more posterior structures) are retensioned

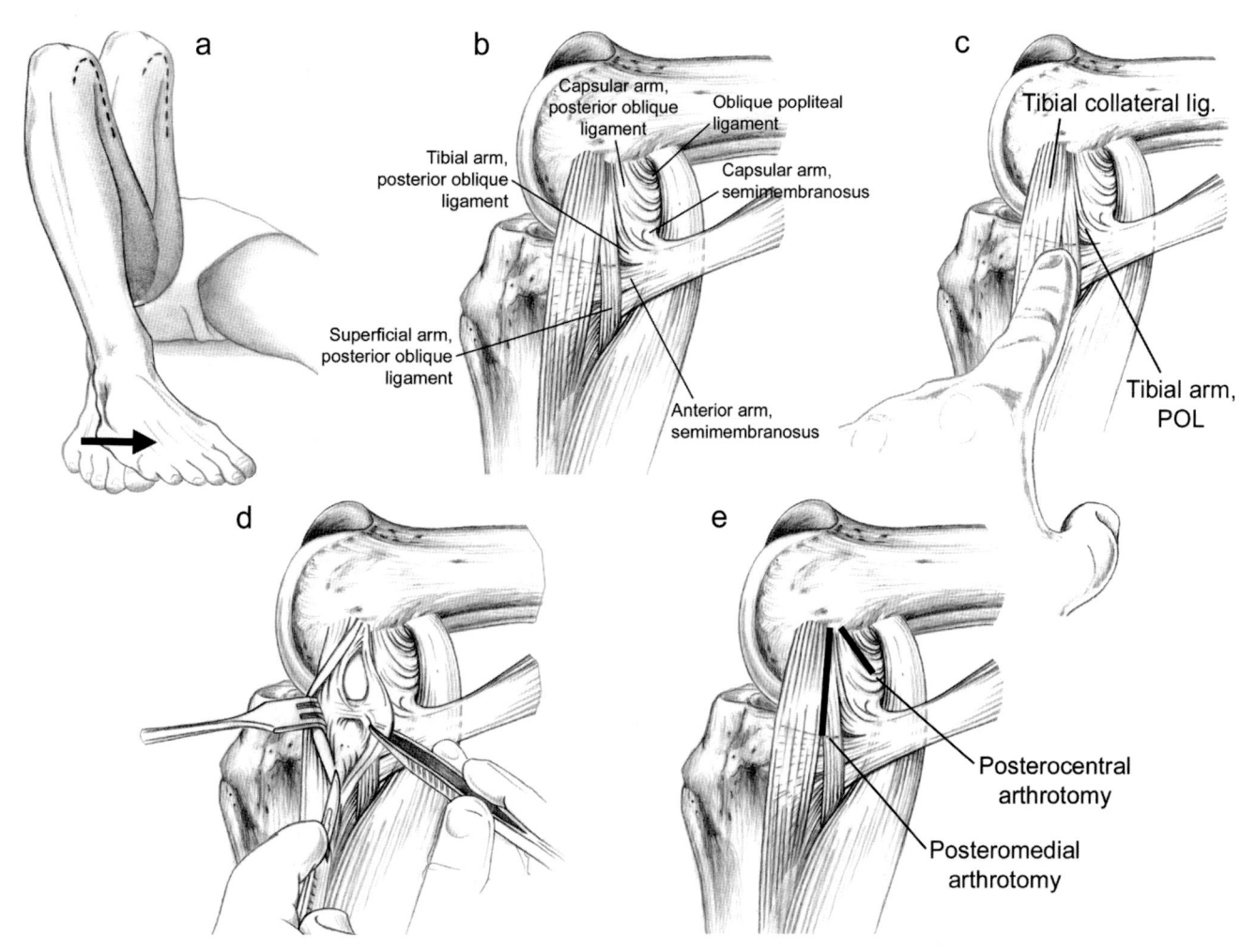

17. Surgical Treatment of Medial Ligament Injuries Associated with PCL Tears

Fig. 17.2. Deeper dissection and arthrotomy incisions. (a) The lateral forefoot is positioned on a sandbag placed under the sterile drapes to internally rotate the tibia and correct anteromedial subluxation. This positions the tibia and femur correctly for retensioning torn or stretched ligaments. (b) Ligamentous structures in the posteromedial corner. (c) A "soft spot" representing a narrow areolar interval between the mid-third capsular ligament and the posterior oblique ligament (POL) can be palpated with a fingertip. It extends distally from the medial epicondyle/adductor tubercle complex or "saddle" when the knee is flexed 90 degrees and serves as the interval for a posteromedial arthrotomy incision. (d) The posteromedial arthrotomy exposes the mid-third capsular ligament deep to the tibial collateral ligament (TCL), the POL, and the posterior and medial attachments of the medial meniscus. It is the interval through which these ligaments are retensioned when injured. (e) Arthrotomy incisions in the posteromedial corner. The posterocentral arthrotomy is used when additional access is needed to the posterocentral attachment of the medial meniscus or to the PCL (for PCL repair in conjunction with capsular repair). Copyright © 2000 by the Hughston Sports Medicine Foundation, Inc.

(Fig. 17.2d). A second utility posterocentral arthrotomy (Fig. 17.2e) can be made for access to the PCL and the posterocentral attachment of the medial meniscus if needed. A medial arthrotomy incision—a utility incision along the anterior border of the tibial collateral ligament (TCL), to assist in repair of the TCL and mid-third capsular ligament—and a medial parapatellar arthrotomy (for access to the anterior joint and intercondylar area) can also be used.

Exploration of Injury Pathology

We advocate a meticulous inventory of all posteromedial structures by a process of inspection, probing, and testing for laxity. Our practice is to

Fig. 17.3. Inspection of the TCL. (a) Interstitial tears are indicated by longitudinal fiber separations or laxity of the ligament as a whole. An interstitially torn ligament takes on a somewhat hemorrhagic and swollen appearance despite remaining in continuity. (b) Exploring the origin from the deep surface. (c) The insertion is obscured by the pes tendons (not shown), so intactness is evaluated by tugging on this attachment with a forceps or meniscal hook retractor. If torn, the mop end will often flop out from under the pes tendons. Copyright © 2000 by the Hughston Sports Medicine Foundation, Inc.

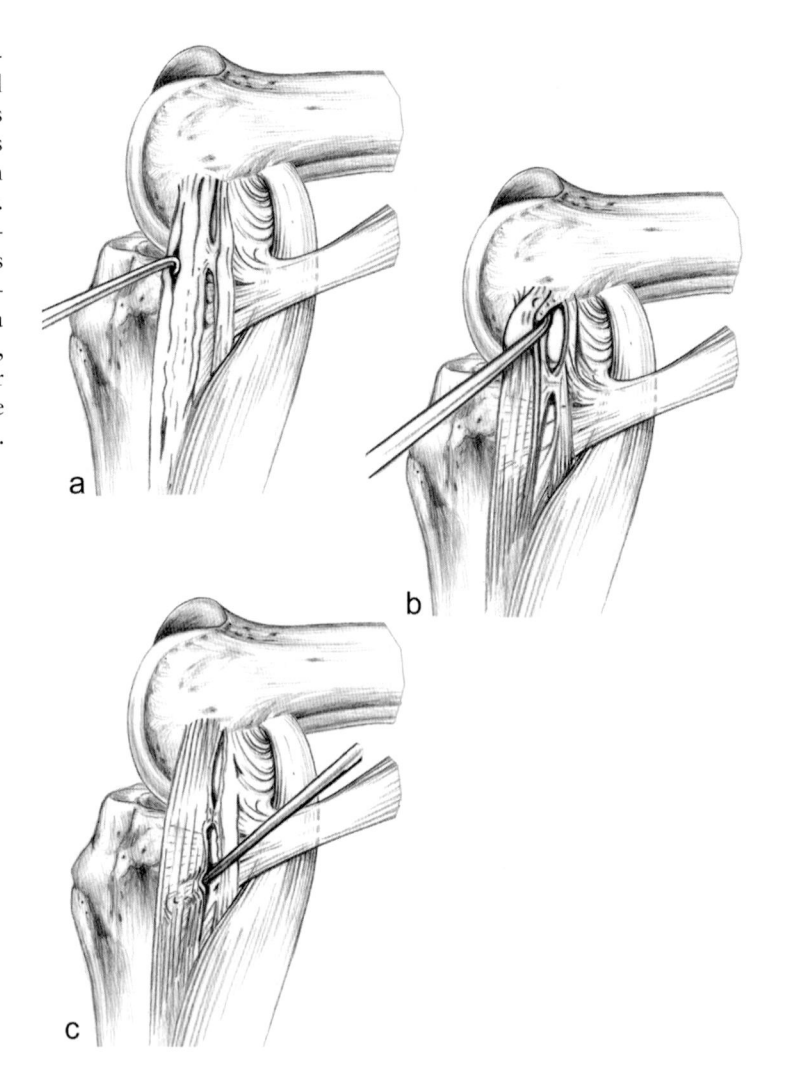

record findings in dictated operative notes, graphically (on diagrams and recorded images), and in databases to support ongoing research. Since our overall operative philosophy is the restoration of anatomy, the operative findings become the blueprint for our surgical repair or reconstruction.

The TCL is evaluated first, since it is the most superficial and thus the first presenting structure. Midsubstance transverse or mop-end tears are obvious, but other subtle patterns will be missed if they are not specifically sought. Probing with a meniscal hook retractor may reveal longitudinal separations of fibers with laxity indicative of interstitial tearing and stretch of the ligament (Fig. 17.3a). A proximal avulsion from the medial epicondyle may be masked by overlying areolar tissue. Exploration from the deep surface of this origin may reveal an unsuspected tear (Fig. 17.3b). Because the distal attachment lies deep to the pes anserinus, it may appear intact, yet an entire mop-end tear may flop out from under the pes tendons if the tendons are tugged on by a forceps or meniscal hook retractor (Fig. 17.3c).

The mid-third capsular ligament lies deep to the TCL; therefore, it is necessary to reflect the TCL away from the arthrotomy incision to inspect this ligament. Techniques of discovery are similar to those outlined above for the TCL (Fig. 17.4a). One of the more commonly missed injuries is an avulsion of the meniscofemoral segment of the mid-third capsular ligament

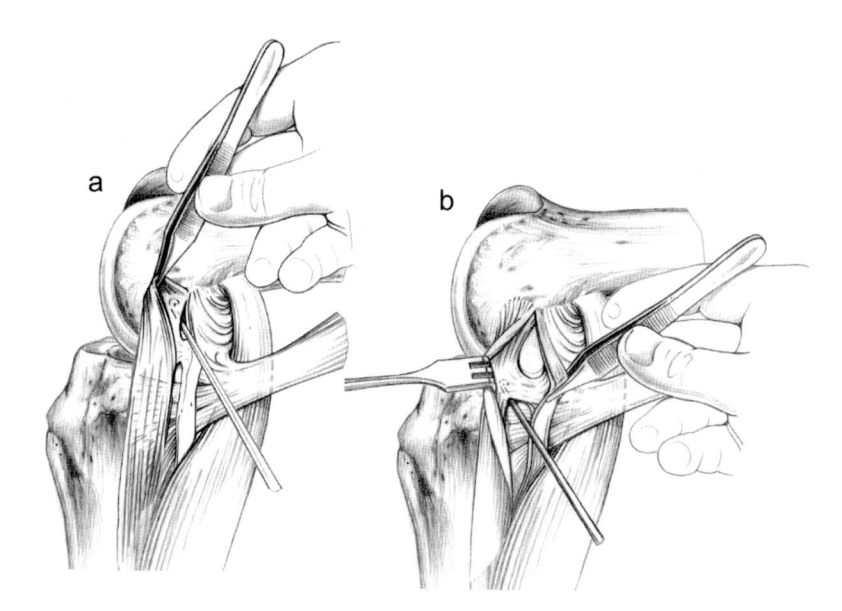

Fig. 17.4. Inspection of the mid-third capsular ligament. (a) The meniscofemoral segment is exposed by retracting the TCL away from the line of the arthrotomy incision. Care should be taken to ensure that there is not a tear of the origin of this segment concealed by the overlying TCL. (b) The meniscotibial segment is evaluated by placing a meniscal hook retractor or tip of a hemostat deep to this ligament and palpating with the instrument from the meniscus to insertion of the ligament on the tibia. The "space" deep to this segment should be just large enough to accept the tip of a small hemostat. If more capacious, it indicates laxity of this segment. Copyright © 2000 by the Hughston Sports Medicine Foundation, Inc.

from the medial epicondyle. This injury is well hidden by the overlying TCL, yet will result, as we have documented,[4–6] in significant instability and disability if not repaired. Retracting the TCL near its origin and probing and tugging on the deeper capsule will reveal this injury, if present (Fig. 17.4b).

The POL is evaluated for interstitial, origin, and insertion tears, as well as for attachment to the periphery of the medial meniscus (Fig. 17.5). A narrow Langenbeck retractor placed in the posteromedial arthrotomy will reveal origin tears from the adductor tubercle. Interstitial tears are indicated by abnormal laxity. This arthrotomy will expose the peripheral attachment of the posterior horn of the meniscus (Fig. 17.5a). Sometimes it is necessary to brush away a thin synovial covering to expose a large and functionally significant subsynovial peripheral detachment (Fig. 17.5b). The tibial arm may avulse distally deep to the tibial arm of the semimembranosus (Fig. 17.5c), which has particular significance for arthroscopic repairs of meniscal tears. Should a peripheral meniscus tear be repaired arthroscopically without confirming the absence of instability, and thus, the absence of a POL tear, the remaining looseness will prevent the semimembranosus from properly retracting the posterior horn of the medial meniscus. Meniscal retear may result.

A tear of the oblique popliteal ligament (OPL) may be detected through the posteromedial arthrotomy incision by palpating the peripheral meniscal posterior horn attachment from its tibial surface (Fig. 17.6a). With the knee flexed to 90 degrees, retraction of the medial head of the gastrocnemius opens the gastrocnemius bursa and reveals the origin of the OPL/capsular arm of the POL (Fig. 17.6b). Excess laxity of this ligament may account for an abnormal anteromedial drawer test.

Continuity of the semimembranosus is determined by either grasping the capsular arm of the tendon immediately posterior to the tibial arm of the POL or grasping the tendon immediately proximal to its capsular arm and retracting it posteriorly (Fig. 17.6c). The posterior horn of the medial meniscus should retract posteriorly with the tendon. If it does not, a disruption of the semimembranosus-POL-medial meniscus complex exists.

Although not directly involved in tibiofemoral stability, the association of tears of the vastus medialis obliquus (VMO) with injuries of the medial

Fig. 17.5. Evaluation of the POL and peripheral meniscal attachment. (a) The posteromedial arthrotomy gives excellent exposure of the POL origin, meniscofemoral segment, and medial meniscus. (b) The peripheral attachment of the medial meniscus is evaluated. An intact overlying synovium may hide a subsynovial detachment that will be more readily seen through the arthrotomy than arthroscopically. (c) Three areas of POL tears commonly encountered are illustrated. The POL tibial arm tear may be hidden by the overlying tibial arm of the semimembranosus. Copyright © 2000 by the Hughston Sports Medicine Foundation, Inc.

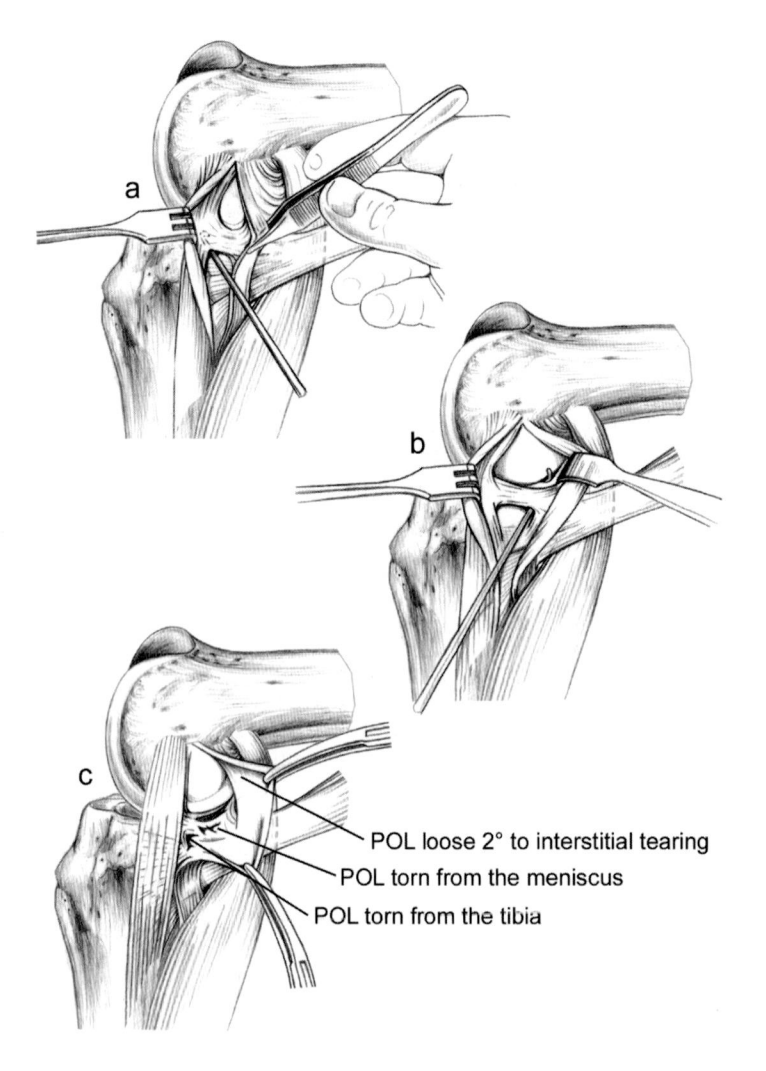

- POL loose 2° to interstitial tearing
- POL torn from the meniscus
- POL torn from the tibia

capsular ligaments has been reported.[22] The most typical pattern is a gross rupture from the adductor tubercle (Fig. 17.7a), but a musculotendinous rupture from the medial retinaculum (Fig. 17.7b) or an insertional avulsion from the patella (Fig. 17.7c) is also possible. Excessive and asymmetric passive lateral hypermobility of the patella in the presence of medial capsular ligament injury should prompt exploration for this lesion.

In chronic instabilities, the findings of hemorrhage, swelling, and obviously disrupted tissue are replaced by tissue in continuity, but it is lax. Tears will "repair" by filling in with areolar tissue. The functional tension of the ligament is not restored; however, in this healing process, the instability of the segment, and the abnormal motion it should prevent, are present. Only avulsions of attachments such as the mid-third capsule from its origin may persist as a discontinuity. While it is important to explore each segment of each ligament, the surgeon should focus on detecting abnormal laxity in a given segment of tissue and possible ways of retensioning that segment. This skill is based on the judgment developed after evaluating many patients with capsular injury.

Acute Injury Repair Strategies

In an acute repair, the operative technique goal, put simply, is to restore normal anatomy, relationships, and dynamic functions. Torn ends of liga-

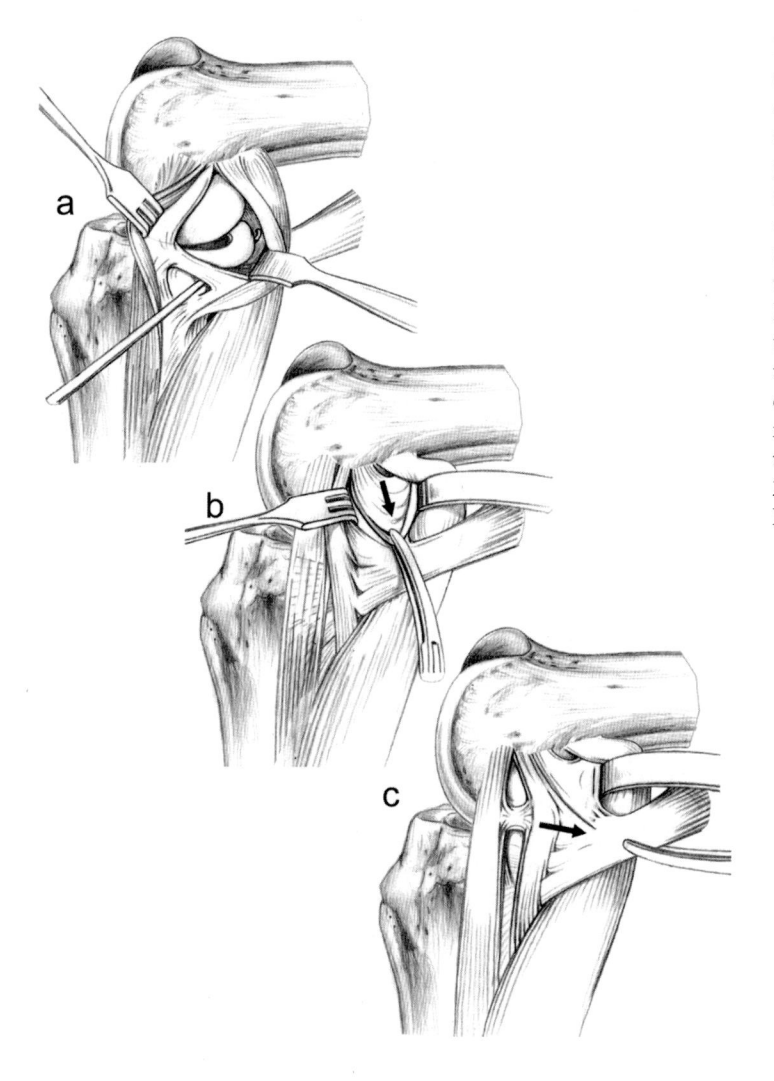

17. Surgical Treatment of Medial Ligament Injuries Associated with PCL Tears

Fig. 17.6. The oblique popliteal ligament (OPL) and the semimembranosus. (a) A tear of the meniscotibial portion of the OPL is revealed when a meniscal hook retractor probing the tibial periphery either emerges (as illustrated) or tents synovium posteriorly at or near the meniscofemoral junction. (b) Reflecting the medial head of the gastrocnemius reveals the origin of the OPL. Laxity of the superior bursal rim may contribute to an abnormal anteromedial drawer test. (c) Testing the semimembranosus tendon (see text). The posterior horn of the medial meniscus should follow when the tendon is retracted posteriorly. Copyright © 2000 by the Hughston Sports Medicine Foundation, Inc.

ments are resewn. In the case of interstitial tears, the ligaments are sutured with one or many horizontal mattress sutures to "reef" redundant tissue upon itself and thereby re-create normal tension in the ligament. Where avulsions have occurred, ligaments or tendons are reattached. If the semi-

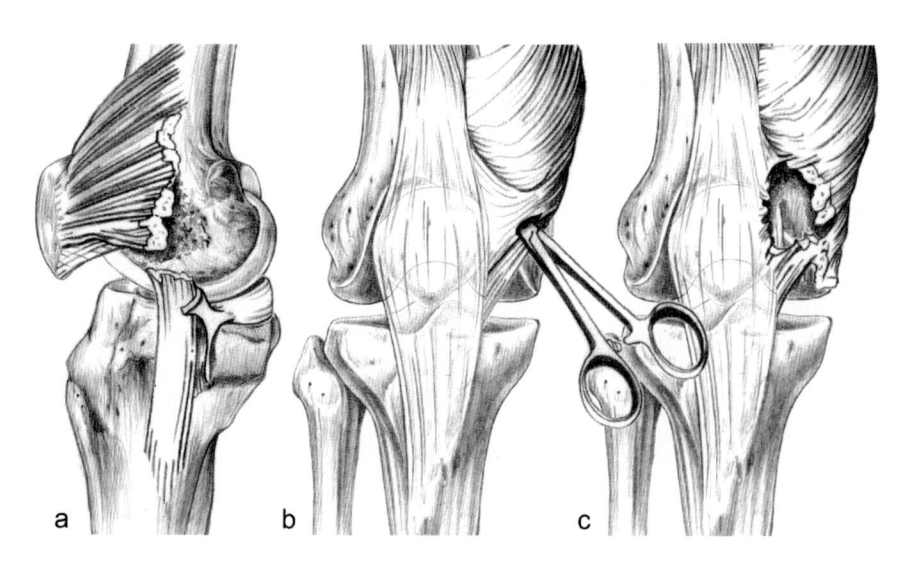

Fig. 17.7. Tears of the vastus medialis obliquus in association with medial capsular injuries. (a) The most common injury is that of a tear at the adductor tubercle. (b) Less commonly encountered is a musculotendinous junction tear. The hemostat inserted deep to the muscle gives complete exposure of the medial metaphysis of the femur. (c) The least frequent injury is an avulsion of the insertion on the patella. Excessive lateral hypermobility of the patella in the face of a medial capsular ligament injury should prompt a search for one of these lesions. Copyright © 2000 by the Hughston Sports Medicine Foundation, Inc.

membranosus-POL-meniscus complex has come uncoupled, dynamic integrity of that complex is restored.

An appropriate order of repair is inherent and necessary in this process, and some of the following steps have already been alluded to:

- Acute repairs are not attempted if the time since injury much exceeds 2 weeks.
- Patient position, as described earlier, provides both ease of access to this area and spatial positioning of the ligamentous attachment points on the femur and tibia to allow optimal retensioning of these ligaments.
- Exploration in a bloodless field and before irrigating fluids obscure pathology by inducing swelling of damaged tissue allows the surgeons to map out segments requiring repair and formulate the final operative tactic.
- Sutures to repair the PCL, if planned, are placed first, but tied last.
- Meniscal body pathology is addressed as appropriate. Peripheral detachments from the capsular ligaments are repaired. Repair of the OPL meniscotibial segment is performed if indicated.
- The mid-third capsular ligament is repaired and retensioned.
- The TCL is repaired and retensioned.
- The POL and OPL origins onto the adductor tubercle of the femur are repaired and advanced.
- The POL is advanced onto the repaired TCL.
- Dynamic input of the capsular arm of the semimembranosus into the POL and OPL is restored.
- Any VMO injury is repaired.
- Dynamic input of the sartorius into the medial retinaculum is restored.
- Arthroscopic procedures follow.

In cases where the PCL is repairable, sutures for this repair must be placed before the capsular ligaments are repaired and retensioned. A posterocentral arthrotomy provides the needed exposure to the PCL insertion in the fovea centralis. The femoral origin is visualized either arthroscopically or through a medial parapatellar arthrotomy incision. Avulsions with or without bone are repaired back to their origin or insertion by a through-bone suture technique, tying the sutures placed over an external (extraarticular/extramedullary) bony cortical bridge.

We have demonstrated that the midsubstance tear of the PCL (unlike the ACL) is repairable because of a different and less vulnerable vascularization of the ligament and the ligament's association with the posterior capsule.[16] The repair technique involves placing reverse-oriented horizontal mattress sutures through the stumps of the ligament, but incorporating posterior capsular tissue with the tibial PCL stump (Fig. 17.8).

Although these sutures are placed initially, they are the last sutures tied (usually through a minimal window left in the wound closure). After these sutures are tied, an anterior drawer stress is maintained until a postoperative brace is applied. The brace maintains a posture that does not stress this repair.

In cases where an arthroscopically assisted intraarticular cruciate ligament reconstruction is planned for an acute injury, capsular surgery should precede the arthroscopic portion of the procedure. If this sequence is not followed, extravasated fluid will cause the torn ligaments to swell, making their identification, repair, and retensioning difficult.

Like the PCL, the medial meniscus, being an intraarticular structure, must be addressed before the arthrotomy windows are closed as the liga-

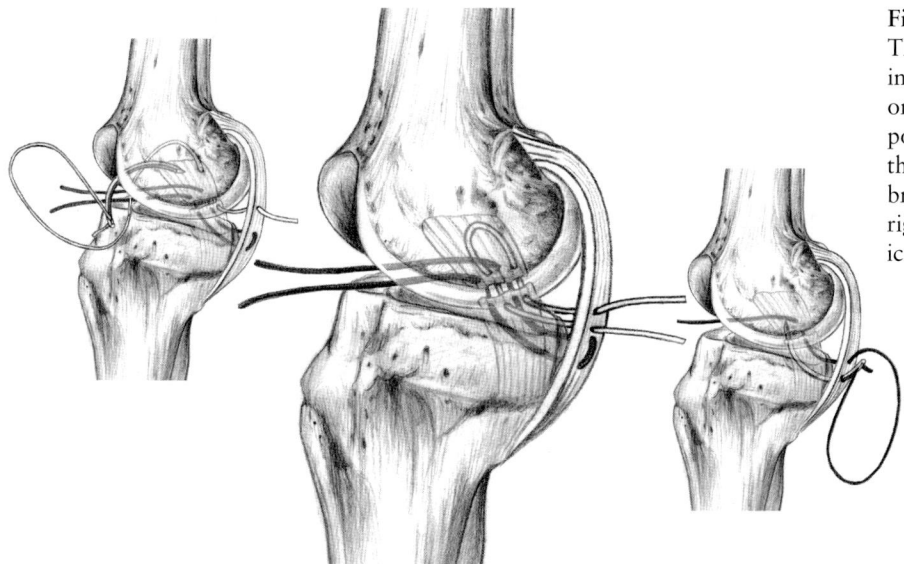

Fig. 17.8. PCL midsubstance acute repair. The two stumps of the PCL are sutured using alternately directed horizontal mattress or modified Kessler suture techniques. The posterior capsule (OPL) is incorporated with the tibial stump. This both augments and brings added vascularity to the repair. Copyright © 2000 by the Hughston Sports Medicine Foundation, Inc.

ments are repaired. Body tears may be trimmed or sutured as needed. (Treatment of a minor inner body flap tear may be deferred if an arthroscopic procedure is planned.) If a tear of the meniscotibial segment of the OPL is detected on examination (see Fig. 17.6a), sutures are placed and tied to repair this ligament insertion (necessary before it is later retensioned on the femur).

If the peripheral tear as illustrated (Fig. 17.9) is unrepaired, the posterior horn is uncoupled from the dynamic function of the semimembranosus (Fig. 17.9b). Repair is accomplished by both exposing the detachment intraarticularly and developing the interval posterior or superficial to the POL and OPL. Using a small stout needle (Davis tonsil), horizontal mattress sutures are placed first through the meniscus into the tear and then through the POL and OPL. As each suture is placed, the POL is grasped and pulled medial and anterior so that when tied, the POL and OPL will be repaired back to the periphery of the meniscus in an advanced position (Fig. 17.9c). These sutures are tagged, but not tied until all ligament repairs are completed, so that these repairs do not affect tensioning of the individual ligaments.

If anterior utility arthrotomy incisions were made, they should be closed before ligament repairs are performed. This is necessary because all subsequent tensioning is based on tensioning of the mid-third capsule and TCL. If these incisions are not closed before ligament repair, tensioning of these two ligaments will not be accurate and more anterior arthrotomies may be more difficult to close.

The mid-third capsular ligament is the first to be repaired. Attachments to the femur, tibia, and meniscus are secured (Fig. 17.10). If any redundancy remains, horizontal mattress sutures are placed in the substance to imbricate the ligament upon itself.

The TCL is repaired next. Again, the principle of suturing gross transverse tears, repairing meniscal detachment, and advancing the origin onto the medial epicondyle and the insertion onto the tibia are followed. If laxity persists, the ligament is further imbricated. Stump avulsions from the origin are repaired using a modified Bunnell or Kessler suture or multiple horizontal mattress sutures. These sutures are anchored in periosteum,

Fig. 17.9. Repair of a peripheral tear of the medial meniscus. (a) Normal coupling of the semimembranosus: the posterior horn retracts posteriorly with contraction of the semimembranosus. (b) Unrepaired, the meniscus does not retract and, as the knee flexes, may become impinged between the femoral and tibial condyles and further damaged. (c) Suture of the meniscus to the POL, OPL, and semimembranosus restores this dynamic function. Copyright © 2000 by the Hughston Sports Medicine Foundation, Inc.

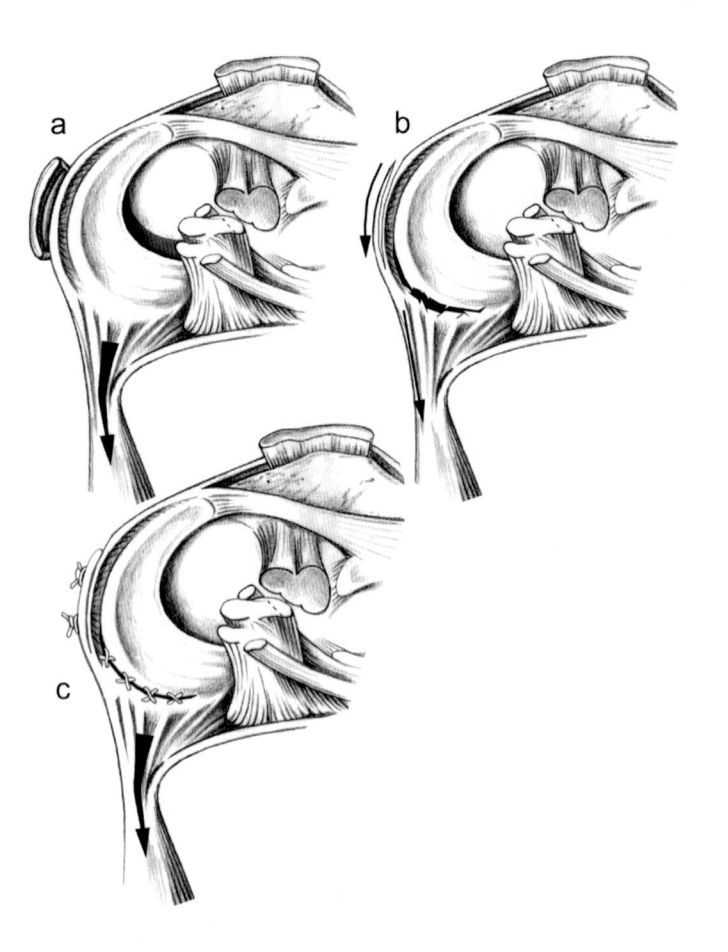

through bone, or with suture anchors (Fig. 17.11a). If both the TCL and mid-third capsular ligament tear from the meniscus, they can be sutured as a single entity. If a reabsorbable suture material is used here, the deep bursa will eventually reform, and dynamic excursion of the TCL in many cases will be reestablished (Fig. 17.11b). If there is an avulsion of the insertion

Fig. 17.10. Repair strategies for the mid-third capsular ligament. (a) Avulsion of the meniscofemoral segment from the medial epicondyle is hidden by the overlying TCL. Repair is accomplished by placing horizontal mattress sutures first in the stump of the ligament and then either through the periosteum or through an intact origin of the overlying TCL. (b) Repair of a meniscofemoral stump to the meniscus. (c) Repair of the meniscotibial segment to the meniscus. (d) Repair of the meniscotibial stump to its insertion. Tactics used in this setting may include suture through periosteum, suture through bone, or suture anchors to accomplish this. Copyright © 2000 by the Hughston Sports Medicine Foundation, Inc.

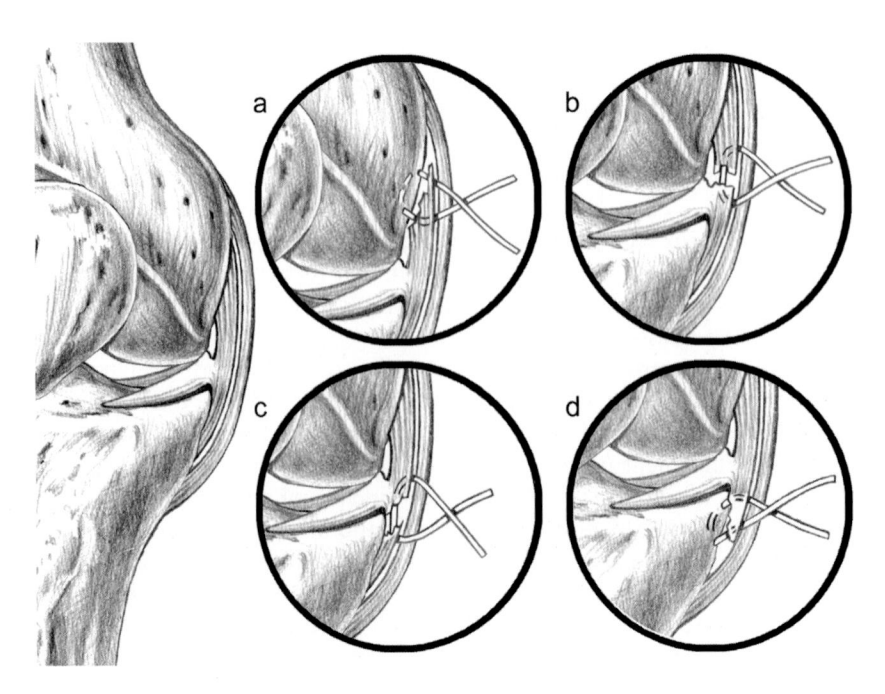

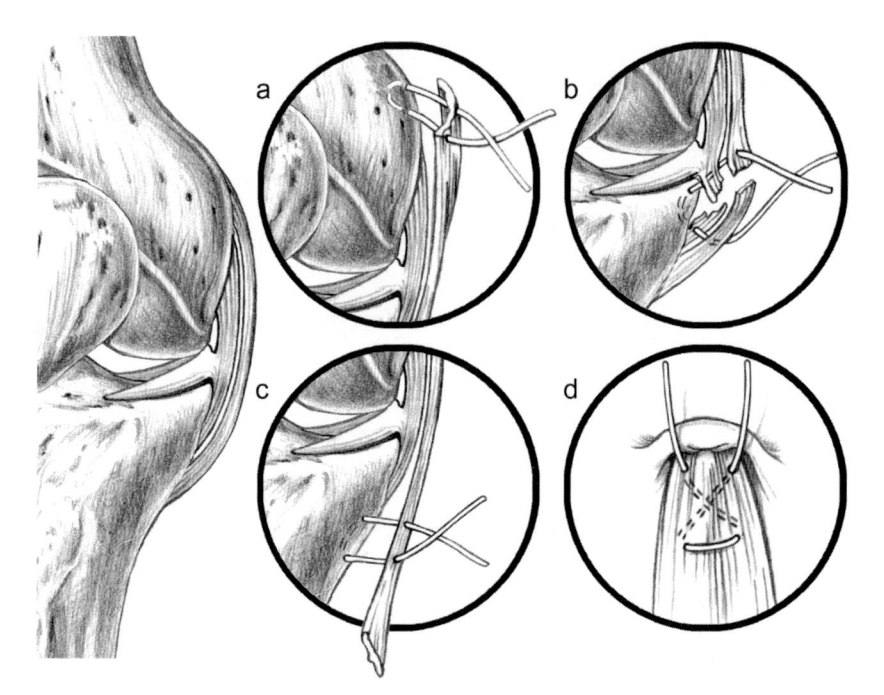

Fig. 17.11. Repair of the TCL. (a) Avulsion of the TCL from the medial epicondyle is repaired using a modified Bunnell suture. (b) Concurrent transverse tears of the TCL and mid-third capsular ligament at the level of the meniscus. The two ligaments may be sutured as one if a reabsorbable material is used. (c) Repair of an insertional avulsion. The ligament is sutured to the tibia proximal to the pes anserinus tendons rather than dissecting through the pes anserinus to find the original insertion point. (d) Placement of horizontal mattress sutures to imbricate or advance the ligament. Sutures are oriented obliquely through substance of the ligament rather than in line with ligament fibers to minimize the potential for pullout. Copyright © 2000 by the Hughston Sports Medicine Foundation, Inc.

or interstitial tearing of the meniscotibial segment, the ligament is anchored back to the tibia at a point just proximal to the pes anserinus tendons rather than to the original insertion deep to the pes anserinus tendons. Our results have shown no adverse outcome from the former, but significant morbidity from the scar resulting from dissecting through the tendons and the deep bursa (Fig. 17.11c).[5]

Several references have been made to imbricating or advancing sutures placed with a horizontal mattress technique, and it is appropriate to digress for a moment to discuss some technical aspects of this method of suture. The anchor, or initial loop, of the suture should be placed in a relatively immovable point such as a point of reattachment to bone. The "legs" are then passed through the tissue to be advanced or imbricated, usually entering from the deep surface so that the knot will be tied on the superficial surface. The legs should enter the ligament at a point that will, when advanced to the anchor, eradicate laxity in the tissue. This point can be determined by grasping the tissue with a forceps and advancing the ligament until the ideal location is determined. The suture should be pulled, held under tension, and tied in the direction the ligament is being advanced. To minimize suture pullout from tissue, the suture legs should pass obliquely across, rather than in line with, the ligament fibers as they traverse the substance of the ligament (Fig. 17.11d). Adherence to these details will allow more effective retensioning of redundant tissue.

With a restored mid-medial complex, attention can be focused on the POL and OPL. Frequently interstitial stretching of the POL is present. In addition, there may be gross transverse tears. After repairing obvious rents, the origin of the POL is advanced back to the "saddle" with a nonabsorbable suture (Fig. 17.12). The anchor of the suture is at the saddle. The legs of the suture enter the POL on its deep (articular) surface 1 to 2 cm distal in a posteromedial direction from the saddle. This places the suture in the capsular arm and effectively retensions the origins of both the POL and the OPL. If the suture is properly placed, the anteromedial drawer should be significantly diminished by this single suture. If not, the suture should be removed and replaced until the anteromedial drawer is dimin-

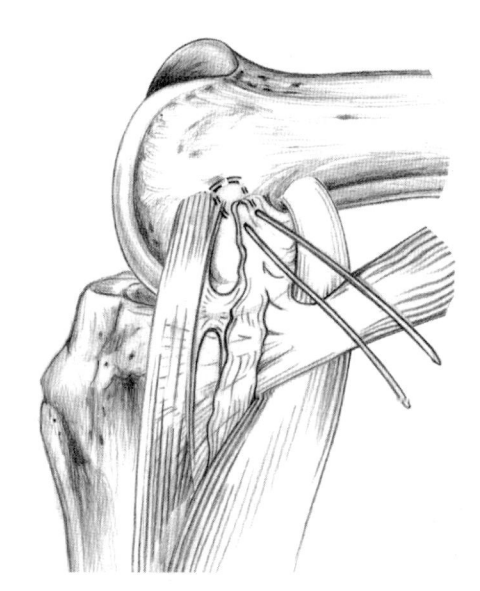

Fig. 17.12. Repair of the POL/OPL. The capsular arm of the POL is advanced back to the saddle (see text). This single suture should significantly diminish the anteromedial drawer. Copyright © 2000 by the Hughston Sports Medicine Foundation, Inc.

ished. The knee should then be briefly extended to ensure that the ligament has not been overtensioned.

The POL margin of the arthrotomy is then advanced over the TCL again with horizontal mattress sutures, thus closing the arthrotomy and further retensioning the posterior capsule and posteromedial corner. The semimembranosus capsular arm is tested and retensioned if needed. Finally, injuries to the VMO and sartorius are addressed as appropriate.

Chronic Injury Reconstruction Strategies

In chronic injuries, transverse and interstitial tears have been replaced by bridging scar. Injured ligaments may be lax and redundant, allowing abnormal subluxations to occur. What may persist as discontinuities are the avulsions and detachment. Ligaments avulsed from origins or insertions may have not healed back to bone. If so, the ligament will be untensioned and will allow gross motion if properly tested. Detachments such as a peripheral medial meniscus tear uncouple the semimembranosus function on the meniscus, leaving it vulnerable to retear. Chronic reconstructions, thus, are designed to reconnect all segments of the overall complex and then restore normal tension to all segments through selective reefing of loose and redundant tissue.

Many of the details of position, exposure, and exploration are as previously described. Care should be taken to explore for an occult avulsion of the origin of the mid-third capsular ligament and peripheral detachment of the meniscus. These tears may have filled in with synovium and areolar tissue. A clue to this is the absence of posterior retraction of the posterior horn of the meniscus when posterior traction is applied to the capsular arm of the semimembranosus.

The basis of the reconstruction then is to advance the POL on a retensioned TCL/mid-third capsular ligament complex. The capsular arm is advanced back onto the medial epicondyle (Fig. 17.13a), taking care not to encroach on the cheek of the femoral condyle (which might alter the axis of rotation). After suturing, the knee should be flexed and extended in a 30- to 40-degree arc from the repair position to ensure that the rotation axis is proper and the ligament has not been overtightened.

The POL margin of the arthrotomy is advanced in "pants-over-vest" fashion onto the TCL. Horizontal mattress sutures are used. This further tensions the POL, OPL, and TCL/mid-third capsule. More distal sutures are placed, each fine-tuning the tension of the POL (Figs. 17.13b–d).

In some cases, eradicating laxity in the capsular ligaments restores normal tension in the capsular arm of the semimembranosus. However, if laxity remains in the semimembranosus after capsular repair, its capsular arm should be advanced onto the capsular repair (Fig. 17.13e). As stated earlier, aside from the essential dynamic functions, the healing process (remodeling in response to stress) will be incomplete if these ligaments are deprived of the normal input of the semimembranosus.

Results

Straight Instabilities

Our series of acute straight medial instabilities consisted of 91 knees in 91 patients treated over a 30-year period.[5,17] The population was predominately male (6:1), with an average age of 24 years. Mechanism of injury

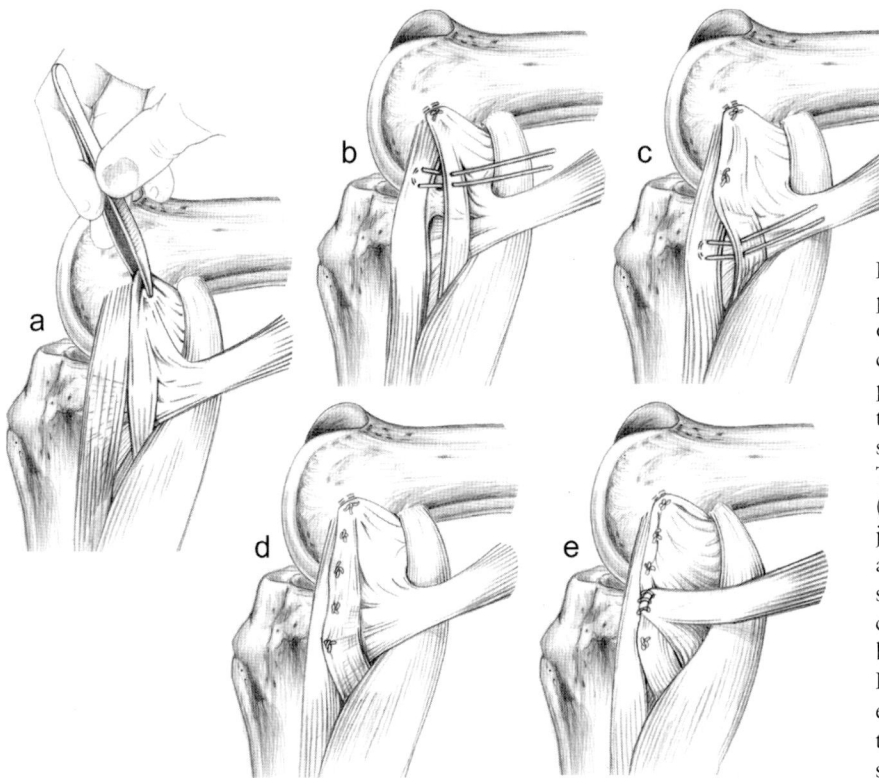

Fig. 17.13. Chronic reconstruction of the posteromedial corner. (a) The capsular arm of the POL is advanced onto the medial epicondyle. (b) A horizontal mattress suture is placed just at the meniscofemoral margin in the TCL/mid-third capsule complex. This suture exits the superficial surface of the TCL and enters the deep surface of the POL. (c) A second suture is placed distal to the joint line advancing the POL onto the TCL and the tibia. (d) The completed reconstruction of the capsular ligaments. (e) The capsular arm of the semimembranosus has been advanced back onto the advanced POL. This restores the dynamic component essential for normal function and healing in this area. Copyright © 2000 by the Hughston Sports Medicine Foundation, Inc.

was sports in 61 patients (39 football), motor vehicle accidents in 17, and other injuries in the remaining 13. At surgery, the PCL and all medial capsular ligaments were torn in all patients. The ACL was torn in 63 of the 91 cases. The medial meniscus was torn in 43 knees (peripheral tears in 38 of the 43 knees), and the lateral meniscus was torn in 23 knees.

Follow-up after 5 years or more was available for 72 patients. Of these, 34 were able to return for reexamination (the remaining patients were followed up via telephone calls). Length of follow-up averaged 11 years, with a range of 5 to 24 years. On follow-up examination, the posterior drawer in neutral tibial rotation was negative in 22 knees, 1+ in seven knees, and 2+ in five knees. The anterior drawer was negative in 20 knees, 1+ in 11 knees, 2+ in three knees, and 3+ in one knee. Five patients who had not undergone concomitant ACL reconstruction had a positive pivot shift test (an additional 20 knees without an ACL had negative pivot shift tests). There was no demonstrable medial instability in 32 knees. The abduction stress test at 30 degrees was 2+ in the remaining two knees. There were no radiographic changes on anteroposterior (AP) films. On weightbearing 30-degree flexed posteroanterior (PA) films, 32% had mild to moderate medial joint space narrowing. No severe osteoarthritic changes were seen in any patient on long-term follow-up. Of these 34 knees, two knees (6%) were considered failures and required further reconstructive procedures.

We have additionally reported a series of knee dislocations in which the principles of acute repair were applied to the medial injury component of the dislocation.[16] Our outcomes must be interpreted in recognition of the fact that in this severe injury, multiple injuries other than those to the medial ligaments were present. These outcomes, therefore, do not reflect the results of the medial ligament repair alone.

Over a 14-year period, 26 acute knee dislocations were treated within 8 days of injury by capsular ligament and PCL repair. One patient was lost to follow-up. Follow-up averaging 8 years was available for the remaining 25 patients. All patients had a normal contralateral knee, which was used as the benchmark for their outcome. Four patients required additional ligament surgery and were considered to have unsatisfactory outcomes. Of the remaining 21 patients, outcomes were as follows: Subjectively, the knees averaged 75% of the normal side. Range of motion was 80% to 95% of the normal side. Stability from examination under anesthesia to last follow-up improved from 60% to 95% of the normal side for all capsular ligament tests. Degenerative changes on radiographs were uncommon; when present, they were mild and did not correlate with outcome.

Rotatory Instabilities

Again, while not in the classification of PCL injuries, repair strategies for acute anteromedial rotatory instability are identical. The report by Hughston[6] in 1994 represents one of the longest follow-up outcome studies of a significant cohort of patients treated by a particular technique. In this study, the outcomes of 41 knees with an average follow-up of 22 years were evaluated. Thirty-eight had good stability and normal range of motion; radiographic changes were slight to nonexistent in all but four knees. Only three knees (7%) were judged as having unsatisfactory outcomes.

In 17 of these knees, no attempt has ever been made to reconstruct the torn ACL. This fact was significant because a large number of satisfactory outcomes were achieved simply by restoring the anatomic and dynamic integrity of the capsular ligaments alone. Although we do not advocate this as a strategy for severe straight instabilities, in properly repaired rotatory instabilities the capsular ligament repairs were sufficiently tenacious as to be able to compensate for the absent cruciate ligament.

References

1. Hughston JC, Eilers AF. The role of the posterior oblique ligament in repairs of acute medial (collateral) ligament tears of the knee. J Bone Joint Surg 1973; 55A:923–940.
2. Hughston JC, Andrews JR, Cross MJ, Moschi A. Classification of knee ligament instabilities. Part I. The medial compartment and cruciate ligaments. J Bone Joint Surg 1976;58A:159–172.
3. Hughston JC, Andrews JR, Cross MJ, Moschi A. Classification of knee ligament instabilities. Part II. The lateral compartment. J Bone Joint Surg 1976;58A: 173–179.
4. Hughston JC, Barrett GR. Acute anteromedial rotatory instability. J Bone Joint Surg 1983;65A:145–153.
5. Hughston JC. Knee Ligaments, Injury and Repair. St. Louis: Mosby, 1993:485.
6. Hughston JC. The importance of the posterior oblique ligament in repairs of acute tears of the medial ligaments in knees with and without an associated rupture of the anterior cruciate ligament: results of long-term follow-up. J Bone Joint Surg 1994;76A:1328–1344.
7. Kannus P. Long-term results of conservatively treated medial collateral ligament injuries of the knee joint. Clin Orthop Rel Res 1988;226:103–112.
8. Price CT, Allen WC. Ligament repair in the knee with preservation of the meniscus. J Bone Joint Surg 1978;60A:61–65.
9. Shapiro MS, Markolf KL, Finerman GAM, Mitchell PW. The effect of section of the medial collateral ligament on force generated in the anterior cruciate ligament. J Bone Joint Surg 1991;73A:248–256.
10. Shoemaker BS, Markolf KL. In vivo rotatory knee stability. J Bone Joint Surg 1982;64A:208–216.
11. Warren LF, Marshall JL, Girgis F. The prime static stabilizer of the medial side of the knee. J Bone Joint Surg 1974;56A:665–674.

12. Ballmer PM, Jacob RP. The nonoperative treatment of isolated complete tears of the medial collateral ligament of the knee: a prospective study. Arch Orthop Trauma Surg 1988;107:273–276.
13. Indelicato PA. Non-operative treatment of complete tears of the medial collateral ligament of the knee. J Bone Joint Surg 1983;65A:323–329.
14. Jokl P, Kaplan N, Stovell P, Keggi K. Non-operative treatment of severe injuries to the medial and anterior cruciate ligaments of the knee. J Bone Joint Surg 1984;66A:741–744.
15. Jones RE, Henley MB, Francis P. Nonoperative management of isolated grade III collateral ligament injury in high school football players. Clin Orthop Rel Res 1986;213:137–140.
16. Flandry FC, Schwartz MG, Hughston JC. Long-term follow up of operatively treated knee dislocations. Orthop Trans 1991;15:805–806.
17. McCluskey LC, Savory C, Hughston JC. Acute straight medial instability of the knee. Orthop Trans 1991;15:613.
18. Butler D, Noyes FR, Grood ES. Ligamentous restraints to anterior-posterior drawer in the human knee: a biomechanical study. J Bone Joint Surg 1980;62A:259–270.
19. Clancy WG, Nelson DA, Reider B, Narechania RG. Anterior cruciate ligament reconstruction using one-third of the patellar ligament, augmented by extra-articular tendon transfers. J Bone Joint Surg 1982;64A:352–359.
20. Nicholas JA. The five-one reconstruction for anteromedial instability of the knee. J Bone Joint Surg 1973;55A:899–922.
21. O'Donoghue DH. Reconstruction for medial instability of the knee. J Bone Joint Surg 1973;55A:941–955.
22. Hunter SC, Marascalco R, Hughston JC. Disruption of the vastus medialis obliquus with medial knee ligament injuries. Am J Sports Med 1983;11:427–431.

Chapter Eighteen

Rehabilitation of Posterior Cruciate Ligament Injuries

Craig J. Edson and Daniel D. Feldmann

History

The natural history of posterior cruciate ligament (PCL) injuries has not been extensively studied, and consequently is not clearly understood. This has resulted in controversy over the most effective treatment for these injuries. Additionally, there is a lack of information of controlled outcome studies in the literature that have focused on effective rehabilitation programs following PCL injury. This is in sharp contrast to the anterior cruciate ligament (ACL), which has amassed an abundance of studies that have outlined surgical management, postoperative rehabilitation, and guidelines for return to sports and functional activities.[1-17]

As with the ACL, the treatment approach to isolated injuries of the PCL initially was conservative. There are still advocates of this philosophy[18-21]; however, there have been a number of studies that have demonstrated significant degenerative changes, especially of the patellofemoral joint, when these injuries are left untreated.[22-25] Unfortunately, these changes may take several years to occur and may also be accompanied by degenerative changes of the medial compartment. The studies that described successful outcomes with a conservative approach often consisted of short-term follow-up and did not delineate specific rehabilitation protocols other than "quadriceps strengthening."[18,20,26] In addition, many studies reported a progressive decline in patient satisfaction with the passing of time.[18-20] Boynton and Tietjens[27] conducted a study that involved isolated PCL injuries and included long-term follow-up. They examined 38 subjects with isolated PCL injuries with a mean of 13.4 years after injury and concluded that patients with isolated PCL injuries have varied results depending on the degree of posterior laxity. Specifically, they found patient satisfaction and function decreased in correlation with the degree of posterior laxity, and radiographic evidence of articular degeneration increased as the length of time from injury increased. If any conclusion can be drawn from the current literature, it appears that patients with isolated PCL injuries and posterior laxity of 2+ or less do well with a conservative program, while patients with 3+ laxity should be considered for surgical reconstruction. In addition, in the presence of associated ligamentous injuries, surgical reconstruction is the best approach.

With this background, it appears that there is the need for specific protocols for both conservative and postsurgical management of PCL injuries. At our institution, we have attempted to establish such guidelines based on clinical examination, arthrometric findings, stress radiographs, and patient function. In the case of the isolated injuries to the PCL, these criteria are essential. With complete ruptures the physical examination often demonstrates an absence of tibial step-off, a positive posterior drawer, corrected arthrometric posterior displacement of 5 mm or greater, and subjective reports of limited function secondary to pain. These patients are considered strong candidates for PCL reconstruction to restore anatomic alignment of the tibia in relation to the femur. When the physical signs are less conspicuous, so too is the optimum treatment. It is quite possible that there are many individuals with grade I or II PCL injuries who are functioning well and will not seek or require medical intervention. These patients may be examined as part of a routine screen following injury, and often present with palpable tibial step-off, grade I or II posterior drawer, arthrometric measurements of 3 mm or less with side-to-side comparison, and no significant radiographic findings of degenerative joint disease. The treatment approach for this patient population is often conservative; however, it is important that they be monitored for any progression in laxity, functional limitations, or pain.

Scientific Basis of Rehabilitation

Rather than discussing specific treatment programs, it may be more appropriate to consider the rationale behind the selection of specific exercises utilized during rehabilitation. The effect of specific exercises and activities on the tibiofemoral joint and consequently on the PCL is the primary concern; however, it is also important to examine the effects of these exercises on the patellofemoral joint. Open- and closed-chain exercises vary in the muscles recruited and forces generated around the knee joint. The body of knowledge as to the effects of these exercises on the tibiofemoral joint mostly stems from research on the knee joint as it relates to rehabilitation of ACL injuries.[3,4,9,10,12,15,17,28,29] It is equally important to discuss the effects of these exercises as they relate to the PCL.

Open- Versus Closed-Chain Exercises

Kinetic chain terminology was originally used to describe linkage analysis in mechanical engineering; however, Steindler[30] was the first to apply this principle to human movement. In general, a closed kinetic chain occurs when the distal segment is fixed. During closed-chain exercises, movement at one joint results in simultaneous movement of all other joints of the kinetic chain in a predictable manner.[31] With the lower extremity, closed kinetic chain exercises incorporate some degree of weight bearing. There are unique physiologic events that occur at the joints when the lower extremity bears weight. With regard to the knee joint specifically, Lutz and associates[32] noted a decrease in shear forces at the tibiofemoral joint during closed kinetic chain exercises as opposed to open kinetic chain exercises. They attributed this decrease to axial orientation of the applied force as well as muscular co-contraction. In addition, Steindler[30] described a phenomenon that he termed the "concurrent shift" of biarticular muscles of the lower extremity. To illustrate, consider the muscles at the hip and knee when one rises from a squatting position. The hamstring muscles shorten as the hip extends and lengthen as the knee extends, thus undergoing a concentric contraction at one joint and an eccentric contraction at the other. Similarly, the rectus femoris lengthens as the hip extends and shortens as the knee extends, again incorporating both concentric and eccentric contractions at the knee and hip, respectively.[32] This co-contraction contributes to joint stability and minimizes shear forces at the tibiofemoral joint.

In contrast to closed-chain exercises, during open-chain exercises the distal segment is free, resulting in isolated motions of flexion and extension of the knee. These motions have varying effects on the tibiofemoral joint and subsequently on the ACL and PCL. It has been shown that open-chain knee extension produces anterior tibial translation between 0 and 60 degrees of flexion.[30,33–35] Beyond 60 degrees, quadriceps activity results in minimal tibial translation until approximately 70 to 75 degrees of flexion. The point at which quadriceps contraction produces neither anterior nor posterior tibial translation has been termed the quadriceps neutral angle.[36] The specific angle at which this occurs encompasses a wide variance within the population. Nonetheless, at flexion angles greater than the quadriceps neutral angle, quadriceps activity will produce posterior translation of the tibia due to the posterior orientation of the quadriceps tendon.[36]

Open-chain knee flexion will have a diametrical effect on translation of the tibia. As knee flexion is initiated, there is a mild degree of posterior force, which has been calculated as less than 1 times body weight (BW) up to 50 degrees of flexion.[37] Beyond 50 degrees, this force increased to 1.7 times body weight. When incorporating resistance, these posterior shear forces increase dramatically. Lutz and associates[32] studied the posterior shear force during maximal isometric knee flexion and found a force of 939 N at 30 degrees, increasing to 1,780 N at 90 degrees. The magnitude of this force can be better appreciated when one considers that the greatest anterior shear force was only 285 N, produced during a maximal isometric contraction of the quadriceps.

Isometric and Eccentric Exercise

Isometric and eccentric exercises are both utilized during the rehabilitative process. Isometrics are usually open chain and therefore follow the same principles as discussed above. They are employed for their efficiency since they have been shown to be second only to eccentric exercises in their ability to generate intramuscular tension.[38] Isometrics are primarily utilized during the early phase of rehabilitation to maintain quadriceps tone. Eccentric movements occur naturally during closed-chain exercises as previously discussed. Open-chain eccentric exercises may also be employed from 0 to 60 degrees, much like open-chain isotonics.

Patellofemoral Joint

At the patellofemoral joint, during closed-chain exercises, there are increased joint reaction forces as the knee flexes. Nonetheless, since this force is distributed over a larger patellofemoral contact area, there is less contact stress per unit area.[39] Conversely, during open-chain knee extension the patellofemoral contact area decreases as the knee extends, resulting in increased contact stress per unit area. Hungerford and Barry[40] found that peak patellofemoral joint reaction forces occurred at 36 degrees with open-chain knee extension against a 9-kg boot.

PCL Stress

Open- and closed-chain exercises have also been shown to create distinct and contrasting effects on ligamentous strain and load. Many studies have illustrated the effects of various exercises and activities on the ACL; however, there have been limited studies regarding their effects on the PCL. Ohkoski and associates[41] did report increased posterior shear forces with increasing angles of trunk flexion secondary to heightened activity of the hamstring muscles. These posterior directed forces were magnified during active knee flexion at all angles beyond 30 degrees. Dahlkuits and colleagues[42] calculated a posterior shear force of 3.0 times BW during squatting activities. Given these findings, it would appear that closed-chain exercises are contraindicated during rehabilitation of the PCL. However, as stated earlier, muscular co-contraction occurs during closed-chain exercises, and this helps to minimize the posterior shear forces and stress on the PCL.

Wilk[43] analyzed the outcomes of several studies and determined that quadriceps and hamstring muscle ratios are similar during the first 60 degrees of knee flexion. In addition, compression of the joint surfaces during weight bearing reduces tibial translation. Other investigators have studied the effects of various activities on posterior tibiofemoral translation and thus the PCL. For example, Ericson and Nisell[44] examined the effect of cycling on the PCL and found a posterior shear force of only 0.05 times BW at 105 degrees of knee flexion. Level walking also produced a relatively low posterior shear force of 0.4 times BW.[45] Conversely, ascending stairs produced a posterior shear force of 1.7 times BW, which occurred at 45 degrees of flexion.[46] This finding is somewhat enigmatic since ascending stairs should involve more quadriceps activity than hamstrings and this force is occurring at a flexion angle (45 degrees) that should produce an anterior shear force. In addition, Smidt[47] found a posterior shear force of only 1.1 times BW with isometric flexion at this same angle of 45 degrees.

There have been numerous reports in the literature about the deleterious effects of open knee extension on the ACL especially, at flexion angles of 60 degrees to full extension. This is secondary to the anterior shear force produced by quadriceps activity. Conversely, in the case of PCL pathology, these exercises can be performed safely since no posterior shear forces are produced. Nonetheless, these exercises should be employed judiciously secondary to the detrimental effects on the articular cartilage of the patellofemoral joint, as previously discussed. Open-chain extension exercises should also be avoided at flexion angles greater than 60 degrees, since, as alluded to earlier, there is a posterior shear force produced by the patellar tendon. Open-chain flexion exercises should be avoided in the case of PCL deficiency and following PCL reconstruction, until such a time that adequate healing has occurred to withstand this force.

Exercise and PCL Grafts

Currently, graft sources consist of allograft and autograft tissues. Common allografts utilized in the reconstruction of ligamentous tissue are cadaver-donated Achilles tendon and bone–patellar tendon–bone grafts. Common autograft sources include bone–patellar tendon–bone and semitendinosus and gracilis looped tissues. There are no known studies on the effects of various exercises with respect to the type of PCL graft employed. For instance, how much force is detrimental to the PCL graft and when is the graft most susceptible to these forces? The type and dimensions of the graft material chosen would certainly play a role in determining the amount of force necessary before interstitial damage occurred. The surgical technique and type of fixation are also factors contributing to the overall stability of the graft.

At our institution, Achilles tendon allograft is utilized for PCL grafts when possible. This allows the surgeon to utilize adequate tissue to simulate the dimensions of the anatomic PCL, without sacrificing any autogenous structures. However, there are inherent dangers with allograft tissues, the most menacing of which is disease transmission.[47] Fortunately, current techniques of tissue preparation and sterilization have diminished this risk to an infinitesimal level.[47] Nonetheless, since the possibility does exist, alternative structures must be available. The bone–patellar tendon–bone autograft is the most studied graft material, and the information available regarding its tensile strength as it relates to ACL substitution[48–50] may be applicable to PCL substitution. However, there is little in the literature to say that this graft acts in a similar fashion as a surrogate for the PCL, especially since it is not possible to take sufficient tissue to mimic the dimensions of the PCL. Recently, consideration has been given to a two-strand graft to better imitate the anterior and posterior fibers of the PCL. A synthetic graft, which has been utilized in Europe but has not been approved for use in the United States, may be of use in the future.

Given the deficiency of controlled studies regarding the effects of exercise and activities on PCL grafts, it is our philosophy to take a cautious approach to rehabilitation following PCL reconstruction. This is especially true when, as is often the case, other ligamentous structures are involved. The conservative and postoperative rehabilitation regimens for isolated and combined PCL injuries will be presented here. These programs have been utilized successfully in those patients with PCL/PLC and ACL/PCL reconstruction as reported in the literature.[51,52]

Nonoperative Treatment

Prior to discussing specific rehabilitation protocols, it is important to review a few general principles that have been discussed by Wilk[43] and have served as the premise for these programs.

- The rehabilitation programs should not be overly regimented, to allow for variability in patient characteristics and goals.
- Programs should be criteria based to establish guidelines for progression from one stage of the rehabilitative process to the next.
- The rehabilitation program is based on current science and fundamental clinical research.
- Healing tissue must not be overstressed.

Additional considerations in conservative management include bracing, gait training, and frequent monitoring/reexamination.

Isolated PCL Injuries

As stated previously, early studies on PCL injuries claimed good results with conservative treatment.[18–21] Those patients who were most satisfied with their outcomes demonstrated superior quadriceps function. Parolie and Bergfeld[20] observed 25 patients with PCL injuries who were managed nonoperatively and found isokinetic quadriceps muscle torque greater than 100% of the uninvolved side in those patients who were able to return to their prior level of function. At our institution, isolated PCL injuries with posterior laxity of 2+ or less are commonly treated with a conservative therapy program (Table 18.1). The primary goals of this program are to enhance dynamic joint stability, improve quadriceps strength, and provide

Table 18.1. Nonoperative treatment for isolated PCL injuries

Acute phase—maximum protection
  Week 1
    Long leg brace—ROM 0 to 60 degrees
    WBAT with crutches
    Modalities to control pain and swelling
    Exercise program
      Quad sets and straight leg raise
      Hip abduction and adduction
      Open chain extension—0 to 60 degrees (min. resistance)
      Wall slides (0 to 45 degrees)
      Isometrics—60 and 40 degrees
  Weeks 2 to 3
    FWB—D/C crutches
    Step-ups (4–6 in. step)
    Leg press—0 to 60 degrees
    Stationary bike for ROM
    Pool program (if available)
    Begin proprioceptive training
Subacute phase—moderate protection
  Weeks 4 to 6
    Discontinue long-leg brace
    Fit for functional PCL brace
    ROM to tolerance
    Add/progress resistance to all above exercises
    Progress proprioception exercises
    Stairmaster, rowing, Nordic Track
    Pool running
    Ongoing evaluation/monitoring of patellofemoral and tibiofemoral joints
Minimum protection phase
  Weeks 6 to 12
    Initiate jogging program
    Progressive strengthening exercises—isolation exercises emphasized
    Progress to cutting, pivoting, and twisting
    Initiate plyometrics
    Return to sports or heavy work given following criteria:
      Functional, pain-free ROM
      Isokinetic test 90% or greater of quadriceps and hamstrings of contralateral side
      Functional tests 90% or greater of contralateral side
      Proprioception equal to contralateral side
      Perform sport-specific or job-specific activities satisfactorily
      Functional PCL brace

D/C, discontinue; FWB, full weight bearing; ROM, range of motion; WBAT, weight bearing as tolerated.

the patient information regarding the extent of the injury, possible sequelae, and aggravating factors.

Weeks 1 to 3

If the patient is seen soon after the injury, cold, compression, and electrical stimulation may be utilized to decrease pain and swelling. Wilk[43] recommends placing the patient in a brace that limits motion from 0 to 60 degrees, to prevent posterior tibial displacement. A long leg brace is preferred to distribute the forces over a larger area and to provide greater leverage. Ambulation with an assistive device and partial weight bearing is initiated, with progression to full weight bearing as tolerated. It is important that the patient maintain an upright posture during gait since flexing at the trunk will elicit increased hamstring activity, resulting in increased posterior tibial displacement. Use of the assistive device may be discontinued when the patient is ambulating without gait deviations and with good quadriceps control.

Isometric exercises for the quadriceps muscles, including "quad" sets and straight leg raises, are adopted immediately in addition to hip strengthening in flexion, abduction, and adduction. Open-chain knee extension is performed between 0 and 60 degrees with light resistance. Heavy-resistance knee extension exercises through a range of 45 to 20 degrees of flexion are avoided as these result in excessive patellofemoral joint reaction forces.[40] Multiangle isometrics can also be employed as long as the same range restrictions are applied. These exercises should be modified if the patient complains of patellofemoral pain or crepitus.

Weeks 3 to 6

As the patient progresses through the acute phase (3 to 6 weeks), closed kinetic chain exercises are instituted. These are performed in a range of motion of 0 to 60 degrees. Stationary cycling is also implemented primarily to assist in improving active range of motion. The closed-chain exercises serve several important functions including proprioception, quadriceps and hamstring strengthening, and dynamic control of the tibia. Use of assistive devices may be discontinued when the patient is ambulating without gait deviations and with good quadriceps control.

By week 6, an aggressive strengthening and conditioning program is instituted including stair-climbing machines, Nordic Track equipment, and resistive stationary cycling. Proprioceptive training is also progressed by implementing a BAPS board, KAT, or similar device. Single-limb balancing on pillows or foam cushions may also be employed.

Weeks 6 to 12

Ongoing evaluation is critical to monitor the tibiofemoral and patellofemoral joints as the exercises are progressed. If severe posterior tibial translation exists, then a functional PCL brace may assist in maintaining the tibia in a more neutral position during exercises and ambulation. As strength and proprioception improve, and in the absence of significant pain and swelling, the patient is gradually progressed to more strenuous activities. These may include jogging, running, sprinting, acceleration/deceleration, cutting, pivoting, and twisting. Plyometrics may also be implemented.

Return to sports and/or heavy labor occurs when the following criteria are met: range of motion is functional and painless, and quadriceps strength is equal to 90% of the contralateral side as assessed via functional tests and isokinetic strength evaluation through a velocity spectrum. Functional tests include single-leg hop test for distance and single-leg, timed hop over a specified distance. A comparative assessment is also performed for proprioception, employing specific balance devices (KAT, BAPS) or on a foam-covered surface. A functional PCL brace is routinely employed to maintain the tibia in a more anterior, anatomic position, protect the posterolateral capsular structures, and enhance proprioception. Athletes are also required to perform sport-specific activities within the brace, at a level that is deemed adequate for the safe return to sports as determined by the medical and coaching staffs. Patients are advised to continue with a regular quadriceps and lower extremity strengthening program indefinitely. In addition, the patient should be followed on a regular basis to monitor laxity, pain, and articular cartilage changes. Serial radiographs or bone scans are utilized to assess the articular cartilage. In addition, periodic stress radiographs (Telos) may be obtained that accurately measure posterior tibial displacement as compared to the contralateral side.

Many of these criteria and assessment tools are utilized for all patients with PCL pathology, be it isolated or associated with other ligamentous injuries. These tests also become essential in our analysis of the outcomes following PCL reconstruction. Arthrometer testing, ligament rating forms, and stress radiographs (Telos) are also employed when compiling the data for the functional outcome studies.

Plyometrics

Plyometric exercises/training incorporate a stretch-shortening cycle and are alleged to increase power of the concentric portion of this cycle as well as decreasing reaction time by enhancing the excitability of the nervous system.[54,55] During the eccentric phase of the stretch-shortening cycle, the series elastic components of the muscles are stretched, storing energy.[33] When this eccentric contraction is immediately followed by a concentric contraction, the stored energy serves to augment this concentric action. The rate at which the individual is able to convert the eccentric contraction into the enhanced concentric contraction is referred to as the amortization phase, and is dependent primarily upon the stretch reflex.[56]

Two types of mechanoreceptors located within the muscle serve to control muscle tension and length via the central nervous system. The muscle spindle lies in parallel with the muscle fibers and provides continuous feedback concerning the length of the muscle. When the muscle is stretched, the primary sensory afferent acts on the alpha motor neuron of the agonist, causing the muscle to contract, thus relieving the stretch on the spindle.[57] This monosynaptic stretch reflex is harnessed during plyometrics to increase agonist concentric force production.[33]

The Golgi tendon organs (GTOs) lie perpendicular to the muscle fibers and respond to increased muscle tension, concentric or eccentric, by inhibiting the agonist and causing relaxation. If, during the eccentric phase of the stretch-shortening cycle, the GTOs are activated, then augmentation of the concentric contraction would be diminished.[33] It is theorized that, through plyometric training, the threshold of the GTOs may be attenuated, thereby allowing more tension to be applied to the system prior to activation.[55] Hence, plyometric training uses eccentric-concentric coupling to load the series elastic components, inhibit the GTOs, and, via the stretch reflex, improve neural efficiency to allow for explosive power.[33] This explosive power is crucial to athletes hoping to return to their sport at a level that is akin to their premorbid status.

Combined Posterior Cruciate Ligament Injuries

As diagnostic schemes and tests have evolved, we are finding that many injuries to the PCL are associated with additional ligamentous pathology. It appears that when there is sufficient force to tear the PCL, additional ligamentous structures are also disrupted. In fact, in our setting, isolated PCL tears are rarely seen.[58,59] Injuries to the posterolateral knee structures, medial collateral ligament, or ACL often accompany injuries to the PCL. It is exceedingly difficult to manage patients with injuries of this severity with a conservative program. There is greater alteration to the normal biomechanics of the knee than those discussed previously. Skyhar et al.[60] have shown that contact pressures of the medial compartment and the patellofemoral joint were exaggerated when the posterolateral complex were sectioned in conjunction with the PCL. The posterolateral complex is composed of the fibular collateral ligament, the popliteus tendon, and the posterolateral capsule.

Postoperative Rehabilitation Programs

PCL Reconstruction

In designing a postsurgical PCL rehabilitation program, the therapist needs to take into account the patient's ultimate goal, the surgeon's preferences, type of graft utilized, and additional ligament reconstruction, as these aspects will all influence the eventual outcome. Although quadriceps strengthening is a key element in the rehabilitative process, this alone is not sufficient to ensure a successful outcome.[24] Protection of the PCL graft may be the most critical component of the rehabilitation program. Ultimately, a successful outcome is dependent on the restoration of muscle strength, proprioception, and function, while minimizing the forces on the healing PCL graft.

Prior to surgery, the patient is scheduled for a preoperative session with the therapist. This includes fitting of the postoperative brace, completion of ligament rating sheets, and an explanation of the postoperative program. We find this session to be extremely beneficial to the patient for addressing any questions or concerns, as well as preparing the patient for the difficult task of the rehabilitation regimen.

Isolated PCL Reconstruction

Weeks 1 to 6

Reconstruction of isolated PCL tears allows for the most latitude within the rehabilitation scheme; however, these programs tend to be less aggressive than those utilized following ACL reconstruction (Table 18.2).

Initially, the knee is braced in full extension, which minimizes tibiofemoral shear force due to the small moment arm of the hamstrings in this position.[31] In addition, this position has also been shown to minimize the stress on the PCL.[61] There are varying opinions regarding weight bearing in the early phase of the postoperative recovery. Noyes[62] and Irrgang[31] recommend non–weight bearing for 3 to 6 weeks. Wilk[43] supports weight bearing of 50% body weight during week 1, with progression to full weight bearing over the next 4 weeks. We utilize a non–weight-bearing gait for the first 6 weeks to provide maximum protection of the graft. More studies are needed to determine the effect of weight bearing on the healing PCL graft.

Immediately after ligamentous surgery, we utilize a continuous cold unit to control pain and hemarthrosis. The pads for these units are incorporated into the postoperative dressings, but do not directly contact the skin. Protective, antibacterial liners are available to minimize the risk of infection and to protect the dressings from becoming damp. Noyes[62] and Wilk[43] both advocate the use of early motion following PCL reconstruction, in a passive range of 0 to 70 degrees. This range minimizes forces on the PCL. This may be accomplished by assisting with the uninvolved extremity, or through a mechanical continuous passive motion (CPM) unit. The brace is opened during ROM exercises but otherwise remains locked in extension; this is especially important at night to avoid the tendency to flex the knee during sleep. Given that the knee is locked in full extension the majority of the time, the patient may develop hamstring pain and tightness. It is important to instruct the patient in hamstring stretching to maintain flexibility, thereby decreasing pain and, more importantly, minimizing posterior

Table 18.2. PCL postoperative rehabilitation

Phase I—0 to 4 weeks
  Goals
    Maximum protection of grafts
    Maintain quadriceps tone
    Maintain patella mobility
    Maintain full passive extension
    Control pain and swelling
  Program
    Non–weight-bearing ambulation
    Brace locked in extension—24 hours/day
    Cryotherapy
    Quad sets—enhance with electrical stimulation and/or biofeedback, progress to SLR
    Patella mobilization
    Ankle pumps—ROM
    Stretching exercises—gastroc-soleus and gentle hamstrings
    Hip abduction and adduction

Phase II—4 to 12 weeks
  Goals
    Increase flexion ROM
    Initiate weight bearing for articular cartilage nourishment (6 weeks)
    Improve quadriceps strength
    Avoid active hamstring contraction
  Program
    At 4 weeks postop—open brace to 70 degrees 2–3 ×/day for passive flexion exercises
    Sleep in locked brace until brace is opened at 6 weeks
    Patella mobilization
    At 6 weeks postop—initiate PWB gait of 25% BW and increase by 25% over next 4 weeks
    Full weight bearing by end of postop week 10
    Prone hangs
    Open chain knee extension 0- to 60-degree range—no resistance
    At 6 weeks postop—open brace 0 to 135 degrees
    Proprioception and weight shift exercises—KAT or BAPS board
    Stationary bike for ROM once brace is opened—passive flexion exercises
    Closed-chain exercises—0 to 60 degrees if quadriceps strength is 3+/5 or greater (no resistance)
    Discontinue brace at end of postop week 12

Phase III—4 to 6 months
  Goals
    Increase ROM
    Maintain full passive extension
    Increase quadriceps and hamstring strength
    Improve proprioception
    Improve functional skills
    Improve cardiovascular endurance
  Program
    4 Months
      Progressive ROM—initiate active hamstring exercises with no resistance
      Single-leg proprioception exercises (KAT, BAPS, mini-trampoline)
      Open chain SAQs—high-speed isokinetics or light-weight isotonics
      Resisted closed-chain exercises—avoid flexion beyond 70 degrees
      Begin active hamstring exercises—no resistance
      Closed-chain conditioning exercises—Stair Climber, ski machine, rapid walking, etc.
      Aggressive ROM—consider manipulation if ROM is <90 degrees by end of month 4
      Hip PREs
      Begin straight-line jogging at end of postop month 4
    5 Months
      Progressive closed-chain strengthening and conditioning exercises
      Initiate low-intensity plyometric exercises
      Resisted hamstring exercises—isokinetic or light-weight isotonics
      Progressive jogging and begin sprints
      Advance proprioception training
      Fit for PCL functional brace
    6 Months
      Progression of all strengthening exercises and plyometrics
      Begin agility drills—carioca, figure 8's, zigzag, slalom running, cutting drills, etc.
      Isokinetic testing at end of postop month 6

Phase IV—7 to 12 months
  Program
    Assess functional strength—single-leg hop for distance, timed hop test, shuttle run, etc.

(continued)

Table 18.2. Continued

Return to sports if the following criteria are met:
  Minimal to no pain and swelling
  Isokinetic *and* functional strength tests equal to 90% or greater of the uninvolved side
  Successful completion of sport-specific drills
  Functional PCL brace

BAPS, biomechanical ankle platform system; BW, body weight; KAT, kinesthetic awareness training; PRE, progressive resistive exercise; PWB, partial weight bearing; ROM, range of motion; SAQ, short-arc quadriceps test; SLR, straight leg raise.

shear forces on the tibiofemoral joint. Patella mobilization is also instituted and the patient is instructed in this technique so that it can be done three to four times per day. Prone hangs are initiated at week 5.

To minimize quadriceps atrophy and inhibition, electrical muscle stimulation is often employed. This modality has been shown to augment quadriceps contraction and facilitate an earlier return of strength.[63] This treatment is employed until such time that the patient demonstrates adequate

Table 18.2a. Rehabilitation summary for PCL reconstruction

	Postoperative weeks					Postoperative months			
	1–2	3–4	5–6	7–8	9–12	4	5	6	7–12
Long-leg brace @ 0 degrees	X	X	X						
Long-leg brace 0–135 degrees				X	X				
Functional brace								X	X
ROM goals (degrees)									
0–70		X	X						
0–90				X					
0–110					X				
0–120						X			
0–130							X		
Procedures									
E-stim	X	X	X						
Patella mob	X	X	X						
Exercises									
Stretching									
Hamstring—gastrocnemius	X	X	X						
Prong hangs			X	X	X	X			
Strengthening:									
Quad sets—SLR	X	X	X						
Knee extension				X	X	X			
Short arc—CKC				X	X	X	X	X	X
Resisted CKC					X	X	X	X	X
Hamstring curls						X	X	X	X
Resisted hamstring curls							X	X	X
Proprioception									
KAT or BAPS				X	X	X	X	X	X
Plyometrics							X	X	X
Conditioning/aerobic									
UBE		X	X						
Stationary bike				X	X	X	X	X	X
Walking						X	X	X	X
Stair climber						X	X	X	X
Ski machine						X	X	X	X
Jogging							X	X	X
Sport specific									
Cutting								X	X
Return to sports									X

CKC, closed kinetic chain; UBE, upper body ergometer.

quadriceps control, as demonstrated by straight-leg raise and multiangle isometrics in the 0 to 60 degree range.

Weeks 7 to 12

At the end of postoperative week 6, the hinged knee brace is open to the maximum flexion allowed and may be removed at night. Range of motion (ROM) goals of 0 to 90 degrees by week 9 and 0 to 120 degrees by week 12 are encouraged. Although the brace allows for full flexion, the patient is advised to avoid aggressive ROM exercises and rapid gains in motion. Inasmuch as the development of arthrofibrosis is a concern, it has been our experience that those patients who make slow, steady gains in motion have less laxity and better outcomes. The patient also begins weight bearing, starting with 25% of body weight and increasing 25% each week until full weight bearing is achieved by week 10. Prone hangs and hamstring stretching exercises continue, so as to avoid hamstring tightness and a flexion contracture.

Aggressive open-chain quadriceps strengthening exercises are implemented at this point, in a range of 0 to 60 degrees. Stationary biking is employed to facilitate motion once the brace is open, and closed-chain exercises are initiated once the patient has achieved full weight-bearing status. Since this does not occur until the end of postoperative week 10, the patient has had adequate time to develop a strong quadriceps contraction to offset the posterior shear forces produced by the hamstrings and hip extensors that occur during closed-chain exercises. The closed-chain exercises are performed in a range of 60 to 0 degrees and are often accompanied by a progression of resistive knee extension from 0 to 60 degrees. As mentioned earlier, open-chain knee extension in this range can be harmful to the patellofemoral joint, so it is important to monitor this joint for crepitus and pain.

Balance and proprioception training activities are initiated as an adjunct to closed-chain exercises. This incorporates single-leg stance activities that the patient can also practice at home in front of a mirror. Pool walking may also be a useful adjunct during the initial phase of weight bearing. It allows for gentle resistance and proprioceptive training in an environment that minimizes the forces of weight bearing.

Months 4 to 6

At the end of postoperative week 12, the hinged brace is discontinued. A goal of 0 to 120 degrees, progressing to full ROM, is encouraged. Flexion limitation of 10 to 15 degrees is not uncommon but may resolve over time. Residual flexion deficits of 10 degrees or less can persist; however, this rarely results in any functional limitation and may result in less stress on the graft since the PCL fibers are most taut in flexion.[64] If active ROM of greater than 90 degrees has not been obtained by the end of postoperative week 12, manipulation under anesthesia and/or possible arthroscopic debridement can be considered.

Closed-chain strengthening exercises are advanced at this time as well as aerobic conditioning exercises, incorporating such devices as a stair climber, rowing machine, and skiing machine. In addition, active flexion exercises are initiated against gravity but without resistance.

At the end of the fourth postoperative month, progressive resistive exercises (PREs) for the hamstrings are initiated in conjunction with a progression of the closed-chain and conditioning exercises previously outlined. Patients also begin straight-line jogging, assuming they are able to demonstrate functional strength that is 70% or greater of the uninvolved extremity. This is assessed through a single-leg hop test during which the patient assumes a single-leg stance, on the uninvolved extremity initially, and jumps forward as far as possible, landing on the same extremity. The patient performs three trials, and the average distance is calculated. The procedure is then repeated for the involved extremity for comparison. Once allowed to run, the patient is advised to begin with a fast walk that is gradually progressed to a mild jog. As tolerance improves, the patient advances to a more consistent jogging pace. The progression of this program is dependent on the patient's goals and the sport or activity that the patient wants to resume. Again, a pool can be beneficial when jogging is initiated by providing resistance and by diminishing the impact-loading forces on the knee.

At the end of the fifth postoperative month, an initial isokinetic assessment is obtained. Higher velocities are used to minimize shear forces.[65] Quadriceps and hamstring deficits of 20% or less are desired before allowing the patient to begin sport-specific activities. The patient is also fitted with a functional PCL brace assuming that thigh circumference is within 2 cm of the contralateral side. If the brace is fitted prematurely, proper fit may be jeopardized as strength, and subsequently thigh circumference, increases. Once braced, more aggressive agility drills and sport-specific activities are permitted. Supervision of these exercises can be coordinated with the athletic training and coaching staffs. During this time, the patient is monitored carefully for signs of pain, swelling, or increased laxity.

Months 7 to 12

At the end of postoperative month 6, an arthrometric evaluation is performed. The KT-1000 arthrometer (Medmetric, San Diego, CA) has been found to be a moderately reliable tool for the assessment of PCL injuries and tibial translation.[66] The KT-1000 is also comparable to stress radiography in determining posterior translation (unpublished data), although Hewett and associates[67] found stress radiography to be more accurate. The quadriceps active displacement (PCL screen) corrected anterior and posterior displacement, and anterior laxity tests are performed on both extremities for side-to-side comparison. These procedures and the subsequent equations to determine the degree of PCL laxity will be discussed in detail at the end of this section.

At the end of the 6th postoperative month, the patient is allowed to return to sports or heavy labor assuming that several criteria are met:

1. Minimal or no pain or swelling.
2. Strength (isokinetic) and functional tests within 90% of contralateral side.
3. Grade I laxity or less with arthrometric testing, physical examination, and/or stress radiography
4. PCL functional brace for sports.

At 1 year from the initial postoperative date, the patient returns for a full reassessment including stress radiography, KT-1000, physical examination, and completion of the ligament rating forms. If PCL laxity is a grade I or less, then the patient is given the option of discontinuing the functional brace for sports. Subsequent appointments are scheduled on an annual basis so that long-term effectiveness can be assessed. The outcomes of those patients who underwent multiple ligamentous reconstruction at our facility have previously been reported in the literature and data collection continues on an ongoing basis.[50,51]

Arthrometric Testing for PCL Laxity

Arthrometric testing of the PCL can be a demanding task for the clinician and requires patient participation. The quadriceps active displacement test is performed with the patient's knee flexed to 90 degrees as described by Daniel et al.[37] The examiner must support the patient's thigh so that the surrounding musculature is relaxed. With the examiner blocking the leg from sliding anteriorly, the patient then performs a gentle quadriceps contraction. During this procedure, a properly functioning PCL will produce posterior tibial translation, due to the posterior shear force generated by the patella tendon. The KT-1000 allows for the measurement of this displacement, in millimeters.

To obtain corrected anterior and posterior displacement measurements, the quadriceps neutral angle is utilized as a reference point. Daniels et al.[37] describe this as the angle of knee flexion that, during an active contraction of the quadriceps, produces neither anterior nor posterior tibial translation. This occurs between 60 and 90 degrees with an average of 71 degrees, and can be determined for each patient through a trial-and-error process. The most effective means of establishing the quadriceps neutral angle is to utilize the uninvolved leg and determine at what point in the range quadriceps neutral occurs. Once identified, the same range of motion is employed on the involved knee. From this point of reference, applying 20 pounds of anteriorly directed and posteriorly directed force, through the KT-1000, determines passive anterior and posterior laxity. To determine the corrected posterior displacement, the amount of displacement that occurs during the quadriceps active test is added to the amount of displacement that is obtained with the 20-pound posterior displacement test. For example, if the tibia translates 2 mm during the quadriceps active displacement test, and there is 2 mm of translation with the 20-pound posterior displacement test, then corrected posterior displacement would be 4 mm.

Corrected anterior displacement is derived in a similar fashion with the exception that the 20-pound active displacement is *subtracted* from the amount of displacement that occurs from the quadriceps active test. If 2 mm of translation occurs with the 20-pound anterior displacement test, then the corrected anterior displacement would be 0 mm, since the 2 mm of translation that occurs with the quadriceps active displacement test is used in both equations. Finally, anterior displacement is derived by the procedure for determining ACL pathology, as originally described by Daniels et al.[68] Although these measurements and the subsequent equations may be tedious and confusing at first, the results provide the clinician and surgeon valuable information regarding the degree of PCL laxity both preoperatively and postoperatively.

Rehabilitation Following PCL/Posterolateral Corner Reconstruction

The theoretical principles behind the protocol for PCL reconstruction serve as the foundation for the programs implemented following multiple ligamentous reconstructive procedures. The principal difference is that, with multiple ligament involvement, protection of the grafts becomes even more critical.

PCL/PLC Postoperative Rehabilitation

Weeks 1 to 6

For the first 6 weeks following PCL/PLC reconstruction, patients are placed in the long-leg hinged brace locked in extension (Table 18.3). Strict non–weight bearing is enforced during this time period. Motion is restricted primarily for protection of the posterolateral corner. A split biceps tendon tenodesis is commonly employed to reconstruct the PLC, anchored by a screw and washer. Repetitive motion could potentially result in splintering or fragmentation secondary to microtrauma, and compromise the integrity of the repair.

Table 18.3. PCL/PLC postoperative rehabilitation

Phase I—0 to 6 weeks
  Goals
    Maximum protection of grafts
    Maintain quadriceps tone/strength
    Maintain patella mobility
    Maintain full passive extension
    Control pain and swelling
  Program
    Non–weight-bearing ambulation with crutches
    Brace locked in extension—24 hours/day
    Cryotherapy
    Quad sets—enhance with electrical stimulation and/or biofeedback, progress to SLR
    Patella mobilization
    Ankle pumps—ROM
    Stretching exercises—gastroc-soleus and gentle hamstrings
    Hip abduction

Phase II—6 to 12 weeks
  Goals
    Increase flexion ROM
    Initiate weight bearing for articular cartilage nourishment
    Increase quadriceps strength
    Improve proprioception
    Avoid isolated active hamstring contraction
  Program
    Open brace to full flexion—continue to wear at night to control varus stress
    Initiate PWB gait of 25% BW and increase by 25% over next 4 weeks
    Full weight bearing by end of postop week 10
    Prone hangs
    Open chain knee extension within 0- to 60-degree range—no resistance
    Proprioception and weight shift exercises—KAT or BAPS board
    Stationary bike to assist ROM—passive flexion exercises
    Initiate closed-chain exercises once FWB and quad strength is 3+/5 or >
    Hip strengthening—no adduction
    Discontinue brace at end of postop week 12

Phase III—4 to 6 months
  Goals
    Increase flexion ROM
    Maintain full passive extension
    Increase quadriceps and hamstring strength
    Improve proprioception
    Improve functional skills
    Improve cardiovascular endurance
  Program
    4 Months
      Progressive flexion ROM—assistive exercises without active hamstring contraction
      Single-leg proprioception exercises (KAT, BAPS, mini-trampoline)
      Open chain, resistive SAQs—high-speed isokinetics or light-weight isotonics
      Resisted closed-chain exercises—avoid flexion beyond 70 degrees
      Begin active hamstring exercises—no resistance
      Closed-chain conditioning exercises—stair climber, ski machine, rapid walking, etc.
      Aggressive ROM—consider manipulation if ROM is <90 degrees by end of month 4
      Hip PREs
      Begin straight-line jogging at end of postop month 4

(continued)

Table 18.3. Continued

- 5 Months
  - Initiate resisted hamstring exercises
  - Initiate low-intensity plyometric exercises
  - Progressive closed-chain strengthening and conditioning exercises
  - Progressive jogging and begin sprints
  - Advance proprioception training
  - Fit for PCL functional brace
- 6 Months
  - Progression of all strengthening exercises and plyometrics
  - Begin agility drills—carioca, figure 8's, zigzag, slalom running, cutting drills, etc.
  - Isokinetic testing at end of postop month 6
- Phase IV—7 to 12 months
  - Program
    - Assess functional strength—single-leg hop for distance, timed hop test, shuttle run, etc.
    - Return to sports if the following criteria are met:
      - Minimal or no pain and swelling
      - Isokinetic *and* functional strength tests equal to 90% or greater of the uninvolved side
      - Successful completion of sports-specific drills
      - PCL functional brace

FWB, full weight bearing.

Table 18.3a. Rehabilitation summary for PCL/PLC reconstruction

	Postoperative weeks					Postoperative months			
	1–2	3–4	5–6	7–8	9–12	4	5	6	7–12
Long-leg brace @ 0 degrees	X	X	X						
Long-leg brace 0–135 degrees				X	X				
Functional brace								X	X
ROM goals (degrees)									
0–70									
0–90					X				
0–110						X			
0–120							X		
0–130								X	
Procedures									
E-stim	X	X	X						
Patella mob	X	X	X						
Exercises									
Stretching									
Hamstring—gastrocnemius	X	X	X						
Prong hangs				X	X	X	X		
Strengtening									
Quad sets—SLR	X	X	X						
Knee extension				X	X	X			
Short arc—CKC				X	X	X	X	X	X
Resisted CKC					X	X	X	X	X
Hamstring curls						X	X	X	X
Resisted hamstring curls							X	X	X
Proprioception									
KAT or BAPS				X	X	X	X	X	X
Plyometrics							X	X	X
Conditioning/aerobic									
UBE		X	X						
Stationary bike				X	X	X	X	X	X
Walking						X	X	X	X
Stair climber						X	X	X	X
Ski machine						X	X	X	X
Jogging							X	X	X
Sport specific									
Cutting								X	X
Return to sports									X

Although there has been some support of early ROM following multiple ligament reconstruction,[64,69] many of these subjects required intervention for motion loss. Others have shown that a more conservative approach prevents undue forces on healing grafts and that an extremely low number of patients required manipulation for loss of motion.[52] Nonetheless, the effects of immobilization are a concern; however, the period of complete immobilization is relatively short (6 weeks) and we have not found any detrimental effects on the articular cartilage with this approach. Arthrofibrosis is also a potential complication of immobilization, although the number of patients who have required manipulation or other intervention due to loss of motion is less than 3%.

During the initial 6 weeks, quadriceps strengthening is encouraged and is accomplished through isometrics and electrical stimulation. In addition, aggressive patella mobilization is instituted to minimize restrictions and any motion loss that could be associated with limited patella mobility. Patients may experience hamstring tightness and posterior knee discomfort when the knee is maintained in extension and no motion is permitted for the entire 6-week period. Therefore, hamstring and gastroc-soleus stretching is encouraged, although vigorous stretching is discouraged to minimize posterior shear forces created by the hamstrings. All patients are advised to avoid resting positions that place varus forces on the knee.

Weeks 7 to 12

At the end of postoperative week 6, ROM is permitted and the patient begins weight bearing of 25% body weight, unless the natural extremity alignment tends toward varus, in which case, non–weight bearing may be extended for an additional 3 to 6 weeks. Assuming adequate alignment, weight bearing continues to increase by 25% body weight in a progressive fashion until the patient is full weight bearing at the end of postoperative week 10. Within this same time frame, passive ROM is also increased, guardedly, since flexion greater than 90 degrees places high forces on PCL grafts.[64] Conversely, full passive extension is encouraged immediately. Studies have shown that full extension places the least amount of force on the PCL.[61]

With the brace opened, active, open-chain quadriceps exercises are initiated in a range of 60 to 0 degrees. Closed-chain exercises are performed in a range of 0 to 60 degrees once the patient has achieved full weight bearing. These exercises are done in the long-leg brace until the brace is discontinued at the end of the 12th postoperative week.

Months 4 to 12

After 12 weeks, the postoperative rehabilitation regimen mirrors that of the PCL. The only variance occurs when the patient's natural alignment has a varus bias. In this event, the patient is immediately placed in a functional brace once the long-leg hinged brace is discontinued. As patients progress through the postoperative regimen, they are closely monitored for varus, posterior, and posterolateral laxity. Several methods have been described for testing of the posterolateral complex and posterolateral instability,[70,71] and many of these have been discussed in previous chapters. At our center, for postoperative assessment, the prone external rotation test is utilized for its ability to assess both the PCL and the posterolateral complex. This test is performed at both 30 and 90 degrees of knee flexion. The feet are then forcefully externally rotated and the degree of external rotation of the foot relative to the axis of the femur is assessed and compared to the contralateral side.[70] If external rotation is increased at 30 degrees but not at 90 degrees, then the posterolateral complex has been compromised. If external rotation is increased at both 30 and 90 degrees, then both the PCL and the posterolateral complex are involved. Posterior drawer testing is also performed routinely to assess PCL integrity.

Progression to jogging, sport-specific activities, and return to sports are based on the achievement of specific strength and ROM criteria outlined in the PCL protocol. Return to unrestricted activity usually occurs some time after the end of the 6th postoperative month, and the patient is encouraged to utilize a functional PCL brace for the next 6 months. Patients are encouraged to return at the 1-year postoperative month for reassessment, and on a yearly basis after that to determine long-term, functional outcomes. At the yearly follow-up, patients are evaluated by stress radiography, KT-1000, and through ligament rating sheets, which include subjective and objective assessments.

Rehabilitation Following ACL/PCL or ACL/PCL/Posterolateral Complex Reconstruction

The postoperative programs following ACL/PCL or ACL/PCL/posterolateral complex reconstruction are identical and deviate very little from the postoperative regimen for PCL/posterolateral complex (Table 18.4). One difference is that, because of involvement of the ACL, open-chain knee extension is not permitted once the brace is unlocked from full extension. Therefore, closed-chain exercises are instituted immediately once the patient has attained full weight-bearing status. If there is difficulty tolerating the closed-chain exercises secondary to quadriceps deficiency, then high-intensity electrical stimulation may be implemented. Since a strong quadriceps contraction is the goal, the knee should be maintained in a position of 60 degrees to minimize anterior tibial translation. This position can be maintained on an isokinetic device by locking the dynamometer at 60 degrees or in a knee extension machine and using the weights and the lever arm to block motion and maintain the 60-degree position. Open-chain knee extension and flexion are initiated at the end of postoperative month 4, without resistance. Resistive exercises begin at the end of postoperative month 5. A combined instability functional brace is utilized for return to sports for the first year.

Conclusion

The knowledge base of the PCL, including injuries, biomechanics, and reconstructive procedures, continues to evolve. This chapter has provided the scientific basis behind the rehabilitation programs utilized for the conservative and postoperative management of PCL and associated ligamentous injuries. Although a period of immobilization is a component of all the rehabilitation protocols, there have been no deleterious effects on the articular cartilage or restoration of motion. It does provide maximum protection of the grafts, which ultimately aids in the restoration of physiologic alignment at the tibiofemoral joint. Utilizing these protocols has allowed the majority of patients treated at out facility to return to their desired level of function.

Table 18.4. ACL/PCL–ACL/PCL/posterolateral complex postoperative rehabilitation

Phase I—0 to 6 weeks
  Goals
    Maximum protection of grafts
    Maintain patella mobility
    Maintain quadriceps tone
    Maintain full passive extension
    Control pain and swelling
  Program
    Non–weight-bearing ambulation with crutches
    Brace locked in extension—24 hours/day
    Cryotherapy
    Quad sets—enhance with low-intensity electrical stimulation or biofeedback
    Patella mobilization
    Ankle pumps—ROM
    Stretching exercises—gastroc-soleus and gentle hamstrings
    Hip abduction

Phase II—6 to 12 weeks
  Goals
    Initiate weight bearing for articular cartilage nourishment
    Increase knee flexion
    Maintain quadriceps tone
    Improve proprioception
    Avoid isolated quadriceps and hamstring contraction
  Program
    Begin PWB gait of 25% BW and increase by 25% over next 4 weeks
    Open brace to full flexion—with posterolateral complex, continue to wear at night
    Prone hangs
    Passive flexion exercises—consider CPM if no involvement of posterolateral corner
    Patella mobilization
    High-intensity E-stim at 60 degrees of knee flexion
    Initiate closed-chain strengthening once FWB and quad strength is 3+/5 or greater
    Stationary bike for ROM assist
    Proprioception and weight shift (KAT or BAPS board)
    Hip strengthening—no adduction if posterolateral corner is involved
    Discontinue brace at end of postop week 12

Phase III—4 to 6 months
  Goals
    Increase knee flexion
    Maintain full passive extension
    Improve quadriceps and hamstring strength
    Improve proprioception
    Improve functional skills
    Increase cardiovascular endurance
  Program
    4 Months
      Closed-chain PREs—avoid flexion beyond 70 degrees
      Isolated quadriceps and hamstring exercises—no resistance
      Single-leg proprioception exercises (KAT, BAPS, mini-trampoline)
      Closed-chain conditioning exercises—Stair Climber, skiing machine, rower, etc.
      Aggressive flexion ROM—consider manipulation if ROM is <90 degrees by end of month 4
      Hip PREs
      Straight-line jogging at end of postop month 4
    5 Months
      Initiate resisted quadriceps and hamstring exercises
      Progressive closed-chain strengthening and conditioning exercises
      Initiate low-intensity plyometrics
      Progressive jogging and begin sprints
      Advance proprioception training
      Fit for ACL/PCL functional brace
    6 Months
      Progression of all strengthening exercises and plyometrics
      Begin agility drills—carioca, figure 8's, zigzag, slalom running, etc., in brace
      Sport-specific drills
      Isokinetic testing at end of postop month 6

Phase IV—7 to 12 months
  Program
    Assess functional strength—single-leg hop for distance, timed hop test, shuttle run, etc.
    Return to sports if the following criteria are met:
      Minimal or no pain or swelling
      Isokinetic *and* functional tests within 10% to 15% of the uninvolved side
      Successful completion of sport-specific drills
      ACL/PCL functional brace

CPM, continuous passive motion.

Table 18.4a. Rehabilitation summary for ACL/PCL and ACL/PCL/posterolateral complex reconstruction

	Postoperative weeks					Postoperative months			
	1–2	3–4	5–6	7–8	9–12	4	5	6	7–12
Long-leg brace @ 0 degrees	X	X	X						
Long-leg brace 0–135 degrees				X	X				
Functional brace								X	X
ROM goals (degrees)									
0–70									
0–90					X				
0–110						X			
0–120							X		
0–130								X	
Procedures									
E-stim	X	X	X						
Patella mob	X	X	X						
Exercises									
Stretching									
Hamstring—gastrocnemius	X	X	X						
Prone hangs				X	X	X	X		
Strengthening									
Quad sets—SLR	X	X	X						
Knee extension						X	X	X	X
Short arc—CKC				X	X	X	X	X	X
Resisted CKC						X	X	X	X
Hamstring curls						X	X	X	X
Resisted hamstring curls							X	X	X
Proprioception									
KAT or BAPS				X	X	X	X	X	X
Plyometrics							X	X	X
Conditioning/aerobic									
UBE		X	X						
Stationary bike				X	X	X	X	X	X
Walking						X	X	X	X
Stair climber						X	X	X	X
Ski machine						X	X	X	X
Jogging							X	X	X
Sport specific									
Cutting								X	X
Return to sports									X

References

1. Bach BR, Tradonsky S, Bojchuk J, Levy ME, Bush-Joseph CA, Khan NH. Arthroscopically assisted anterior cruciate ligament reconstruction using patellar tendon autograft. Five- to nine-year follow-up evaluation. Am J Sports Med 1998;26:20–29.
2. Barber SD, Noyes FR, Mangine R, DeMaio M. Rehabilitation after ACL reconstruction: function testing. Orthopedics 1992;15:969–974.
3. Beynnon BD, Johnson RJ. Anterior cruciate ligament injury rehabilitation in athletes. Biomechanical considerations. Sports Med 1996;22:54–64.
4. Bynum EB, Barrack RL, Alexander AH. Open versus closed chain kinetic exercises after anterior cruciate ligament reconstruction. A prospective randomized study. Am J Sports Med 1995;23:401–406.
5. Dandy DJ. Historical overview of operations for anterior cruciate ligament rupture. Knee Surg Sports Trauma Arthrosc 1996;3:256–261.
6. Frndak PA, Berasi CC. Rehabilitation concerns following anterior cruciate ligament reconstruction. Sports Med 1991;12:338–346.
7. Harner CD, Olson E, Irrgang JJ, Silverstein S, Fu FH, Silbey M. Allograft versus autograft anterior cruciate ligament reconstruction: 3- to 5-year outcome. Clin Orthop 1996;324:134–144.

8. Howell SM, Taylor MA. Brace-free rehabilitation with early return for knees reconstructed with double-looped semitendinosus and gracilis graft. J Bone Joint Surg 1996;78A:814–825.
9. Irrgang JJ. Modern trends in anterior cruciate ligament rehabilitation: nonoperative and postoperative management. Clin Sports Med 1993;12:797–813.
10. Lysholm M, Messner K. Sagittal plane translation of the tibia in anterior cruciate ligament-deficient knees during commonly used rehabilitation exercises. Scand J Med Sci Sports 1995;5:49–56.
11. Noyes FR, Barber-Westin SD. Reconstruction of the anterior cruciate ligament with human allograft. Comparison of early and later results. J Bone Joint Surg 1996;78A:524–537.
12. Parker MG. Biomechanical and histological concepts in the rehabilitation of patients with anterior cruciate ligament reconstruction. J Orthop Sports Phys Ther 1994;20:44–50.
13. Sailors ME, Keskula DR, Perrin DH. Effect of running on anterior knee laxity in collegiate-level female athletes after anterior cruciate ligament reconstruction. J Orthop Sports Phys Ther 1995;21:233–239.
14. Shelbourne KD, Klootwyk TE, Wilckens JH, DeCarlo MS. Ligament stability two to six years after anterior cruciate ligament reconstruction with autogenous patellar tendon graft and participation in an accelerated rehabilitation program. Am J Sports Med 1995;23:575–579.
15. Shelbourne KD, Wilckens JH. Current concepts in anterior cruciate ligament rehabilitation. Orthop Rev 1990;19:957–964.
16. Veltri DM. Arthroscopic anterior cruciate ligament reconstruction. Clin Sports Med 1997;16:123–144.
17. Yack HJ, Collins CE, Whieldon TJ. Comparison of closed and open kinetic chain exercise in the anterior cruciate ligament-deficient knee. Am J Sports Med 1993;21:49–54.
18. Cross MJ, Powell JF. Long-term followup of posterior cruciate ligament rupture: a study of 116 cases. Am J Sports Med 1984;12:292–297.
19. Dandy DJ, Pusey RJ. The long-term results of unrepaired tears of the posterior cruciate ligament. J Bone Joint Surg 1982;64:92–94.
20. Parolie JM, Bergfield JA. Long-term results of nonoperative treatment of isolated posterior cruciate ligament injuries in the athlete. Am J Sports Med 1986;14:35–38.
21. Shino K, Horibe S, Nakata K, Maeda A, Hamada M, Nakamura N. Conservative treatment of isolated injuries to the posterior cruciate ligament in athletes. J Bone Joint Surg 1995;77B:895–900.
22. Clancy WG, Shelbourne KD, Zoellner GB, Keene JS, Reider B, Rosenberg TD. Treatment of knee joint instability secondary to rupture of the posterior cruciate ligament. J Bone Joint Surg 1983;65A:310–322.
23. Geissler WB, Whipple TL. Intraarticular abnormalities in association with posterior cruciate ligament injuries. Am J Sports Med 1993;21:846–849.
24. Keller PM, Shelbourne KD, McCarroll JR, Rettig AC. Nonoperatively treated isolated posterior cruciate ligament injuries. Am J Sports Med 1993;21:132–136.
25. Torg JS, Barton TM, Pavlov H, Stine R. Natural history of the posterior cruciate ligament-deficient knee. Clin Orthop 1989;246:208–215.
26. Fowler PJ, Messieh SS. Isolated posterior cruciate ligament injuries in athletes. Am J Sports Med 1987;15:553–557.
27. Boynton MB, Tietjens BR. Long-term follow-up of the untreated isolated posterior cruciate ligament-deficient knee. Am J Sports Med 1996;24:306–310.
28. Lutz GE, Stuart MJ, Sim FH, Scott SG. Rehabilitative techniques for athletes after reconstruction of the anterior cruciate ligament. Mayo Clin Proc 1990;65:1322–1329.
29. Rennison M. Open versus closed chain kinetic exercises after anterior cruciate ligament reconstruction. A prospective randomized study. Am J Sports Med 1996;24:125.
30. Steindler A. Kinesiology of the Human Body Under Normal and Pathological Conditions. Springfield, IL: Charles C Thomas, 1970.
31. Irrgang JJ. Rehabilitation for nonoperative and operative management of knee injuries. In: Fu FA, Harner CD, Vince KG, eds. Knee Surgery. Baltimore: Williams & Wilkins; 1994:485–502.

32. Lutz GE, Palmitier RA, An KN, Chao YS. Comparison of tibiofemoral joint forces during open-kinetic-chain and closed-kinetic-chain exercises. J Bone Joint Surg 1993;75A:732–739.
33. Snyder-Mackler L. Scientific rationale and physiological basis for the use of closed kinetic chain exercises in the lower extremity. J Sports Rehabil 1996;5:2–12.
34. Arms SW, Pope MH, Johnson RJ, Fischer RA, Arvidsson I, Eriksson E. The biomechanics of anterior cruciate ligament rehabilitation and reconstruction. Am J Sports Med 1984;12:8–18.
35. Grood ES, Suntay WJ, Noyes FR, Butler DL. Biomechanics of the knee-extension exercise: effect of cutting the anterior cruciate ligament. J Bone Joint Surg 1984;66A:725–734.
36. Renstrom P, Arms SW, Stanwyck TS, Johnson RJ, Pope MH. Strain within the anterior cruciate ligament during hamstring and quadriceps activity. Am J Sports Med 1986;14:83–87.
37. Daniel DM, Stone ML, Barnett P, Sachs R. Use of the quadriceps active test to diagnose posterior cruciate-ligament disruption and measure posterior laxity of the knee. J Bone Joint Surg 1988;70A:387–391.
38. Davies GJ. A comparison of isokinetics. In: Clinical Usage of Isokinetics, 3rd ed. La Crosse, WI: S&S, 1987:4–28.
39. Kaufman KR, An KN, Litchy WJ, Morrey BF, Chao EYS. Dynamic joint forces during knee isokinetic exercise. Am J Sports Med 1991;19:305–316.
40. Hungerford DS, Barry M. Biomechanics of the patellofemoral joint. Clin Orthop 1979;144:9–15.
41. Ohkoshi Y, Yasuda K, Kaneda K, Wada T, Yamanaka M. Biomechanical analysis of rehabilitation in the standing position. Am J Sports Med 1991;19:605–611.
42. Dahlkuits NJ, Mago P, Seedholm BB. Forces during squatting and rising from a deep squat. Eng Med 1982;11:69–76.
43. Wilk KE. Rehabilitation of isolated and combined posterior cruciate ligament injuries. Clin Sports Med 1994;13:649–677.
44. Ericson MO, Nisell R. Tibiofemoral joint forces during ergometer cycling. Am J Sports Med 1986;14:285–290.
45. Morrison JB. The biomechanics of the knee joint in relation to normal walking. J Biomech 1970;3:51–60.
46. Morrison JB. Function of the knee joint in various activities. Biomech Eng 1969;4:573–580.
47. Smidt GL. Biomechanical analysis of knee extension and flexion. J Biomech 1973;6:79–92.
48. Fanelli GC, Feldmann DD. The use of allograft tissue in knee ligament reconstruction. In: Parisien JS, ed. Current Techniques in Arthroscopy, 3rd ed. New York: Thieme; 1998:47–55.
49. Krishna KV, Sagar JV. In vitro biomechanical evaluation of anterior cruciate ligament graft substitutions. Indian J Med Res 1994;100:295–298.
50. Steiner ME, Hecker AT, Brown CH, Hayes WC. Anterior cruciate ligament graft fixation: comparison of hamstring and patellar tendon grafts. Am J Sports Med 1994;22:240–246.
51. Graf BK, Rothenberg M, Vanderby R. Anterior cruciate ligament reconstruction with patellar tendon: an ex vivo study of wear-related damage and failure at the femoral tunnel. Am J Sports Med 1994;22:131–135.
52. Fanelli GC, Gianotti BF, Edson CJ. Arthroscopically assisted combined posterior cruciate ligament/posterior lateral complex reconstruction. Arthroscopy 1996;12:521–530.
53. Fanelli GC, Gianotti BF, Edson CJ. Arthroscopically assisted combined anterior and posterior cruciate ligament reconstruction. Arthroscopy 1996;12:5–14.
54. Belli A, Bosco C. Influence of stretch-shortening cycle on mechanical behavior of triceps surae during hopping. Acta Physiol Scand 1992;144:401–408.
55. Bosco C, Komi PV. Potentiation of the mechanical behavior of human skeletal muscle through prestretching. Acta Physiol Scand 1979;106:467–472.
56. Wilk KE, Voight MA, Keirns MA, Gambetta V, Andrews JR, Dillman CJ. Stretch-shortening drills for the upper extremities: theory and clinical application. J Orthop Sports Phys Ther 1993;17:225–239.
57. Gordon J, Ghez C. Muscle receptors and spinal reflexes: the stretch reflex. In:

Kandel ER, Schwartz JH, Jessel TM, eds. Principles of Neural Science, 3rd ed. New York: Elsevier, 1991:564–580.
58. Fanelli GC. PCL injuries in trauma patients. Arthroscopy 1993;9:291–294.
59. Fanelli GC, Edson CJ. PCL injuries in trauma patients. Part II. Arthroscopy 1995;11:526–529.
60. Skyhar MJ, Warren RF, Ortiz GJ, Schwartz E, Otis JC. The effects of sectioning of the posterior cruciate ligament and the posterolateral complex on the articular contact pressure within the knee. J Bone Joint Surg 1993;75:694–699.
61. Ogata K, McCarthy JA. Measurements of length and tension patterns during reconstruction of the posterior cruciate ligament. Am J Sports Med 1992;20:351–355.
62. Noyes FR. Rehabilitation protocols following PCL reconstruction. Paper presented at the Cincinnati Sportsmedicine 1997 Advances on the Knee and Shoulder, Hilton Head Island, SC, May 1997.
63. Snyder-Mackler L, Ladin Z, Schepsis AA, Young JC. Electrical stimulation of the thigh muscles after reconstruction of the anterior cruciate ligament: effects of electrically elicited contraction of the quadriceps femoris and hamstring muscles on gait and on strength of the thigh muscles. J Bone Joint Surg 1991;73:1025–1036.
64. Noyes FR, Barber-Westin SD. Reconstruction of the anterior and posterior cruciate ligaments after knee dislocation: use of early protected postoperative motion to decrease arthrofibrosis. Am J Sports Med 1997;25:769–778.
65. Kaufman KR, An K-N, Litchy WJ, Morrey BF, Chao EYS. Dynamic joint forces during knee isokinetic exercise. Am J Sports Med 1991;19:305–315.
66. Huber FE, Irrgang JJ, Harner C, Lephart S. Intratester and intertester reliability of the KT-1000 arthrometer in the assessment of posterior laxity of the knee. Am J Sports Med 1997;25:479–485.
67. Hewett TE, Noyes FR, Lee MD. Diagnosis of complete and partial posterior cruciate ligament ruptures: stress radiography compared with KT-1000 arthrometer and posterior drawer testing. Am J Sports Med 1997;25:648–655.
68. Daniel DM, Malcolm LL, Losse G, Stone ML, Sachs R, Burks R. Instrumented measurement of anterior laxity of the knee. J Bone Joint Surg 1985;67:720–726.
69. Shapiro MS, Freedman EL. Allograft reconstruction of the anterior and posterior cruciate ligaments after traumatic knee dislocation. Am J Sports Med 1995;23:580–587.
70. Veltri DM, Warren RF. Anatomy, biomechanics, and physical findings in posterolateral knee instability. Clin Sports Med 1994;3:599–614.
71. Hughston J, Norwood L. The posterolateral drawer test and external recurvatum test for posterolateral rotatory instability of the knee. Clinical Orthopedic and Related Research 1980;82:147.

Chapter Nineteen

Complications in Posterior Cruciate Ligament Surgery

Gregory C. Fanelli and Timothy J. Monahan

Anterior cruciate ligament (ACL) surgery is a frequently performed orthopedic procedure in the United States. Posterior cruciate ligament (PCL) injuries occur less frequently in this country, and the experience of the orthopedic surgeon is correspondingly less for PCL examination, diagnosis, and surgical reconstructive procedures. Studies indicate that acute PCL injuries are related to geographic region, frequency of blunt trauma, and the population density of orthopedic surgeons.[1,2] Fanelli[1,2] reports a 38% incidence of PCL tears in acute knee injuries at his tertiary care regional referral center. The frequency of PCL related injuries in this study is 38.3% of acute knee injuries in a study population of 222 knees; 48 (56.5%) were multiple trauma related, while 28 (32.9%) were sports related injuries. There were only three isolated PCL injuries (3.5%), and 82 (96.5%) PCL combined with one or more injured ligaments. The most frequently encountered PCL/combined knee ligament injuries were PCL/posterolateral corner (35/85, 41.2%), and combined ACL/PCL tears (39/85, 45.9%).

It is estimated that relatively few orthopedic surgeons perform PCL surgery when compared to ACL surgery, and complications may result from lack of experience in diagnosis, surgical techniques, and postoperative care.

This chapter discusses the nature and type of each complication related to PCL surgery, how the particular complication could have been prevented, and how to treat, manage, or salvage the problem. Our goal is to provide practical solutions to problems that may arise in conjunction with PCL surgery. Table 19.1 summarizes the complications associated with PCL surgery. Each complication is discussed in detail in the following sections.

Failure to Recognize Associated Ligament Injuries

Failure to recognize associated ligament injuries is possibly the most common reason for poor results following PCL surgery. Studies have shown that the isolated PCL tear is an uncommon injury, and combined PCL/posterolateral and combined ACL/PCL instabilities are the most common injury patterns involving the PCL.[1,2] Failure to recognize posterolateral instability that is combined with PCL instability is a common error, and is the most likely reason for poor results following PCL reconstruction. PCL reconstruction without addressing or achieving good static posterolateral stability will yield poor results. Failure to correct the posterolateral laxity results in excessive external rotation at both 30 and 90 degrees of knee

Table 19.1. Complications associated with PCL surgery

Failure to recognize associated ligament injuries
Neurovascular complications
Persistent posterior sag
Osteonecrosis
Knee motion loss
Anterior knee pain
Fractures

flexion. This causes the posterior tibial insertion of the PCL to rotate medially and anteriorly, shortening the distance to its insertion on the medial femoral condyle and producing a functionally lax PCL.[3]

The best way to determine if there are associated ligament injuries in addition to the PCL injury is with careful history and physical examination, magnetic resonance imaging (MRI) studies in selected cases, and examination under anesthesia and diagnostic arthroscopy.[4] Physical examination findings suggestive of multiple ligament injuries include opening on varus/valgus stress on full extension, grade III posterior drawer test, negative tibial step-off test, positive dial test, increased external rotation thigh foot angle test of 10 degrees greater than the normal side, and a positive posterolateral drawer test.[4] If the integrity of other ligaments or capsular structures is in doubt based on physical examination, appropriate diagnostic studies should be performed. Examination under anesthesia and diagnostic arthroscopy are our preferred methods for accurate evaluation; however, MRI may be useful if the above are impractical. The most predictable way to salvage knees with multiple ligament injuries is with the correct preoperative diagnosis, and appropriate treatment of all injured ligaments.

Neurovascular Complications

The neurovascular structures most at risk during PCL reconstruction are the popliteal artery and the tibial nerve. The neurovascular bundle is located posterior to the posterior horn of the lateral meniscus, with the posterior capsule of the knee joint interposed between the two structures[5] (Fig. 19.1).

Reconstruction of the PCL using the transtibial tunnel technique involves passing a guide wire from anterior to posterior through the proximal tibia. The guide wire emerges at the inferior lateral aspect of the PCL fossa-tibial ridge area. The tunnel is subsequently created by reaming over the guide wire with a cannulated reamer. Arterial and/or nerve injury can occur with overpenetration of the guide wire during placement, or with inadvertent guide wire advancement during tibial tunnel reaming. It is also possible to cause neurovascular injury with the drill bit during reaming, ei-

Fig. 19.1. Axial section at the level of the posterior cruciate ligament (PCL) insertion demonstrating the relationship of the neurovascular structures and their proximity to the exit point of the drill bit. (From Jackson et al.,[6] with permission.)

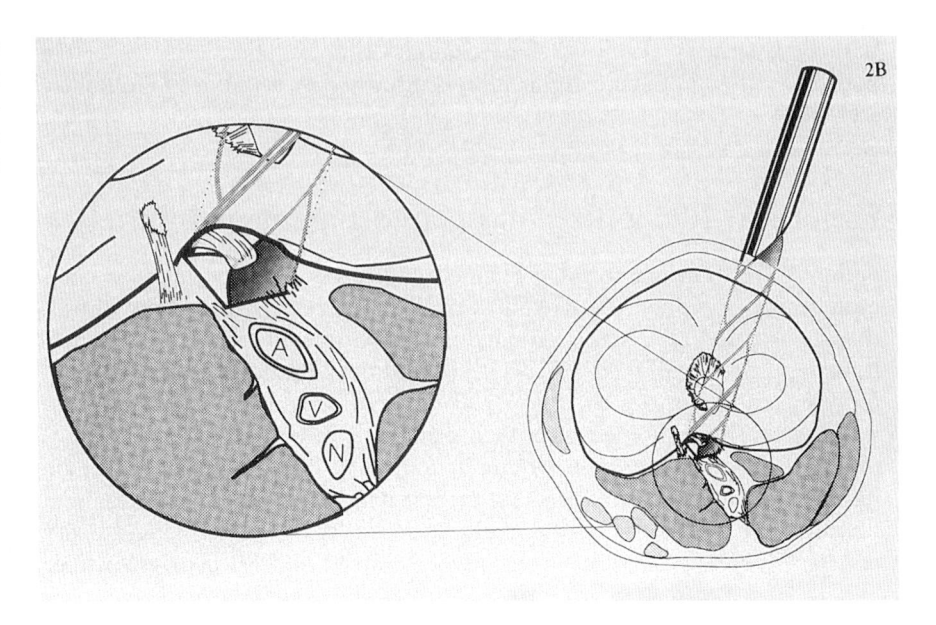

19. Complications in Posterior Cruciate Ligament Surgery

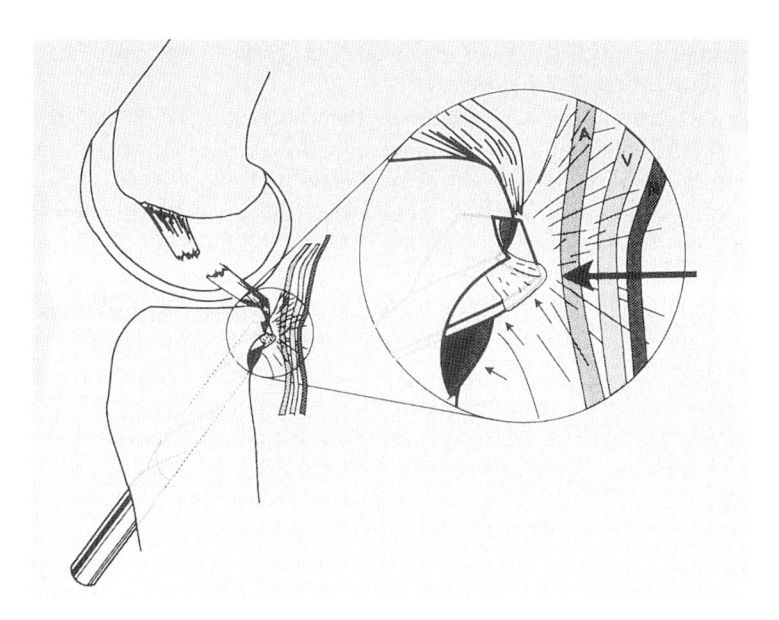

Fig. 19.2. Sagittal section demonstrating the relationship of the neurovascular structures to the drill bit that has prematurely exited the posterior cortex. (From Jackson et al.,[6] with permission.)

ther by direct contact of the drill bit or by "winding up" of the neurovascular structures (Fig. 19.2).

There are several methods to make transtibial tunnel creation safer. The use of a spade-tip guide wire instead of a trochar-tip guide wire, the use of an oscillating drill, and the use of a tapered instead of a square drill bit will reduce the likelihood of penetration, or winding up the neurovascular structures.[6]

Intraoperative radiographs and image intensification/fluoroscopy help to confirm guide wire placement, and monitor guide wire and reamer position during drilling. The use of intraoperative C-arm image intensifier is cumbersome, and care must be taken not to compromise the sterile field.

The best method of neurovascular injury prevention during transtibial drilling is the use of a posteromedial safety incision[4,7,8] (Fig. 19.3). The posteromedial safety incision is a 2-cm incision made just inferior to the posterior medial joint line. The crural fascia is carefully incised, and the interval is developed between the posterior capsule of the tibiofemoral joint anteriorly and the medial head of the gastrocnemius muscle and the neurovascular structures posteriorly. The surgeon can then place his finger in this extracapsular position to monitor the position of tools, guide wires,

Fig. 19.3. (a) The posteromedial safety incision is made on the posteromedial aspect of the proximal tibia area just inferior to the medial joint line. (b) This allows the surgeon's gloved finger to assume an extraarticular position posterior to the capsule of the tibiofemoral joint, and anterior to the neurovascular structures. This facilitates guide wire placement and correct tibial tunnel positioning, and protects the neurovascular structures.

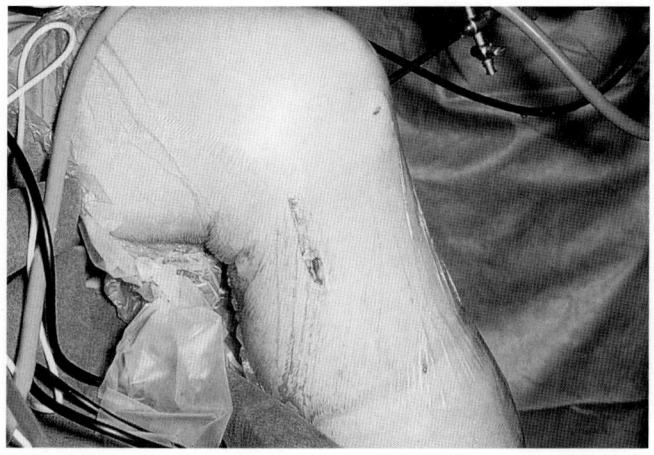

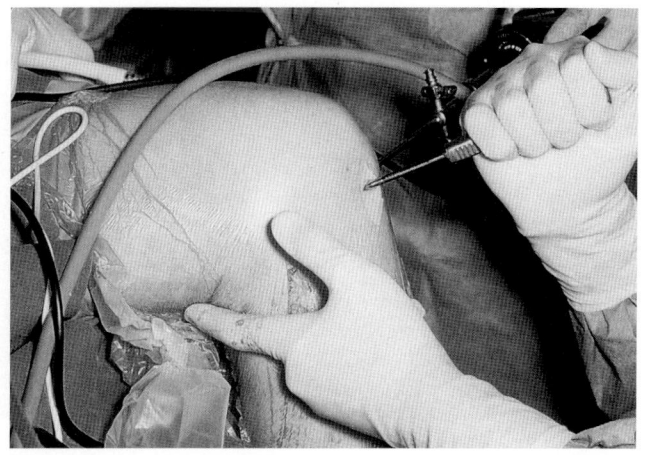

a   b

reamers, and drill bits in the posterior aspect of the knee, thus protecting the neurovascular structures.

Two additional advantages of the posteromedial safety incision are to provide better guide wire placement in the medial-lateral and superior-inferior planes, and to allow an escape pathway for extravasating arthroscopic irrigation fluid if a capsular tear occurs. This posteromedial safety incision will reduce or eliminate the likelihood of a posterior calf compartment syndrome due to arthroscopic irrigation fluid extravasation. We have reported our results in 21 arthroscopically assisted combined PCL/posterolateral complex reconstructions, and 20 arthroscopically assisted combined ACL/PCL reconstructions,[7,8] and have performed a total of 107 PCL reconstructions using the transtibial tunnel technique. Spade-tip guide wires, tapered reamers, and the posteromedial safety incision were used in each case, and no neurovascular injuries or compartment syndromes occurred.

Particular caution should be exercised in the case of previously PCL-operated knees with posterior scarring due to the prior surgical procedures. In these cases it is possible to have aberrant positions of the neurovascular structures. Inadvertent damage may occur during routine PCL surgery due to the transposed neurovascular bundle (Fig. 19.4). Arthroscopic application of vascular clips and immediate vascular repair are required.

The only salvage procedure of an intraoperative vascular injury is recognition of the injury and arterial repair by a vascular surgeon. The orthopedic surgeon should always have a safety-first attitude about PCL surgery.

Fig. 19.4. (a) Posterior view of the capsule demonstrating a septum-like structure, which is actually the popliteal artery transposed after prior PCL surgery. (b) Lacerated popliteal artery from arthroscopic shaver. (c) Arthroscopically applied vascular clamps applied to the popliteal artery before arterial repair. (Photographs courtesy of M.M. Malek, M.D.)

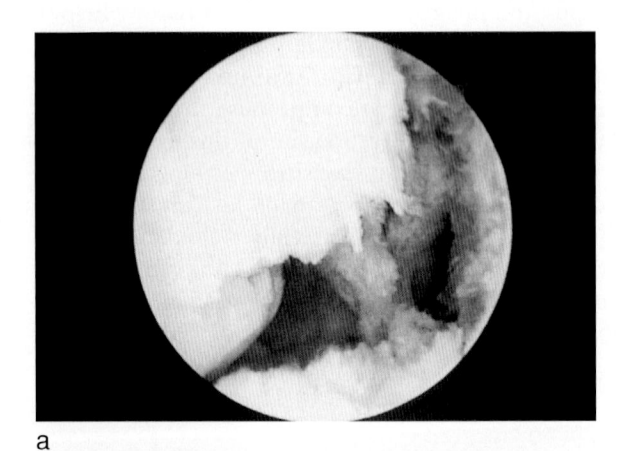

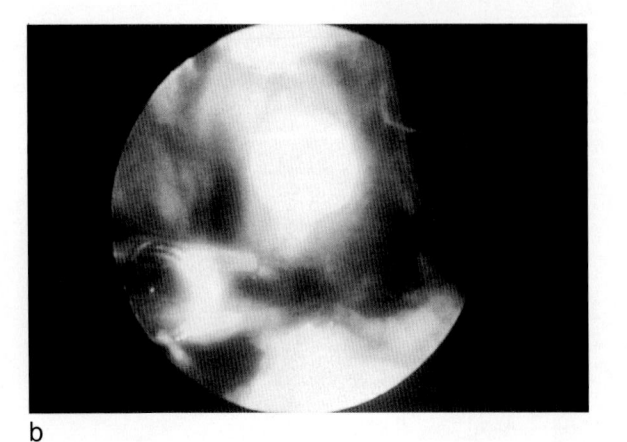

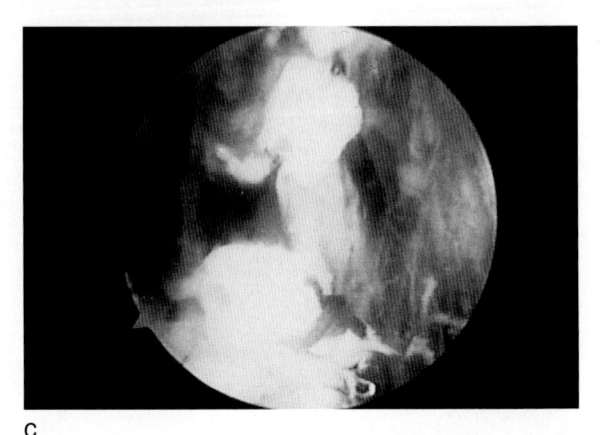

Persistent Posterior Sag

Persistent posterior sag following PCL surgery produces a suboptimal result of PCL reconstruction. Factors contributing to persistent posterior sag include failure to address posterolateral or posteromedial instability, early weight bearing in flexion, a PCL substitute of insufficient mechanical strength, improper graft placement, improper graft tensioning, poor graft fixation, or early initiation of open chain hamstring strengthening during rehabilitation.[9-14]

Failure to recognize associated ligament injuries is possibly the most common reason for poor results following PCL surgery. We have shown that the isolated PCL tear is an uncommon injury, and that combined PCL/posterolateral, and combined ACL/PCL instabilities are the most common injury patterns involving the PCL.[1,2] The best way to determine if there are associated ligament injuries in addition to the PCL injury is with careful history and physical examination, MRI studies in selected cases, and examination under anesthesia and diagnostic arthroscopy.[4] The most predictable way to salvage this situation is with the correct preoperative diagnosis and appropriate treatment of all injured ligaments.

The correct diagnosis of ligament injuries occurring in association with the PCL injury is critical, and plans must be developed to treat these associated injuries. While it is preferable to address these acute knee injuries within a 3- to 6-week time period of the index injury, this may be impossible in severe multisystem-injured patients. We have reported excellent results with delayed surgery provided the knee is protected in the preoperative period.[7-9]

Technical considerations to prevent persistent posterior sag include the use of strong graft tissue, correct tunnel placement, correct graft tensioning, and secure graft fixation. Regardless of the graft material chosen, incorrect tunnel placement will doom even the strongest graft material to failure. The most predictable PCL reconstructive procedures occur with reconstruction of the anterior lateral fiber bundle region of the PCL. The femoral tunnel should be positioned to reproduce the anterolateral band of the PCL (Fig. 19.5). This is distal and anterior to the PCL isometric point, and allows the graft to tighten in flexion.[11]

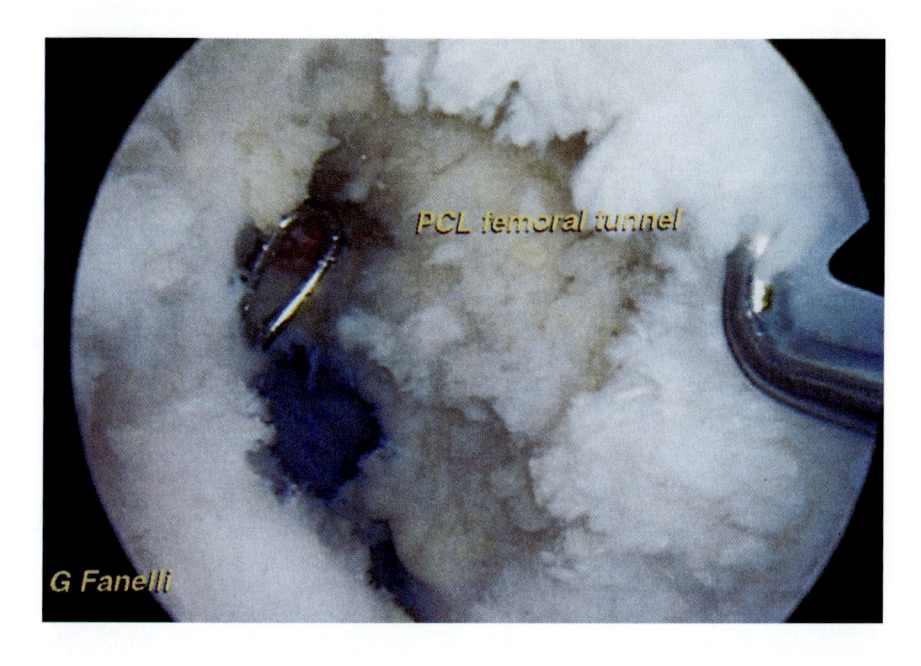

Fig. 19.5. The femoral tunnel in PCL reconstruction is optimally placed in the center of the stump of the anterolateral fiber bundle region of the native PCL. This provides the most successful and reproducible results.

The ideal tibial tunnel position is emerging inferior to the apex of the tibial ridge in the lateral aspect of the PCL anatomic insertion site. This will allow the PCL substitute to turn a series of gentle 45-degree turns around the posterior proximal tibia with no acute 90-degree angle turns that may compromise graft integrity.[7,8]

Given proper tunnel placement, graft tensioning should be performed in 70 to 90 degrees of knee flexion.[11] This will allow for achievement of the tibial step-off without restricting range of motion into flexion. Tensioning of the PCL substitute in full extension will cause excessive graft tension as the knee progresses into flexion with subsequent graft rupture, or range of motion loss in flexion.[11]

Secure graft fixation is critical to the success of PCL reconstruction. Direct fixation of the graft material to bone with interference fixation, screw and spiked ligament washer, or a combination of fixation methods is preferable. Suture fixation of grafts to a post and washer assembly is a very successful method of fixation; however, postoperative care must be modified by the security of the graft fixation.

Inappropriate postoperative rehabilitation can adversely affect PCL reconstruction results, and lead to a persistent posterior sag.[12–14] The PCL is subject to higher forces than the ACL, so a slower postoperative course than with ACL postoperative rehabilitation is warranted. Posterior tibial sag will recur with early weight bearing in flexion, or open chain hamstring activity, which will increase forces in the PCL. Postoperative immobilization should be in full extension for a period of 4 to 6 weeks. This allows for preliminary biologic fixation to occur of the graft material in the tunnels. If early mobilization is utilized, the proximal tibia must be supported.

Osteonecrosis

Avascular necrosis of the medial femoral condyle has been reported by Athanasian et al.[15] as complication of PCL reconstructive surgery. The patients present with medial knee pain and tenderness over the medial femoral condyle. Radiographic findings include flattened articular surface of the medial femoral condyle and radiolucency. The osteonecrosis is thought to occur from increased pressure in the bone causing vascular insufficiency. Etiologic factors include drilling the femoral tunnel too close to the articular surface, which may disturb the single nutrient vessel providing the intraosseus blood supply to the medial femoral condyle,[16] and extensive soft tissue dissection over the medial femoral condyle during surgery, which may compromise the vascular supply to the medial femoral condyle.

This complication is prevented with accurate femoral tunnel placement, leaving an adequate bone bridge between the femoral tunnel and the medial femoral condyle articular surface. Our preferred method for femoral tunnel creation is the outside-in method using an ACL/PCL drill guide[4,7,8,17] (Fig. 19.6). The knee is flexed 90 degrees, and the guide wire is positioned externally with the help of the ACL/PCL drill guide at a point halfway between the medial femoral epicondyle and the articular surface of the medial femoral condyle at the level of the superior one-third of the patella with the knee in 90 degrees of flexion.[17] The guide wire emerges through the center of the femoral insertion of the anterior lateral bundle of the PCL. These guidelines result in accurate and reproducible PCL femoral tunnel placement. Our series had no complications of osteonecrosis.[7,8] When an Achilles tendon allograft is used for PCL reconstruction, we prefer to contour the calcaneal bone plug into the shape of a wedge, and to press fit this portion of the Achilles tendon allograft into the femoral tunnel. This tech-

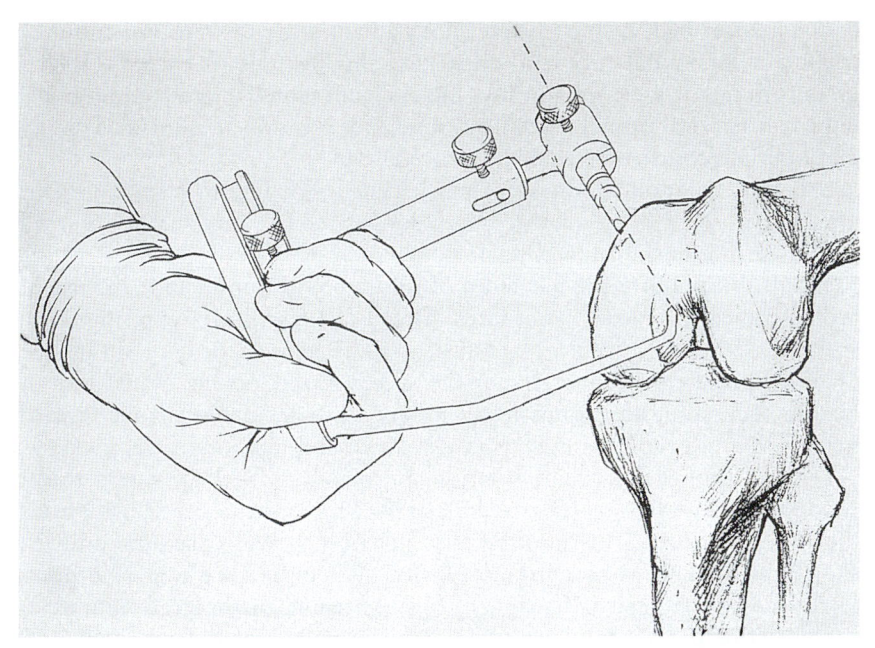

Fig. 19.6. (a) The ACL/PCL drill guide positioned for creation of femoral tunnel. (b) Interoperative photograph demonstrating drill guide position.

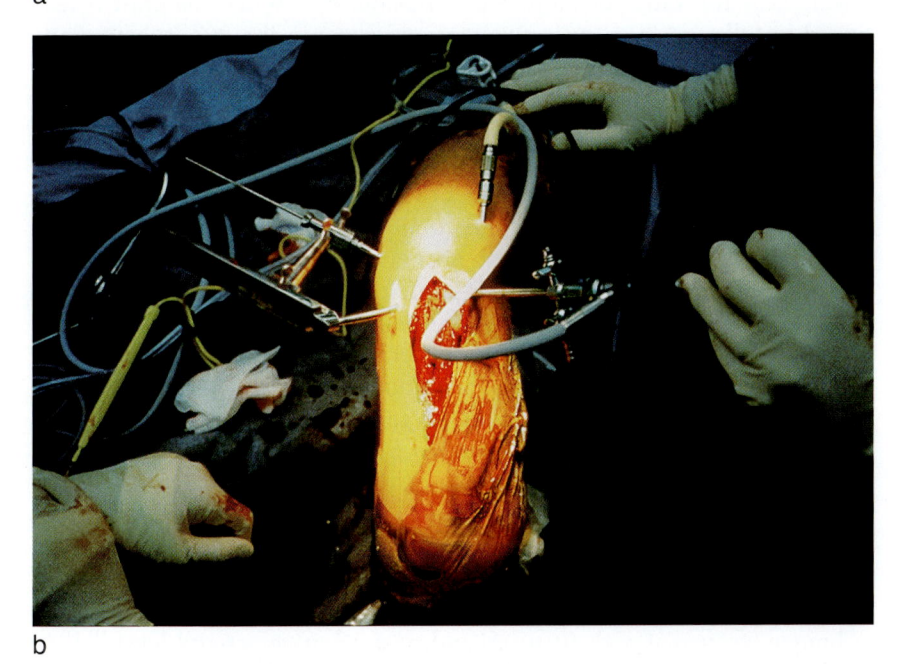

nique bone grafts the femoral tunnel and provides secure fixation for the Achilles tendon allograft in PCL reconstruction.

Salvage of this complication lies in early detection of the osteonecrosis. A high index of suspicion should be maintained in a postoperative patient with persistent medial femoral condylar pain. Clinical examination and imaging studies will confirm the diagnosis. Non–weight bearing and bone grafting of the tunnel defect may salvage the problem.[15]

Knee Motion Loss

Knee motion loss may occur after PCL reconstruction, and may present as loss of extension or loss of flexion.[18] Loss of flexion appears to be more common after PCL reconstructive surgery than loss of extension, and may become a functional problem when knee flexion fails to exceed 110 degrees. Causes of knee flexion loss include suprapatellar pouch adhesions, improper femoral tunnel placement, improper position of graft tensioning, and multiple concurrent ligament procedures.

Suprapatellar pouch adhesions may occur following an arthrotomy, or after harvesting a bone-quadriceps tendon graft.[7,8,10] Any surgery that induces scarring in the suprapatellar pouch area predisposes to flexion loss. Methods to minimize the incidence of this complication include the use of arthroscopically assisted procedures instead of an arthrotomy to minimize scarring. When a quadriceps tendon graft is harvested, care should be taken to preserve the integrity of the suprapatellar pouch. When loss of flexion occurs due to suprapatellar pouch adhesions, arthroscopic lysis of adhesions and gentle manipulation is a successful method of treatment.[7,8]

Femoral tunnel placement anterior and distal in the PCL stump to reproduce the anterolateral fiber region of the PCL is the desired position for the PCL substitute. This nonisometric positioning will cause the graft to shorten as the knee is brought into flexion. When the knee is in 70 degrees of flexion as the graft is tensioned, this tightening of the graft eliminates the posterior drawer, and maintains the normal tibial step-off, while allowing near full range of motion into flexion.[7,8,11] When the graft is tensioned in full extension, excessive tension occurs in the PCL substitute as flexion progresses, resulting in flexion range of motion loss or graft failure through tension overload.

Medial collateral ligament repair combined with PCL reconstruction may be associated with loss of motion. The large surgical dissection required, combined with plication types of procedures, increases the risk for postoperative motion loss. Direct anatomic repair is preferable to plication procedures in acute cases. When augmentation grafts are used, these must reproduce anatomic structures. This is preferred to capsular plication procedures, which may be nonanatomic and nonisometric.

It should be noted that some loss of terminal flexion is to be expected with PCL reconstruction surgery. This is due to the nonisometric nature of PCL reconstructions, and their failure to reproduce the complex PCL fiber bundle regions. Our series demonstrated a mean 10-degree terminal flexion range of motion loss.[7,8] This has not proven to be a functional problem for these patients. Rapid return of full flexion after PCL reconstructive surgery should be avoided as it may compromise the graft integrity and lead to persistent posterior sag.

Graft rupture, or failure through elongation, may be related to surgical technique, graft material, fixation, and postoperative activity level.[10,19,20] Artificial ligaments have been associated with failure due to abrasion of the graft around tunnel edges. Technical considerations that may reduce or eliminate this complication include smooth, well-chamfered tunnel edges, and anatomic positioning of the PCL reconstruction tunnels so the graft material does not make any acute angle turns or experience severe tension overloads.

The area that has the most acute bend for the graft is the proximal portion of the tibial tunnel in the transtibial reconstructive technique. To avoid this problem, the guide wire should exit inferior to the apex of the tibial ridge in the lateral aspect of the PCL anatomic insertion site. This will allow the PCL substitute to make two 45-degree turns, instead of a 90-degree bend on the posterior aspect of the proximal tibia. This applies to autograft, allograft, and synthetic ligament PCL surgical reconstructive techniques.

Another technique to avoid acute-angle graft bending at the proximal tibial tunnel edge is to position the bone plug tendon interface at the edge of the tunnel opening.[21] This will allow the graft's soft tissue portion to hinge on its own anatomic insertion into the bone plug, and not the tunnel edge. The tibial inlay technique of PCL reconstruction is performed with direct posterior exposure of the proximal tibia, and fixation of a bone–patellar tendon–bone graft in a posterior tibial trough.[22] This eliminates acute graft bending around the corner of a tibial tunnel.

Anterior Knee Pain

Anterior knee pain following PCL reconstruction may occur as a result of persistent posterior sag with resultant patella baja and increased patellofemoral forces.[13] This increased patellofemoral force may cause patellofemoral degenerative joint disease in long-term follow-up.[3] The solution to this problem is a technically sound PCL reconstruction with strong graft material to restore the anatomic tibial step-off, and return patellofemoral forces to a more normal level.

Other causes of anterior knee pain are prominent hardware, graft harvest site pain, and postoperative synovitis. Painful hardware can be removed in the postoperative period; however, low-profile fixation devices may reduce the need for this second procedure. Anterior knee pain associated with the harvest of an autogenous bone–patellar tendon–bone (BTB) graft may be reduced with the following technical considerations.[23] The harvested BTB autograft should be no wider than one-third the width of the patellar tendon measured at the tibial tubercle insertion. Harvesting the patella bone plug should be done with an oscillating saw, not osteotomes. This minimizes concussive effects to the articular cartilage. The depth of the patella bone plug harvest should not exceed one-third the thickness of the patella at the median ridge, usually about 8 mm. The superior one-third of the anterior cortex of the patella should be left intact. This will reduce stress riser effects to the patella. Bone grafting of the patella harvest defect may decrease the stress riser effect, but we have found no difference in the incidence of anterior knee pain with or without bone grafting of the patella donor defect.[23]

Fractures

Tibial fracture has been reported as an intraoperative complication during PCL reconstruction.[24] This can occur as a result of large-diameter tunnels in the femur and tibia, as well as the convergence of two tibial tunnels in combined ACL/PCL reconstruction.[8] During combined ACL/PCL reconstructions, the ACL and PCL tibial tunnels must diverge. This will prevent proximal tibial fracture. Tibial fracture has also been associated with hammering staples into the tibia for graft fixation. Our recommendation is the use of screws and spiked ligament washers or interference screws for graft fixation. If staples are to be used, predrilling the holes may avoid tibial fractures associated with hammering. As mentioned earlier, femoral tunnels should be positioned to allow adequate bone stock between the femoral tunnel and the medial femoral condyle articular surface. This will decrease the likelihood of subchondral bone collapse and/or osteonecrosis as outlined above.

Fanelli Sports Injury Clinic Experience

We have performed 92 PCL reconstructions at our center from August 1990 through March 1998. These reconstructions have included ACL/PCL reconstructions, PCL/posterolateral corner reconstructions, ACL/PCL/posterolateral reconstructions, ACL/PCL/medial collateral ligament (MCL) reconstructions, and PCL/MCL reconstructions. Our complications include loss of flexion requiring arthroscopic lysis of adhesions and manipulation (five cases), removal of painful prominent hardware used for graft fixation (21 cases), and superficial suture abscess (one case). We have had no incidence of nerve injuries, vascular injuries, deep infections, fractures, osteonecrosis, or skin slough complications. Our clinical results evaluated with Lysholm, Tegner, and Hospital for Special Surgery knee ligament rating scales, and KT-1000 arthrometer have been previously published.[7,8]

Analysis of our patients with loss of flexion requiring arthroscopic lysis of adhesions and manipulation are as follows. The first patient had ACL, PCL, MCL, posterolateral complex tears, and an acutely dislocated patella. Repair/reconstruction of all four ligaments and extensor mechanism repair was performed as an open procedure. Adhesions developed requiring arthroscopic adhesolysis and manipulation. The second patient had right knee ACL/PCL/MCL tears and an ipsilateral tibial shaft fracture. The tibial shaft fracture was treated with open reduction and internal fixation using plate and screws. The MCL was treated in a brace for 6 weeks, and healed with normal valgus stability. Arthroscopic combined ACL/PCL reconstruction was then performed using double-loop semitendinosus/gracillis hamstring autograft, and bone–patellar tendon–bone autograft. The patient developed suprapatellar pouch adhesions postoperatively requiring arthroscopic adhesolysis and manipulation.

The third patient was a 20-year-old with a right PCL/posterolateral complex tear combined with a hip fracture and multiple fractures of the foot. The fractures were addressed with open reduction and internal fixation, followed by arthroscopic PCL/posterolateral complex reconstruction with bone–patellar tendon–bone autograft and biceps tendon transfer. The patient developed postoperative suprapatellar pouch adhesions that responded well to arthroscopic lysis of adhesions and manipulation.

Patients four and five were a 15-year-old girl and a 22-year-old woman with PCL/posterolateral complex tears (one acute, one chronic) who underwent arthroscopic PCL/posterolateral reconstruction with Achilles tendon allograft and Clancy biceps tendon tenodesis in one case, and bone–patellar tendon–bone autograft and split-biceps tendon transfer in the second case. Both patients developed suprapatellar pouch adhesions postoperatively requiring arthroscopic lysis of adhesions and manipulation. Both patients returned to full unrestricted activity with stable knees.

We were not able to develop any trends to predict which patients may develop motion loss postoperative with the exception of arthrotomy versus arthroscopy. Arthrotomy seems more likely to have postoperative range of motion problems. Only five patients of 92 (5.4%) developed motion loss requiring surgical intervention and manipulation, indicating a low rate of this complication.

We attribute our low rate of complications in our series of PCL reconstructions to our accurate preoperative evaluations, surgical technique, and postoperative rehabilitation. Our procedures are described below.

All but one patient in our series of 92 PCL reconstructions had arthroscopically assisted procedures. The acute cases had their reconstruction performed approximately 3 to 6 weeks postinjury to allow for restoration of motion and for other injuries to stabilize. The chronic cases had their surgery performed from 6 months to 16 years postinjury. Our surgical technique has been described elsewhere.[4,7,8] We describe here specific technical aspects of arthroscopically assisted PCL/posterior lateral corner reconstruction.

When allograft tissue is selected, it is prepared prior to bringing the patient into the operating room. When autograft tissue is used, it is harvested and prepared at the beginning of the surgical procedure. The arthroscopic instruments are then inserted into the knee through the standard portals, and gravity inflow is used.

An extracapsular posterior medial safety incision is made that allows the surgeon's gloved finger to be positioned between the posterior capsule and the neurovascular structures. This allows the surgeon to feel through the capsule all instruments working in the back of the knee joint. The PCL stumps are debrided, and the posterior capsule is elevated from the posterior tibial ridge with the special PCL curved instruments.

The PCL tibial and femoral tunnels are then drilled with the help of an ACL/PCL drill guide, and the PCL graft is positioned. The PCL graft is anchored on the femoral side first, and left free on the tibial side. The posterior lateral reconstruction is then performed. We used the biceps femoris tendon tenodesis described by Clancy in our reported series.[3,7]

The posterior lateral reconstruction is tensioned with the knee in 30 degrees of flexion. With the knee in 70 degrees of flexion, an anterior drawer force is applied to the proximal tibia to restore the normal tibial step-off, and the PCL graft is then secured on the tibial side. Restoration of the normal tibial step-off at 70 degrees of flexion has provided the most reproducible method of establishing the appropriate tibia-femoral relationship in our experience. Wound closure is performed in the standard fashion, and the knee is immobilized in full extension.

The knee is kept in full extension for 5 to 6 weeks, with progressive weight bearing using crutches. Progressive range of motion occurs after week 6. The brace is unlocked at the end of 6 weeks, and the crutches are discontinued. Progressive closed kinetic chain strength training and continued motion exercises are performed. The brace is discontinued after the 10th postoperative week. Return to sports and heavy labor occurs after the 9th postoperative month, when sufficient strength and range of motion have returned.

Conclusion

Posterior cruciate ligament reconstruction is technically demanding surgery. Complications encountered with this surgical procedure include failure to recognize associated ligament injuries, neurovascular complications, persistent posterior sag, osteonecrosis, knee motion loss, anterior knee pain, and fractures. A comprehensive preoperative evaluation, including an accurate diagnosis, a well-planned and carefully executed surgical procedure, and a supervised postoperative rehabilitation program will help to reduce the incidence of these complications.

References

1. Fanelli GC. PCL injuries in trauma patients. Arthroscopy 1993;9(3):291–294.
2. Fanelli GC, Edson CJ. PCL injuries in trauma patients. Part II. Arthroscopy 1995;11:526–529.

3. Clancy WG. Repair and reconstruction of the posterior cruciate ligament. In: Chapman M, ed. Operative Orthopaedics. Philadelphia: JB Lippincott, 1988:1651–1665.
4. Fanelli GC, Giannotti BF, Edson CJ. Current concepts review. The posterior cruciate ligament arthroscopic evaluation and treatment. Arthroscopy 1994;10(6):673–688.
5. Muller W: The Knee: Form, Function, and Ligamentous Reconstruction. New York: Springer-Verlag, 1983.
6. Jackson DW, Proctor CS, Simon TM. Arthroscopic assisted PCL reconstruction: a technical note on potential neurovascular injury related to drill bit configuration. Arthroscopy 1993;9(2):224–227.
7. Fanelli GC, Giannotti BF, Edson CJ. Arthroscopically assisted combined posterior cruciate ligament/posterior lateral complex reconstruction. Arthroscopy 1996;12(5):521–530.
8. Fanelli GC, Giannotti BF, Edson CJ. Arthroscopically assisted combined anterior and posterior cruciate ligament reconstruction. Arthroscopy 1996;12(1):5–14.
9. Thomann YR, Gaechter A. Dorsal approach for reconstruction of the PCL. Arch Orthop Trauma Surg 1994;113:142–148.
10. Noyes FR, Barber-Westin SD. Posterior cruciate ligament allograft reconstruction with and without a ligament augmentation device. Arthroscopy 1994;10(4):371–382.
11. Burns WC, Draganich LF, Pyevich M, Reider B. The effect of femoral tunnel position and graft tensioning technique on posterior laxity of the posterior ligament-reconstructed knee. Am J Sports Med 1995;23(4):424–430.
12. Ogata K. Posterior cruciate ligament reconstruction: a comparative study of two different methods. Bull Hosp Jt Dis Orthop Inst 1991;5(2):186–198.
13. Paulos LE, Wnorowski DC, Beck CL. Rehabilitation following knee surgery. Recommendations. Sports Med 1991;11(4):257–275.
14. Ogata K, McCarthy JA. Measurements of length and tension patterns during reconstruction of the posterior cruciate ligament. Am J Sports Med 1992;20(3):351–355.
15. Athanasian EA, Wickiewicz TL, Warren RF. Osteonecrosis of the femoral condyle after arthroscopic reconstruction of a cruciate ligament. J Bone Joint Surg 1995;77A(9):1418–1422.
16. Reddy AS, Frederick RW. Evaluation of the intraosseous and extraosseous blood supply to the distal femoral condyles. Am J Sports Med 1998;26(3):415–419.
17. Rosenberg TD, Paulos LE, Abbot PJ. Arthroscopic cruciate repair and reconstruction: an overview and description of technique. In: Feagin JA, ed. The Crucial Ligaments. New York: Churchill Livingstone, 1988:409–423.
18. Irrgang JJ, Harner CD. Loss of motion following knee ligament reconstruction. Sports Med 1995;19(2):150–159.
19. Bosch U, Kasperczyk WJ. Healing of the patellar tendon autograft after posterior cruciate ligament reconstruction—a process of ligamentization? An experimental study in a sheep model. Am J Sports Med 1992;20(5):558–566.
20. Sterling JC, Meyers MC, Calvo RD. Allograft failure in cruciate ligament reconstruction. Am J Sports Med 1995;23(2):173–178.
21. Miller MD, Harner CD, Koshwaguchi S. Acute posterior cruciate ligament injuries. In: Fu FH, Harner CD, Vince KG, eds. Knee Surgery. Baltimore: Williams & Wilkins, 1994:749–767.
22. Berg EE. Posterior cruciate ligament tibial inlay reconstruction. Arthroscopy 1995;11(1):69–76.
23. Fanelli GC, Desai BM, Cummings PD, Hanks GA, Kalenak A. Divergent alignment of the femoral interference screw in single incision endoscopic reconstruction of the anterior cruciate ligament. Contemp Orthop 1994;28(1):21–25.
24. Malek MM, Fanelli GC. Technique of arthroscopically assisted posterior cruciate ligament reconstruction. Orthopedics 1993;16(9):961–966.

Index

Page numbers in *italic* type refer to illustrations.

ABC ligament, 189
Achilles tendon graft, for posterior cruciate ligament reconstruction, 138, *138*, 143, *143*, 224, *224*, 271, 296–297
ACL. *See* Anterior cruciate ligament
AC 30 RA ligament, 195, *195*
ACTOR 10 ligament, 194, *195*
AL. *See* Anterolateral (AL) bundle
Anterior cruciate ligament (ACL), 3, *5*, 6, *7*
   athletic injuries, 67
   combined ligament injuries, 222–234, *224–230*
   and LARS, 206–207
   original replacement of, 189
   rehabilitation following reconstruction, 284, *285–286*
   straight instabilities, 263
   surgery, 142, 291
Anterior drawer test, *58*
Anterior knee pain, 299
Anterior-posterior stabilization, 41–42, *42*
Anterior third capsular ligament, 49
Anteroinferior fascicle, 37, *37*
Anterolateral (AL) bundle, 4, *4*, *5*, 7–9, *8*, 15, 200, *200*, 202
Anteroposterior translation, 160
Arthrography, 79
Arthroscopy, 95–105, *99–105*, 203
   and combined injury reconstruction, 222–234
   and double femoral tunnel technique, 182–183
   and tibial inlay technique, 162
   and transtibial tunnel technique, 141–155
Athletic injuries, 66, 67, 126
Automobile accidents. *See* Motor vehicle accidents

Biceps femoris, 24, 27, 242
   long head of, 33, *33–34*
   short head of, 33–35, *34–35*
Biceps tenodesis, 242, *242*
Biomechanics
   changes and compensatory mechanisms in posterior cruciate ligament-deficient knees, 123–124
   of closed kinetic chain exercises, 122–123
   of open kinetic chain exercises, 122

   of posterior cruciate ligament, 7–16, 237–238
   of posterolateral aspect of knee, 40
   of posteromedial aspect of knee, 54–55, *56*
   of rehabilitation, 120–122
Bone-patellar tendon-bone graft, 162–163, *163–164*, 271

Canulated interference screws, 199
Capsular ligaments, 47–50, *48–49*, 57, *59*, 251
Carbon fiber ligaments, 189
CKCEs. *See* Closed kinetic chain exercises
Closed kinetic chain exercises (CKCEs), 121, 122–123, 125, 168–169, 268–269
Complications
   anterior knee pain, 299
   failure to recognize associated injuries, 291–292
   fractures, 299
   knee motion loss, 297–299
   in multiple-ligament-injured knee, 232
   neurovascular, 292–294, *292–294*
   osteonecrosis, 296–297
   persistent posterior sag, 295–296, *295*
   surgical, 291–301
Computed tomography (CT), 78
"Concurrent shift" of biarticular muscles, 268
Coronary ligament of lateral meniscus, 39, *39*
CT. *See* Computed tomography

Dacron mesh ligaments, 189
Deep mid-third capsular ligament, 49–50
Dial test, 44, *44*, 88, *88*, 240, *240*
Dimple sign, 219
Double femoral tunnel technique, 175–186
   alternate inside-out fixation on femur, 184–185
   creation of femoral tunnel, 183
   creation of tibial tunnel, 183
   femoral tunnel placement, 180–181, *180–181*
   graft choices, 177–178
   graft passage, 184
   hamstring tendon autografts, 182–185, *182*, *185*

   operative principles of, 177–182
   optimizing tunnel placement, 179–181, *179–181*
   tensioning and tibial fixation, 184
   tensioning of grafts, 181–182
   tibial tunnel placement, 179, *179*
Drill guides, 225, *227*, *228*, 296, *297*
Dynamic posterior shift test, 72

Eccentric exercise, 269
Epicondyle, 239, *239*, 241, *242*, 243
External rotation recurvatum test, 73, *73*

Fabellofibular ligament, 26, 39–40, *40*
Fanelli drill guide, 225, *227*, *228*
Femoral fixation, 184–185, 202, 243–244, *244*
Femoral tunnels, 201, *201*, 203, *203*, 227–228
   creation, 296, *297*
   double femoral tunnel technique, 175–186, *178–182*, *185*
   placement, 295, *295*, 298
Fibular collateral ligament. *See* Lateral collateral ligament
Fibular head, 239, *239*, 240, *241*
   drill hole, 242, *242*, 243, *244*
Fractures, 299

Gastrocnemius muscle, 25, 40
Godfrey's test, 71
Golgi tendon organs, 274
Gore-Tex fiber ligaments, 190, *190*
Gracilis graft, 136–137
Graft(s)
   bone-patellar tendon-bone, 162–163, *163–164*, 271
   choices for reconstruction, 136–138
   double femoral tunnel technique, 175–186
   and exercise, 270–271
   hamstring, 136, *136*, 182–185, *182*, *185*
   patella tendon, 137, *137*
   properties of ideal, 135–136
   quadriceps tendon, 137–138, *137*
   rupture, 298
   selection, 135–139, 143, 177–178, 224
   semitendinosus, 136–137, 242–244, *242*, *244*

Graft(s) (*continued*)
 tensioning, 13–14, *14*, 150–151, 167, 181–182, 184, 230–231
 tibial inlay technique, 157, 160–171
 transtibial tunnel technique, 141–155
 tunnel placement, 13, *13*, 179–181, *179–181*
Gravity sign near extension test, 71

Hamstring grafts, 136, *136*, 182–185, *182*, *185*
High-energy mechanisms, 66–67
Hockey-stick incision, 229, *230*, 232, *232*

Iliotibial band, 32, *32*
Iliotibial tract, 25–26, 29
Imaging, 77–84, 220, *220*
 See also specific techniques
In situ forces, 10–12, *12*, *13*
Isometric exercise, 269, 272
Isometry, 177, *177–178*, 238–239, *239*

Joint contact forces, 10

Killer tunnel angle, 190, *190*, *191*
Kinematics, 9–10, 12
KLT. *See* Knee laxity test
Knee
 absolute surgical indications, 221
 anatomic structures at risk in dislocated, 218–219
 anatomy of, 215–216
 anterior pain, 299
 classification of dislocated, 216–217
 combined injuries, 110–111, 222–224
 definitive surgical management, 222–223
 imaging studies, 220, *220*
 mechanism of injury, 217
 motion loss, 297–299
 physical examination, 219
 unreduced dislocated, 220–221
 vascular injuries, 221
 See also specific ligaments and aspects
Knee laxity test (KLT), 91–92
KT-1000 arthrometer, 79, 88–90, *89*, 119, 279, 280

Laminae, 29
LARS (Ligament Advanced Reinforcement System) synthetic ligament, 191, *191*, 192–212
 autogenous option, 204
 biological environment, 193
 complications, 208
 development and design, 192–197
 histological studies, 195–197
 material, 207
 mechanical stresses, 192–193
 posterolateral laxities, 205–206
 postoperative care, 204–205
 results, 208–210, *208–210*
 surgical technique, 199–207
Lateral collateral ligament (LCL), 5
 anatomy of, 23–25, 27, 35–36, *36*
 and anterior-posterior stabilization, 41–42, *42*

dial test, 44, *44*
and internal/external tibial rotation, 42–43
isometry of, 238–239, *239*
and posterolateral reconstruction, 226, *229*
reconstruction using semitendinosus free graft, 243–244
and varus stabilization, 40–41
Lateral femoral epicondyle, 241, *241*, 243
Lateral gastrocnemius complex, 39–40, *40*
Lateral meniscus stability, 43
LCL. *See* Lateral collateral ligament
Leeds-Keio ligament, 189–190
Ligament Advanced Reinforcement System. *See* LARS
Ligament augmentation device, 189
Ligament of Humphry, 4, 7
Ligament of Wrisberg, 4, *5*, 38–39, *39*
Ligament synthesis, 203
Load-elongation curves, 8, *8*

Magnetic resonance imaging (MRI), 38, *38*, 80–84, *81–84*, 119–120
MCL. *See* Medial collateral ligament
Measurement, 87–93
Medial capsular ligaments, 48–49
Medial collateral ligament (MCL), 215
 anatomy of, 216
 brace treatment of, 142, 223
 posterior lateral corner injury, 222
 reconstruction, 224, 229–230
 repair and loss of motion, 298
Medial femoral condyle, *13*
Medial hockey-stick incision. *See* Hockey-stick incision
Medial ligament injuries. *See* Posteromedial aspect of knee
Medial meniscus, 52, 252, *253*, 255, *256*, 258–259, *260*
Medial retinaculum. *See* Anterior third capsular ligament
Meniscofemoral ligaments (MFLs), 4–8, *5*, 7–9
MFLs. *See* Meniscofemoral ligaments
Midsubstance tear, 59, 258, *259*
Mid-third capsular ligament, 36, *36*, 254, *255*, 259, 260, *263*
Motion loss, 297–299
Motor vehicle accidents, 66–67
MRI. *See* Magnetic resonance imaging
Musculotendinous meniscocapsular aponeurosis, 53–54, *53–54*

Neurovascular complications, 292–294, *292–294*
Nuclear medicine, 79

Oblique popliteal ligament (OPL)
 anatomy of, 38, *38*, 50–51, *50*, *52*
 repair of, 261, *261*
 tear of, 255, *257*, 259
Ogata's points, 201, *202*
OKCEs. *See* Open kinetic chain exercises
Open kinetic chain exercises (OKCEs), 121–122, 125, 168–169, 268–269
OPL. *See* Oblique popliteal ligament
Osteonecrosis, 296–297

Pain, 299
Patella tendon graft, 137, *137*
Patellofemoral joint, 269
PCL. *See* Posterior cruciate ligament
Peroneal nerve, 218–219
Persistent posterior sag, 295–296, *295*
Physical examination
 in polytrauma patient, 219
 of posterior cruciate ligament, 68–74, 158
 for posterior cruciate ligament insufficiency, 175–176
 of posterolateral instability, 240
PLS. *See* Posterolateral structures
Plyometrics, 274
POL. *See* Posterior oblique ligament
Polytrauma, 219, *220*
Popliteal artery, 218, *294*
Popliteal hiatus, 28, 30
Popliteofibular fascicle, 30
Popliteofibular ligament, 226, *229*, 238–239, *239*, 242, 243–244
Popliteomeniscal fascicles, 43
Popliteus bypass, 242, *242*
Popliteus complex, 5, 37–38, *37–38*
Popliteus muscle, *13*, 25–28, 30
Popliteus tendon, 30, 37, *37*, 238, *239*
Posterior cruciate ligament (PCL)
 anatomy of, 3–7, 176–177, *176*
 arthrometric testing for laxity, 260
 arthroscopic evaluation of, 95–105
 biomechanics of, 3, 7–16, 120–124, 237–238
 clinical examination of, 65–75
 clinical management of, 15
 combined ligament injuries, 3, 110–111, *110–111*, 222–224, *224–230*, 274
 cross-sectional area, 5, *6*
 diagnosis of injuries, 117–119
 history of, 65–67
 imaging of, 77–84
 insertion site anatomy, 5–6
 isolated injuries, 3, 111–114, *121*, 271–274, 275–279
 measurement of, 87–93
 mechanism of injury, 141
 midsubstance repair, 59, 258, *259*
 natural history of, 109–114
 nonoperative treatment of injuries, 117–127, *124*, 271–274
 physical examination, 68–74, 158, 175–176
 rehabilitation of injuries, 267–286
 stress, 269–270
 surgical complications, 291–301
 tears associated with medial ligament injuries, 249–264
 three-zone concept, 96, *97*
 ultrastructure, 6–7
 use of LARS in, 192–212
 See also Graft(s); Reconstruction; Surgery
Posterior drawer test, 69, *69*, 70–71, *71*, 87, 88, 158–159, *159*
Posterior oblique ligament (POL), 58
 anatomy of, 50–51, *50–52*
 evaluation of, 255, *256*
 and posterior cruciate ligament tears, 259

Index

and posteromedial arthrotomy, 252, *253*
repair of, 261–262, *261*, *263*
Posterior sag test, 69, *69*
Posterolateral aspect of knee
  anatomy of, 31–43, *238*, *239*, 274
  anterior-posterior stabilization and, 41–42, *42*
  arthroscopic evaluation of, 95–105
  biomechanics of, 40
  early studies on anatomy of, 23–31
  injury to, 43–44, 237, 239
  internal and external tibial rotation and, 42–43
  measurement of, 87–93
  motion testing, *44*
  reconstruction, 226, *229*
  rehabilitation following reconstruction, 280–284, *281–282*
  torque on tibia with anterior and posterior force, *43*
  varus stabilization and, 40–41
  See also Posterolateral instability
Posterolateral drawer test, 70
Posterolateral instability, 72, 237
  advancement or recession of lateral femoral epicondyle, 241, *241*
  options for treating, 241–245
  physical examination of, 240
  reconstructions, 240–244, *240–244*
  symptoms of, 239–240
  tibial osteotomy, 244–245, *245*
Posterolateral structures (PLS), 3, 10, 14, 15, *15*
Posterolateral tibia, 240, *241*
Posteromedial arthrotomy, 252, 253, *253*
Posteromedial aspect of knee, 47–60
  acute injury repair strategies, 256–262
  anatomy of, 47–54
  biomechanics of, 54–55, *56*
  chronic injury reconstruction strategies, 262, *263*
  exploration of injury pathology, 253–256
  injuries associated with posterior cruciate ligament tears, 249–264
  ligaments, 48–52, 55–59, 249–264
  meniscus, 52, *252*, *253*, 255, *256*, *258–259*, *260*
  muscular input, 52
  musculotendinous meniscocapsular aponeurosis, 53–54, *53–54*
  operative versus conservative management, 249–250
  reconstruction, 229–230, *230*
  rotatory instabilities, 264
  straight instabilities, 262–264
  surgical approaches, 250–253, *252*
Posteromedial bundle, 4, *4*, 7–9, *8*
Posteromedial drawer test, 70, 73
Posteromedial pivot-shift test, 73, *74*
Posteromedial safety incision, 293–294, *293*
Pretibial trauma, 66
Prone posterior drawer test, 70, *70*
Proximal tibia, abrasions of, 95, *96*

Quadriceps active drawer test, 72, *72*
Quadriceps tendon-bone graft, 162–163, *163–164*
Quadriceps tendon graft, 137–138, *137*

Radiography, 119
  See also Stress radiography
Reconstruction, 12–14, *14*
  combined injury, 222–214, *224–230*
  double femoral tunnel technique, 175–186, *176–182*, *185*
  posterolateral, 240–244, *240–244*
  posteromedial, 262, *263*
  synthetic, 189–192
  tibial inlay technique, 157, 160–171, *163–166*
  transtibial tunnel technique, 141–155, *145–154*
Rehabilitation
  biomechanics of, 120–122
  following anterior cruciate ligament reconstruction, 284, *285*
  following posterior cruciate ligament/posterolateral corner reconstruction, 280–284, *285–286*
  of posterior cruciate ligament injuries, 267–286
  postoperative combined injury, 231–232
  postoperative posterior cruciate reconstruction, 168–171, 275–280, *276–277*
  scientific basis of, 268–271
Retrograde fiber tracks, 29–30
Reverse pivot shift test of Jakob, 73, *73*
Robotics, 11, *11*, 12
Roentgenographic imaging, 77–79, *77*, *78*
Rotational laxiometer, 92–93, *92*

Semimembranosus muscle, *47*, 52, *52–53*, 255
Semitendinosus graft, 136–137, 242–244, *242*, *244*
Short lateral ligament, 26
Single photon emission computed tomography (SPECT), 79
SPECT. *See* Single photon emission computed tomography
Split-biceps tendon transfer, 226, *229*
Sports injuries. *See* Athletic injuries
"Static" ligaments, 47
Stress radiography, 74–75, *75*, 79, 90–91, *90–91*, 119
Suprapatellar pouch adhesions, 298
Surgery
  absolute indications, 221
  and arthroscopic evaluation, 96–98, *97–98*
  complications, 291–301
  cruciate ligament reconstruction with synthetics, 189–192
  definitive management, 222–223
  on lateral side of knee joint, 24–25
  for torn ligament, 59–60
  See also Graft(s); Reconstruction; *specific techniques*
Synthetic grafts, 138
Synthetic replacements, 189–192

TCL. *See* Tibial collateral ligament
Telescopic tubes, 198, *198*
Telos stress radiography, 90–91, *90–91*
Thigh-foot angle test, 73, *74*
Tibial collateral ligament (TCL), 51–52, 253–255, *253–255*, 259–261, *260–261*, 263
Tibial external rotation test, 73, *74*
Tibial fixation, 202, *203*
Tibial fracture, 299
Tibial guide, 198, *198*
Tibial inlay technique, 157, 160–171
  bone-patellar tendon-bone graft, 162–163, *163–164*
  diagnostic/operative arthroscopy, 162
  femoral tunnel selection, 163–164, *164–165*
  graft tensioning and femoral fixation, 167
  patient education, 161
  patient positioning and preparation, 161–162
  posterior approach to tibia, 165–167, *165–166*
  postoperative care, 167–168, *168*
  postoperative reconstruction rehabilitation, 168–171
  quadriceps tendon-bone graft, 162–163, *163–164*
  scientific rationale for, 160–161, *161*
Tibial nerve, 218
Tibial osteotomy, 244–245, *245*
Tibial rotation, 42–43
Tibial step-off test, 69
Tibial tunnel, 179, *179*, 183, 199, 226, *227*, 231, *232*, *233*
Tomography, 78
Transtibial tunnel technique, 141–155, 292
  associated ligament surgery, 151–153, *153–154*
  capsule elevation, 145, *145*
  drill guide positioning, 146–147, *146–148*
  femoral tunnel drilling, 149–150, *149–150*
  graft tensioning and fixation, 150–151, *153*
  patient positioning and preparation, 144
  posteromedial safety incision, 144–145, *144*
  postoperative rehabilitation, 153
  tibial tunnel drilling, 148, *149*
  tunnel preparation and graft passage, 150, *151–153*
Trevira ligament, 189
Two-strand graft, 271

Ultrasound, 80, *80*
Universal Force-Moment Sensor, 11, *11*, 12

Valgus stress test, *56*
Varus stabilization, 40–41
Varus/valgus stresses, 215–216
Varus/valgus tests, 159
Vascular injuries, 221
Vastus medialis obliquus (VMO), 255–256, *257*, 262
VMO. *See* Vastus medialis obliquus
Voluntary evoked posterolateral drawer sign, 73

ISBN 0-387-98573-5